ECOLOGICAL ASPECTS OF THE MINERAL NUTRITION OF PLANTS

THE BRITISH ECOLOGICAL SOCIETY

SYMPOSIUM NUMBER NINE

FRONTISPIECE. The Park Grass Experiment, Rothamsted (12.8.68)

Above, plot 11/2 complete fertilizer and lime treatment, mainly *Alopecurus pratensis* (not flowering) and *Arrhenatherum elatius*.

Below, plot 12 unmanured, showing original diversity of species

For details see J.M. Thurston in this volume.

ECOLOGICAL ASPECTS OF THE MINERAL NUTRITION OF PLANTS

A Symposium of

THE BRITISH ECOLOGICAL SOCIETY

Sheffield 1–5 April 1968

Edited by

I. H. RORISON

University of Sheffield

with the assistance of

A. D. BRADSHAW
University of Liverpool

M. J. CHADWICK
University of York

R. L. JEFFERIES
University of East Anglia

D. H. JENNINGS
University of Liverpool

P. B. TINKER
University of Oxford

BLACKWELL SCIENTIFIC PUBLICATIONS
OXFORD AND EDINBURGH

SBN 632 05660 6

FIRST PUBLISHED 1969

Printed in Great Britain by
SPOTTISWOODE, BALLANTYNE AND CO LTD
LONDON AND COLCHESTER
and bound by
THE KEMP HALL BINDERY, OXFORD

CONTENTS

SECTION TWO · MINERAL NUTRIENT SUPPLY FROM SOILS

SECTION THREE · MINERAL NUTRITION OF THE WHOLE PLANT SYSTEM

SECTION FIVE · SUMMATION

PREFACE

The aim of the symposium was to assess current knowledge of the mineral nutrient factors which influence plant distribution and of the physiological mechanisms associated with the adaptation of plants to their edaphic environments. Papers were read by ecologists, geneticists, physiologists and soil chemists. They each gave an interpretation of concepts and methods from the viewpoint of their own discipline and an attempt was made to present an integrated appraisal of some fundamental problems.

The papers were grouped in five sections. The first dealt with mineral nutrition in relation to plant distribution with examples from specific habitats. One paper appears in this section which was not announced in the programme. It records a short illustrated talk on the Park Grass Experiment at Rothamsted given one evening by Miss J.M.Thurston. It made a strong impression on those present, and since the experiment is producing a wealth of data of which too few people are aware, it was decided to publish a brief account together with references to detailed results.

The second section which dealt with mineral nutrient supply from soils, aimed to give a comprehensive and detailed picture of one part of soil chemistry. Most of the time was therefore given to reviews of recent work on the properties of ions in the soil as they affect uptake by plants. The value of models in stimulating precise research into complex systems was clearly demonstrated.

In the third section, on mineral nutrition of the whole plant system, the problems of growing plants in isolation in the laboratory were considered. Some of the complicating factors which impede the interpretation of field situations through extrapolation of laboratory findings were then discussed.

The fourth section dealt with mechanisms of mineral nutrition and in it a search was made for links between whole plant, tissue, and cellular studies. The problems were discussed with reference to experiments in which ions of ecological interest were studied and their contribution to the understanding of ion uptake and toxicity mechanisms was considered.

In the fifth and final section an ecologist, a physiologist and a soil chemist were given a free hand to emphasize what they considered had been the main problems raised or omitted by previous speakers, to criticize hypotheses and to summarize conclusions for a final discussion.

Points which were made in discussion and which were not covered in the text of the published papers have been summarized at the end of each section.

The symposium was illuminating in a number of ways. In particular it showed that biochemists, physiologists, and to some extent soil chemists, are seeking basic mechanisms and systems which might apply to all plants. Ecologists on the other hand are looking for *specific* mechanisms of adaptation which as yet they are unable to define with sufficient precision to enlist the co-operation of researchers at the metabolic level. The inter-disciplinary nature of the meeting, and the presence of many guests from the Society of Experimental Biologists and the British Society of Soil Science, stimulated widespread discussion and opened new avenues of thought for many people. In a few years the Society may perhaps sponsor a symposium on *Mechanisms of plant adaptation to extreme environments and their mode of evolution*. Such a symposium will surely demonstrate that further inter-disciplinary co-operation has meanwhile been achieved.

I.H.Rorison

February 1969

ACKNOWLEDGMENTS

The symposium was held at Sheffield University by invitation of the Vice-Chancellor and Professor A.R.Clapham F.R.S., and the warmth of the hospitality provided by the University staff was a major factor contributing to its success.

Professor A.D.Bradshaw, Dr M.J.Chadwick, Dr R.L.Jefferies, Professor D.H.Jennings and Dr P.B.Tinker assisted with the planning and editing throughout. They were joined as chairmen of sessions by Dr R.K. Cunningham, Professor Emanuel Epstein and Professor J.L.Harley F.R.S., and together with Dr D.Gunary, J.F.Handley, C.P.Harding, Dr C.Marshall and Dr.R.W.Snaydon, they recorded the discussions. Dr T.T.Elkington organized the demonstrations, Dr J.P.Grime led a very successful field excursion, Dr P.S.Lloyd ensured the smooth running of domestic arrangements, and Mrs Jean Hampshire carried the entire secretarial burden. To all these staunch friends the editor is deeply grateful.

Acknowledgment is also made to the D. Van Nostrand Co Inc for permission to reproduce Fig. 3, p. 42 and Fig. 4, p. 43 from Fruchter's *Introduction to Factor Analysis*; to the Editor, *Journal of Experimental Botany* for Fig. 1, p. 328 and Fig. 2, p. 330; to the Director, Rothamsted Experimental Station for Fig. 1, p. 416; and to Yale University Press and Duke University Press for Fig. 4, p. 423.

INTRODUCTION

A. R. Clapham

Department of Botany, University of Sheffield

It is first my very great pleasure to welcome members of the British Ecological Society and their guests to this Symposium Meeting in the University of Sheffield. The Southern Pennines have been a special attraction to ecologists since the earliest days of British ecological studies, the days of Moss, Woodhead, Adamson and others. Professor W.H. Pearsall was appointed to the Chair of Botany in this University in 1938 and Dr Verona Conway was a member of his staff when she began her important investigations into the history of the local vegetation as revealed in the peat of Ringinglow Bog, just to the west of the city. And perhaps I may be allowed to recall that in 1945 there was a field meeting of the Society here in Sheffield and Professor Tansley, although well into his seventies, reached the top of Kinderscout with the rest of us. His main worry during his stay in Sheffield, as I remember well, was that there were no adequate facilities for securing alcoholic drinks with meals at Crewe Hall. He complained that he could not digest his dinner without alcohol, and he dined with my wife and me on all nights after the first.

I have been asked to give this introductory paper to your Symposium on Ecological Aspects of the Mineral Nutrition of Plants. Now this is a Society of Ecologists but you have many guests here today who are themselves ecologists only in the very broadest sense of that term. It is therefore, I trust, not inappropriate for me to explain that a plant ecologist is a botanist with a special interest in where plants of a given kind grow and what company they keep. His taste and training may make him content to observe and record the facts of species distribution and to attempt to formulate generalizations based on his findings. But most plant ecologists wish to know why a given plant grows more or less well or not at all in a given place, and it may be surmised that most who are here today have been attracted through their interest in explanations of species distribution. That matters of mineral nutrition are relevant to this can hardly be denied.

It seems to me that, from the standpoint of the plant ecologist of this kind, three aspects of the study of the mineral nutrition of plants are of special importance. These are:

1. *The specification of the whole plant-soil system*, in the context of mineral ion uptake, transport and utilization and *within the range of naturally*

occurring conditions. This involves identification of the appropriate parameters and selection and refinement of techniques for measuring them.

2. The identification, specification and measurement of *differential behaviour*, both between and within species, in respect of the uptake, transport and utilization of ions.

3. The identification, specification and measurement of *competitive effects*, both between and within species, in respect of the uptake of ions.

In all three fields of investigation there are areas that concern the plant physiologist, the biophysicist or the biochemist more closely than the ecologist. I have particularly in mind the detailed elucidation of mechanisms of uptake, transport and utilization of ions, which may take the research worker very far indeed from the study of the plant in the field. Nevertheless the ecologist needs to become familiar with the outcome of such studies, since an understanding of mechanisms must deepen his understanding of the ecological situation and guide his ecological research. The same is true of the detailed study of soils. Much of this again lies outside the immediate field of the ecologist, but here too he must be familiar with the progress of pedological research because he cannot afford to be ignorant of it.

We are fortunate in having as contributors to this Symposium not only a large number of ecologists *sensu stricto*, from this and from several other countries, but also many distinguished pedologists, plant physiologists, biophysicists and others whom we cordially welcome because we know in how valuable a way they will help us in our discussions and bring us up-to-date in our awareness of what is going on in neighbouring worlds of scientific research.

I should like now to turn back to my three main areas of concern for ecologists in the general field of the mineral nutrition of plants.

I referred first to the *specification of the plant-soil system* in the context of ion uptake, transport and utilization and within the range of naturally occurring conditions, and to the identification and measurement of the relevant parameters. I mean by this the construction of a sufficiently comprehensive and detailed picture of the plant/soil system to enable us predict the results of changes in any components of the system. We are reminded by the title of the 3rd Section of this Symposium that the ecologist is primarily interested in the behaviour of the whole intact plant, not in isolated bits and pieces; and of the plant rooted in natural soil, not in simpler and more readily comprehensible substrata. This means that the required specification of the formidably complex system will certainly not be achieved quickly and is unlikely to be achieved by ecologists working on their own. We must edge our way in the general direction of the goal. The important point is to learn more than we know at present about comparative magnitudes. What components of the system can safely be neglected in the early rounds of model-making? How do we avoid being seriously misled about the true facts of plant life when we experiment with simplified systems? This is a major problem of ecological experimentation and many of our speakers will deal with aspects of it.

There are two other points about the soil as a source of mineral nutrients and other ions to which I should like to make some passing reference. There is first the problem, and it may well be insoluble, of finding readily measurable analytical parameters whose utility extends over the whole range of ecological substrata: sands, clays, peats; acid and alkaline, dry and wet. Then, second, there is the problem of making the necessary distinction between the *rate of supply* and uptake of an ion and its *maintained level* in the soil solution: between the scale of the process and the pattern of the time-slice afforded by analyses at points of time. This is, of course, a long-recognized difficulty, but one that is still often overlooked. We have amongst our speakers members of teams working on the specification of soil phosphate status, where the problem is specially acute. Some headway has been made, as we shall learn, in attempts to apply kinetic considerations to the selection of suitable procedures for defining the phosphate status of natural soils. But the supplying process may be neither single nor simple even from the purely chemical standpoint, and other than chemical factors will commonly be operating. We are to be told, for instance, about the importance of the uptake of phosphate by micro-organisms, especially at low phosphate levels; and about root-surface enzyme systems that may release inorganic phosphate from compounds at the absorbing surface. In all the complex circumstances it might seem very doubtful whether any analytical procedure could be as satisfactory as measuring the rate of uptake of phosphorus by individuals of one or more standard plant species sampled and analysed after successive time-intervals. But then we have to decide what standard species to choose, and by what criteria. We must find means, further, of determining the appropriate stage of life, previous treatment, length of experimental period and conditions to be maintained during its course. These again are matters about which we shall hear from one or more of our speakers.

Other problems concern the variations in mineral nutrient supply in time and in space. There are certainly marked seasonal changes in availability. When I was at Rothamsted, long years ago, I assisted L.R. Bishop in a comparison of a soil famous for the frequency with which it, and the man who farmed it, produced prize-winning malting barleys with one notorious for not doing so. We grew barley from seed and estimated the rate of supply of nitrate by incubating soil samples for given periods and by analysing periodic samples of the barley seedlings. The outstanding difference between the two soils was that the 'good' soil supplied nitrate rapidly only for a relatively short period at the beginning of the growing season, while the 'bad' fen soil supplied it rapidly throughout the season.

Changes in space include the depth-stratification of available ions in relation to the position of absorbing system in the soil, and I am glad to see that this is to be the subject of one of our contributions.

I am not going to say any more about the soil as a source of nutrient ions: we shall be prevented by our speakers from overlooking the fact of its complexity or the fact that all soils are different in relevant ways. The soil and its roots are indeed extraordinarily inaccessible to observation and experiment. But the whole living plant rooted in the soil is certainly no less complex a system, and none of its activities can safely be thought irrelevant until the necessary investigation has been undertaken. Professor Weatherley has something to say about this.

My second point was the identification, specification and measurement of *differential behaviour*, both between and within species. The crop physiologist and ecologist share an interest in the different requirements, tolerances and intolerances of different species with regard to the levels and availability of ions present in the soil; but for the ecologist this has always been a central issue. There is a long history of recognition of correlations between species-distribution in the field and chemical features of the substratum. Terms such as calcicole, calcifuge, halophyte bear witness to these distributional findings and inferences. During recent decades there has been an increased understanding of the bearing of pH and HCO_3^- and Ca^{++} concentrations on other factors affecting the mineral nutrition of plants. We know, for instance, a good deal more than formerly about lime chlorosis and ionic interactions, about aluminium and manganese toxicity. Yet the picture is very far from clear. We rarely know unequivocally what features of the edaphic situation are basically responsible for the facts of differential distribution—or even that they are matters of soil chemistry at all in any very direct sense.

There are many data showing that species do differ quite markedly in the relative proportions in which various mineral ions are taken up. Collander, in particular, has supplied a valuable store of information of this kind. It appears that some species tend to accumulate one ion, some another. There are interspecific and intraspecific differences in the proportions of the absorbed ions which are transported from root to shoot and in the effectiveness of their uptake and transport in bringing about the growth of the plant. Rates of uptake per unit dry weight of root or of whole plant presumably depend upon the areas of absorbing surface and their absorptive efficiencies and also upon the extent to which the concentration gradients of the various ions are maintained either by the growth of roots into previously unoccupied volumes of soil or by mass flow of soil solution consequent upon transpiration. But differential growth-behaviour seems often to arise not from differential ion-absorption but from differential effectiveness of assimilation, transport and utilization within the plant. That seedlings of *Deschampsia flexuosa*, for example, show a much smaller change in growth-rate with decreasing phosphate concentration than do those of *Urtica dioica*, and that they continue to grow at concentrations much lower than the minimum for *Urtica*, seems inexplicable in terms of absorptive efficiency; and the same is apparently true of the uptake of iron by certain species of ecological races differing in their susceptibility to lime chlorosis. Differential susceptibility to aluminium and manganese toxicity is no more readily accounted for by differential uptake.

Where there is real differential *uptake* the explanation may still not lie in differential absorptive efficiencies in a straightforward sense. Micro-organisms rendering certain ions available may be differently numerous or differently active in the rhizospheres of different species; and mycorrhizas may exaggerate specific differences. Root-surface enzyme systems, such as those splitting inorganic phosphate from organic compounds, may or may not be present or may be differentially active. We are to hear about enzyme systems of this kind and their possible differential ecological significance.

Much has been written about the cation-exchange capacities of roots and their possible role in differential absorption; and there are numerous other facts and theories that may or may not have a bearing on specific differences in mineral nutrition. Ecologists would much like to hear views from specialists on some of these matters, for it is the differences between species in their plant/soil relations that he wishes to be able to explain.

Finally there is the question of *competition* for mineral nutrients, which may be interspecific and intraspecific. We know that with an adequate water-supply, and with shoot-competition excluded, plants of different species sharing root-space may behave as more or less successful competitors. We know very little, however, about the mechanisms of such root-competition. It may be merely a matter of the comparative dimensions of root-systems equally effective per unit area of ion-absorbing surface, though it would not be easy to demonstrate that this is so. Or it might be a matter of different ion-absorbing efficiencies of the two root-systems, for whatever reason. There could be said to be competition only if the superior capacity or intensity of absorption by the one root-system resulted in a deprivation of the other. Deprivation of the kind required to explain root-competition could readily be envisaged as operating through the superior capacity or intensity (or both) of one root-system as a diffusion-sink for ions, or as a more literal sink in mass-flow actuated by transpiration, or again through the more rapid invasion by one of the root-systems of volumes of soil previously unoccupied by either but later to be occupied by both—this involves hypotheses about rates of replacement of absorbed ions. There are clearly other possibilities: root-exudates from one root-system, for example, may affect the capacity of the other to absorb ions. This is an aspect of mineral nutrition that is of very great ecological importance, but we know little about it and, I am sorry to see, are to hear little about it during the Symposium.

Ecologists, then, find themselves faced by the full and daunting complexity of the plant/soil system when they set out to understand what part mineral nutrition plays in determining the facts of species distribution. They have to make up their minds what should be their role as ecologists in the total scientific attack on this and comparable questions. Two extreme attitudes would be either that ecologists should be content to record the facts of species distribution in the field and leave it to plant physiologists, soil scientists, biophysicists, biochemists and the rest to elucidate them; or that they should recognize the problems as essentially their own and assume the full responsibility for solving them. I think neither of these extremes is desirable, and I think the latter impracticable. There are, however, middle ways. I would hope on the one hand that a man who calls himself a plant ecologist would always be a student of natural vegetation. As a general rule, on the other hand, he ought to be an experimenter as well.

The essential first steps are in any case his to take. He must observe and analyse the field situation. He must adopt observational methods that enable him to specify features of plant distribution in quantitative terms and correlate his findings with correspondingly quantitative specifications of environmental factors. This is the initial 'survey' stage of an ecological investigation, and it poses plenty of methodological problems. The grassland survey being undertaken by Drs Rorison, Grime and Lloyd in the Botany Department here, some of the results of which are being exhibited in the Department, is an example of this initial stage.

From the correlations obtained in the survey stage the ecologist will be led to infer or suspect causal relationships, and these he should himself be prepared to test in a second stage of investigation, which is one of *experimentation*. Experiments may be carried out first in the field (as in the interesting field-fertilizer trials undertaken by Dr Willis at Braunton Burrows†), but sooner or later they will take place in the garden, greenhouse or growth chamber according to the closeness of the required control of environmental variables. Methods are now much the same as those of the whole-plant-physiologist, and problems of control of variability of the plant material and rooting medium and problems of sampling and analysis will need to be solved.

If his initial inferences were well-founded and his experiments well-designed—and if he has been generally fortunate—he may now be ready for the third stage of investigation and the second of experimentation: that of the reconstruction of the field situation, simulating it in as detailed a way as seems necessary for the immediate experimental purpose.

The whole process may have to be repeated for each factor suspected of influencing the ecological situation to an appreciable degree. Eventually, however, the experimental ecologist should be in a position to announce his findings in the form of a statement of acceptably inferred cause-and-effect. He may be able to say that species A is more tolerant of phosphate deficiency, in a specifiable sense of the term, than is B; that this explains why in certain types of substratum A thrives better than B and may even oust B from the plant-assemblage by this or that means; and why if phosphate is added to the substratum in the field B can thrive as well as A and even reverse the competitive situation. At this point, I think, the true ecologist, however experimental in inclination and habits, should be content to hand over the problems of mechanism to the appropriate laboratory scientists.

† Willis A. J. (1963). *J. Ecol.* **51**, 353–374.

SECTION ONE

MINERAL NUTRITION AND PLANT DISTRIBUTION

THE EFFECT OF LIMING AND FERTILIZERS ON THE BOTANICAL COMPOSITION OF PERMANENT GRASSLAND, AND ON THE YIELD OF HAY

J. M. THURSTON
Rothamsted Experimental Station, Harpenden, Herts.

INTRODUCTION

The Park Grass experiment on the effect of fertilizers on yield and quality of hay from permanent grassland began on Rothamsted farm in Hertfordshire in 1856 and still continues. It started as an agricultural investigation, but 25 years later Lawes and Gilbert (1880) concluded that it 'afforded results of more interest to the botanist, the vegetable physiologist and the chemist than to the farmer'. This is still true. Some treatments were changed and a few plots were added in the early years of the experiment in the light of experience gained (Laws Agricultural Trust 1966) but 15 plots retain their original treatments. Two plots whose treatments were changed in the early years, were dropped from the main experiment in 1965 and are now used for small-plot tests of treatments, rates and combinations not represented on the large plots (Fig. 1, Plots 5 and 6). The vegetation on plot-sections still receiving their original treatments alters very little between years but on those receiving the new liming rates it is now changing gradually. The plant communities differ greatly according to treatment, and as the soil-type, aspect, weather and management are the same for all plots, these differences can result only from applying fertilizers and lime.

METHODS

The 8-acre field is level and was under permanent grassland consisting of indigenous species and strains of plants for several hundred years before the experiment began. The soil is silt-loam overlying stiff yellow-red clay upon chalk. Drainage is good and agricultural drains are unnecessary. The mixed vegetation of Gramineae, Leguminosae and other families appeared fairly uniform over the whole area at the start.

The plan of the experiment as it exists in 1968 is shown in Fig. 1. The

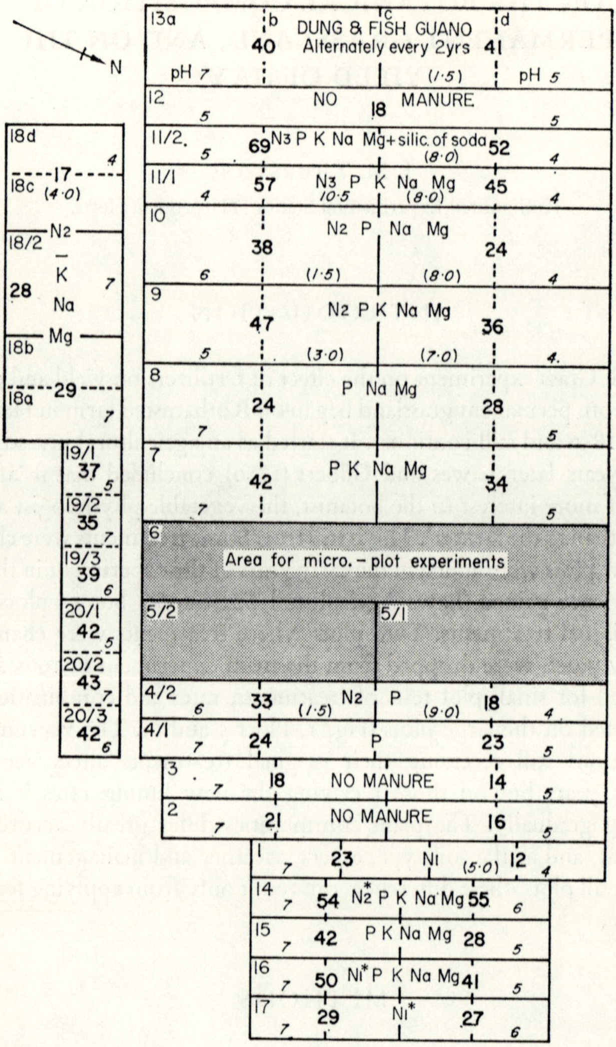

FIG. 1. Diagram of the Park Grass plots. The plots are numbered in the top left-hand corner. Numbers in bold type = mean annual yield of dry matter cwt/acre (1952–59). Numbers in italic type = pH at present. Numbers in brackets = ground chalk tons/acre applied to 'b' and 'c' subplots 1965–68.

Inorganic treatments

Nitrogen (applied annually in spring)

N_1, N_2, N_3 ammonium sulphate supplying 43, 86, 129 lb N.

plots differ in shape and are from 0·5–0·12 acres. Mean yields of hay from the first cut in cwt/acre are marked in bold figures on half-plots. Approximate pH values are shown by small italic numbers. Plots 1–13 are the oldest. Plots 14–17 were added to compare the effect of nitrogen as sodium nitrate with ammonium sulphate used on the older plots, and plots 18–20 compare unlimed and two different liming-rates for three treatments not represented in plots 1–17. The shape of the field and the presence of trees determined the lay-out of the experiment.

There are two completely unmanured plots at opposite ends of the experiment (plots 3 and 12) but other treatments are neither replicated nor randomized. The grass is cut in June every year and made into hay on the plot on which it grew. At first the plots were grazed by sheep after the hay was carted, but since 1873 a second cut has been taken in autumn, recently by forage-harvester, and the field is never grazed now.

Liming of half-plots began in 1903 because of acidity developed on plots receiving ammonium sulphate, but no major changes were made between 1903 and 1964. The liming scheme was revised in 1965, and five of the most acid plots are now divided into four sub-sections, one unlimed and three with different amounts of lime, aimed at producing four pH values for each fertilizer treatment. Three others have light liming on half of the old unlimed end.

FIG. 1 caption continued

N_1*, N_2* sodium nitrate supplying 43, 86 lb N.
(86 lb. N is equivalent to 412 lb ammonium sulphate or 550 lb sodium nitrate).

Minerals (*applied annually in winter*)

P 363 lb superphosphate supplying 30 lb P.
K 500 lb potassium sulphate supplying 200 lb K.
Na 100 lb sodium sulphate supplying 14 lb Na.
Mg 100 lb magnesium sulphate supplying 10 lb Mg.
Sodium silicate 400 lb of water-soluble powder.

Lime

1903–1964
2000 lb/acre to appropriate half-plots every fourth year.
From 1965
Will be applied as necessary to obtain and maintain pH values of approximately 4, 5, 6 and 7 on quarter-plots of the more acid treatments. Amounts applied in 1965–8 on these sub-plots are shown in brackets.

Organic treatments (applied every 4th year)

Dung 14 tons farmyard manure (bullocks).
Fish meal to supply 86 lb N (about 6 cwt meal).

Fertilizer treatments are in three main groups: (1) No nitrogen, (2) Nitrogen as ammonium sulphate, (3) Nitrogen as sodium nitrate. Within groups there are plots with and without phosphorus (given as superphosphate) and with or without potassium, sodium and magnesium sulphates. Three plots have dung every 4th year; one has sodium nitrate, phosphorus and potassium in the other 3 years and another has fish meal once in 4 years alternating with dung. These treatments with partly-rotted dung carted out from the cattle-yards and spread as evenly as possible are not the same as dung left by grazing animals, and are neither associated with grazing nor with the deposition of fresh urine.

The herbage on the plots began to differ in botanical composition in the second year of treatment. Differences increased with time but eventually the plant communities stabilized and altered very little between 1919 (Brenchley 1924) and 1948 (Brenchley and Warington 1958). By 1903 the soil on plots receiving ammonium sulphate was very acid so half of each plot was limed then and every following 4th year. After 112 years the boundaries between contrasting plant communities on adjacent plots are straight and distinct, confirming that treatments have been applied accurately and that neither soil nor fertilizers have moved between plots.

Hay yields are recorded annually for all plots and visual botanical surveys are made twice yearly, before cutting. Botanical analyses of hay samples (Lawes, Gilbert and Masters 1882; Brenchley 1924; Brenchley and Warington 1958) and chemical analyses of produce and soil (Lawes and Gilbert 1900; Warren and Johnston 1964) have been made at intervals and soil pH has also been studied.

RESULTS

Only selected results and main conclusions can be given here. Botanical and chemical details and yields are published in the references cited.

The effect of chemical treatments on floristic composition and yield

The unmanured plots probably resemble the original vegetation. They have about 60 species of higher plants (Brenchley and Warington 1958). Every plant found on the other plots also occurs on the unmanured ones and some grow only on the unmanured. Fertilizers and liming encourage some species and discourage others, but do not cause invasion by species not previously present. Gaps in the sward occur only on very acid plots after severe winters. There is no dominant species on the unmanured plots. Gramineae, Leguminosae and other families are all present, including some plants characteristic of poor land, e.g. *Briza media*, *Primula veris*. The vegetation is short and hay yield is small. Some plants rare in the locality occur here, e.g. *Dactylorchis fuchsii*, *Listera ovata* and the small fern *Ophioglossum vulgatum*. *Dactylorchis* was recorded for the first time in 1963 and flowered again in 1964. It does not occur elsewhere on Rothamsted farm and may have been there from the beginning, possibly not flowering.

The addition of phosphorus, potassium, sodium and magnesium but no nitrogen encourages the Leguminosae (*Lathyrus pratensis*, *Trifolium pratense* and *T. repens*, *Lotus corniculatus*) but the vegetation is still short and contains many species. The yield is double that of an unmanured plot, and liming increases it further.

Annual applications of ammonium sulphate without other fertilizers or lime eliminate all the Leguminosae and all except the most acid-tolerant grasses and weeds. The vegetation is chiefly *Agrostic tenuis*, with some *Festuca rubra* and occasional plants of *Rumex acetosa* and *Potentilla reptans*. The very acid plots develop a layer of peat above the soil. The grasses tend to root in this and are then easily killed by frost or drought, especially where vigorous harrowing has pulled the peat away from the underlying soil. The mean yield is less than an unmanured, unlimed plot.

The addition of phosphorus as well as ammonium sulphate but without lime, produces a sward very similar to ammonium sulphate alone, but with some *Anthoxanthum odoratum* and less *R. acetosa*.

Where potassium is given in addition to phosphorus, sodium and magnesium, plus the highest rate of ammonium sulphate, but where acidity is not corrected by liming, *Holcus lanatus* is dominant, with a mat of undecayed vegetation below it. The shorter grasses have disappeared. After a severe winter, the *Holcus* dies in patches, but when a mild winter is followed by a moist spring, the soil is completely covered and no other species is present. A few plants of *Chamaenerion angustifolium* may invade the bare areas, but do not persist, being discouraged by cutting and ousted by *Holcus* regenerated from seed. Liming overcomes the mat-formation and encourages agriculturally-useful grasses, especially *Alopecurus pratensis*, and also some less desirable species, e.g. *Taraxacum officinale*, but Leguminosae cannot survive under the dense cover of tall grasses. This plot, and its neighbour with sodium silicate in addition, yield the most in the experiment, and produce good-quality hay.

Where nitrogen is applied as sodium nitrate, lime is unnecessary. With complete fertilizers including sodium nitrate, *A. pratensis* and *Dactylis glomerata* flourish and *Arrhenatherum elatius* is also abundant. The hay crop is heavier than on the limed half of a plot receiving complete fertilizers, including the same amount of nitrogen as ammonium sulphate, and there is no peat layer.

The first effect of lime added in 1965 to previously very acid quarter-plots was seen in the spring immediately after application, when the dead vegetation rotted more quickly on the newly-limed areas. Liming also inhibited flowering of *Holcus* right from the start, and this effect showed more clearly in 1967 than in the preceding two years. *Agrostis* and *Anthoxanthum* showed the same effect in 1967. In the spring, the new growth on the very acid quarter-plots is obscured by the yellowish stubble from the previous year's flower-stalks but where flowering was inhibited by the new liming the quarter-plots look green. By 1966 a few plants of *Trifolium pratense* had established on the newly-limed Section C of plot 11–2, which was previously too acid for clover. In 1968 they are still there, and also a few on 11–1 C. In contrast, seedlings of *Taraxacum officinale* which germinated on bare patches of the same section in 1966 did not develop beyond the 2–3 leaf stage, and died during the summer. Occasional seedlings from 1967 have survived into 1968. *Festuca ovina* is increasing on the newly-limed acid areas and in 1967 *Cerastium holosteoides* occurred on nearly all the recently limed areas of acid plots, but not on the unlimed parts. More species were present on the newly-limed areas in 1967 than ever before. Increasing the lime on quarter-plots at the previously limed end has not noticeably affected the vegetation.

Effects on individual species

This experiment indicates tolerances or preferences of individual species for some combinations of fertilizers and soil acidity, but competition between species must also affect the results. Experiments on isolated plants, as described at the Symposium by Goodman (this volume) for *Lolium perenne* and demonstrated by Snaydon for *Anthoxanthum odoratum* plants obtained from the Park Grass plots are necessary to determine the factors involved and their relative importance. Snaydon (unpublished) found that with strains of *Anthoxanthum* differing in height, response to nutrients and susceptibility to disease occurred on plots receiving different fertilizers, and that they preserved these characteristics when grown in pots under uniform conditions. The effect of treatment on some species has already been indicated in discussing the plant communities, e.g. all grasses are encouraged by nitrogen, *Agrostis tenuis* tolerates acidity and also potassium and phosphorus deficiency, *Holcus lanatus* also tolerates acidity but requires phosphate and potash and benefits from much nitrogen, and *Anthoxanthum odoratum* requires phosphorus but tolerates acidity and potassium deficiency. *Alopecurus pratensis* requires complete fertilizers and will not grow on very acid soil. Leguminosae do not need nitrogenous fertilizers but benefit from phosphorus and potassium.

Taraxacum officinale is very abundant only on plots that are less acid than pH 6·0 and receive potassium. Occasional plants occur on no-potassium or moderately acid plots, but it is not known whether these individuals are more tolerant than the species as a whole, or whether they are in spots where soil conditions are more favourable to them than on the rest of the plot.

Rumex acetosa tolerates phosphorus deficiency in very acid soil. After a drought or severe frost has killed most of the *Agrostis* on such a plot, *R. acetosa* may be the only living plant. Probably depth of rooting is responsible for the difference between *Agrostis* and *Rumex*.

Effects on the distribution of fauna

Changes in flora, caused by treatments, affect the fauna of the plots. Acid soil is unfavourable to earthworms, as may be seen from the distribution of worm-casts. When moles invade the field they throw up mole-hills only on the plots where earthworms are abundant. Swallows tend to fly mostly over plots where Umbelliferae are flowering abundantly, because the inflorescences attract flies.

REFERENCES

Brenchley W.E. (1924). Manuring of grassland for hay. *The Rothamsted Monographs on Agricultural Science*. London.

Brenchley W.E. (revised by Warington K.) (1958). *The Park Grass plots at Rothamsted* 1856–1949. Rothamsted Experimental Station, Harpenden, Herts.

Lawes Agricultural Trust. (1966). *Details of the classical and long-term experiments up to* 1962. Rothamsted Experimental Station, Harpenden, Herts.

Lawes J.B. and Gilbert J.H. (1880). Agricultural, botanical and chemical results of experiments on the mixed herbage of permanent grassland, conducted for more than twenty years in succession on the same land. Part I. The agricultural results. *Phil. Trans. A & B.* **171**, 289–416.

Lawes J.B. and Gilbert J.H. (1900). Agricultural, botanical and chemical results of experiments on the mixed herbage of permanent grassland, conducted for many years in succession on the same land. Part III. The chemical results. *Phil. Trans. B.* **192**, 139–210.

LAWES J.B., GILBERT J.H. and MASTERS M.T. (1882). Agricultural, botanical and chemical results of experiments on the mixed herbage of permanent grassland, conducted for more than twenty years in succession on the same land. Part II. The botanical results. *Phil. Trans. A & B.* **173**, 1181–1413.

WARREN R.G. and JOHNSTON A.E. (1964). *The Park Grass Experiment.* Rep. Rothamsted exp. Stn for 1963, 240–262.

See also yields and occasional botanical or chemical notes in Rothamsted Experimental Station Annual Report, and Numerical Results of Experiments, both published annually.

Note: The Latin names of species in this paper are those used in Clapham, Tutin and Warburg (1962) *Flora of the British Isles* (2nd edition) Cambridge University Press and are not always the same as those used in the papers and publications listed above.

DISTRIBUTION OF PASTURE PLANTS IN RELATION TO CHEMICAL PROPERTIES OF THE SOIL

J. P. VAN DEN BERGH

Institute for Biological and Chemical Research on Field Crops and Herbage, Wageningen, the Netherlands

INTRODUCTION

Certain plant species are to be found more abundantly on soils with certain chemical properties than on other soils. It has been shown that plant species and even strains of the same species may differ in their response to particular mineral levels. In extreme situations the causal relationship between the specific response to certain elements and plant distribution appears obvious. In most of these cases some species only are persistent in a particular environment, while the other species, even in monoculture, are not persistent and may frequently show toxicity symptoms.

In less extreme environments almost all species may grow in monoculture, but in mixed stands a great number will disappear because they are replaced by better adapted ones. It is often supposed that the vigour or competitive ability of the species is more or less proportional to their productivity and that the effect of a treatment on the result of competition may be deduced from the effect of this treatment on the yields of monocultures. In other cases, however, the least productive species gains. This phenomenon is called the Montgomery effect (Montgomery 1912; Gustafsson 1951). It has been emphasized by Bradshaw, Chadwick, Jowett and Snaydon (1964) that under conditions of low environmental potential, low yield may be of selective advantage.

Besides vegetative reproduction, reproduction by seed is very common, even in perennials. In particular the soil environment may affect germination and seedling establishment. As it is well known that germination is very sensitive to pH special attention should be paid to this aspect.

The results of some experiments with perennial grasses in monoculture and in mixtures will be discussed and special attention will be paid to conditions in which growth is limited by the mineral supply.

MATHEMATICAL TREATMENT AND EXPERIMENTAL PROCEDURE

Studying competition between perennial grass species de Wit and van den Bergh (1965) introduced the relative yield total (RYT), which is the sum of the relative yields of the components of the mixture and in many cases about equal to 1:

$$\mathrm{RYT} = r_a + r_b = O_a/M_a + O_b/M_b = 1 \tag{1}$$

O and M represent the yield of a component in the mixture and monoculture resp. Furthermore the relative replacement rate (ρ) of species a with respect to species b at the nth harvest with respect to the mth harvest was defined by (de Wit and van den Bergh 1965):

$$^{nm}\rho_{ab} = \frac{^{n}r_a/^{m}r_a}{^{n}r_b/^{m}r_b} \tag{2}$$

From equations (1) and (2) it follows that the relative yields at the nth harvest are equal to:

$$^{n}r_a = \frac{^{nm}\rho_{ab}\ ^{m}r_a}{^{nm}\rho_{ab}\ ^{m}r_a + ^{m}r_b} \quad \text{and} \quad ^{n}r_b = \frac{^{m}r_b}{^{nm}\rho_{ab}\ ^{m}r_a + ^{m}r_b} \tag{3}$$

During the experiment RYT of the successive cuts may deviate somewhat from 1. To examine whether this deviation is systematic or not, it is preferred to substitute k_{ab} for ρ_{ab} and k_{ba} for ρ_{ab}^{-1}:

$$^{n}r_a = \frac{^{nm}k_{ab}\ ^{m}r_a}{^{nm}k_{ab}\ ^{m}r_a + ^{m}r_b} \quad \text{and} \quad ^{n}r_b = \frac{^{nm}k_{ba}\ ^{m}r_b}{^{nm}k_{ba}\ ^{m}r_b + ^{m}r_a} \tag{4}$$

in which k_{ab} and k_{ba} are the relative crowding coefficients of species a with respect to species b and of species b with respect to species a respectively (de Wit 1960). When k_{ab} is the reciprocal value of k_{ba}, RYT will be equal to 1 and $k_{ab} \equiv \rho_{ab}$; in this case it is said that the species exclude each other.

If the product of the k-values deviates systematically from 1, the species do not exclude each other and are mutually stimulating (i.e. $k_{ab} \times k_{ba} > 1$ and RYT>1; see de Wit 1960; Bakhuis and Kleter 1965; de Wit, Tow and Ennik 1966) or conversely mutually suppressing (i.e. $k_{ab} \times k_{ba} < 1$ and RYT<1; see de Wit 1960; van den Bergh and Elberse 1962).

Figure 1 shows the results of a pot experiment in which tillers of *Lolium perenne* L. (perennial ryegrass) and *Anthoxanthum odoratum* L. (sweet vernal grass) were planted in four different ratios and in monoculture according to the replacement principle (de Wit 1960). Pots with a 20-cm diameter and a volume of 8 kg soil were used. To prevent the crop from over-hanging the pot, PVC-cylinders (8 cm high) were placed on the pots, one on top of the other as growth proceeded. The plants were cut every 3 weeks at a height of 5 cm and the dry-matter yields of the separate species were

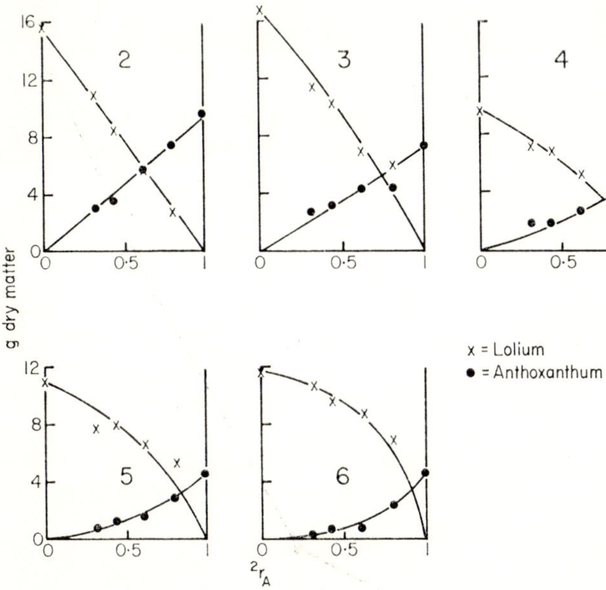

FIG. 1. Replacement diagrams of a competition experiment with *Lolium perenne* (L) and *Anthoxanthum odoratum* (A). The relative yields of *Anthoxanthum* at the 2nd harvest (^2r_A) on the horizontal axis and the yields of 5 successive harvests on the vertical axis.

determined. The regrowth of five successive periods are plotted against the relative yields of *Anthoxanthum odoratum* at the second harvest (replacement diagrams). The curves are drawn according to equation (3). The observations are not at variance with these curves, so ρ is independent of the composition of the mixture and RYT = 1.

The curves indicate the degree to which *A. odoratum* (A) is replaced by *Lolium perenne* (L). This process is better expressed by plotting the ρ-values of the successive cuts on a logarithmic scale against time (Fig. 2). The angle of the line (de Wit and van den Bergh 1965) with the horizontal is a measure of the rate at which *L. perenne* replaces *A. odoratum*. The logarithmic scale is used to enable comparison between the replacement rates of mixtures from various treatments; parallel lines mean the same replacement rate.

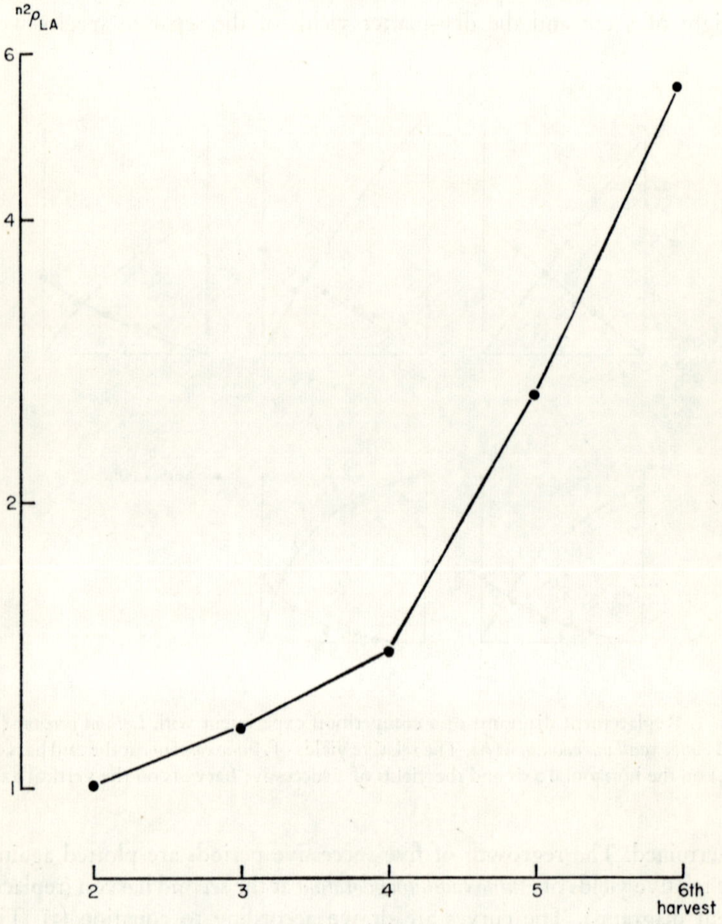

FIG. 2. Relative replacement rates with respect to the 2nd harvest presented as a course line for *Lolium* (L) with respect to *Anthoxanthum* (A) during 5 successive harvests.

NKP-SUPPLY AND pH OF THE SOIL

In plant ecology the pH of the soil is often considered an important factor in determining the botanical composition of vegetation. However, in nature the pH is closely bound to many other factors which affect growth (fertility, soil temperature and structure etc.), so that the causal relationship between the pH and the vegetation, if any, is not clearly understood.

To examine this phenomenon more closely, replacement series of *Agrostis tenuis* Sibth. (common bent) and *Dactylis glomerata* L. (cocksfoot) were sown in pots filled with a very poor and acid sandy soil (pH=4·2).

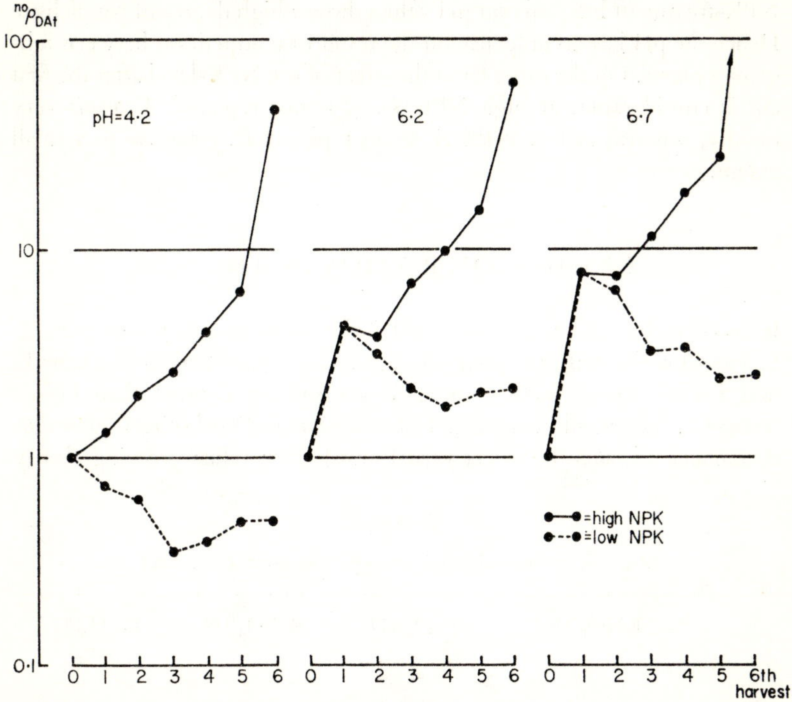

FIG. 3. Relative replacement rates of *Dactylis glomerata* (D) with respect to *Agrostis tenuis* (At) at 3 pH's and 2 NKP-levels during 6 successive harvests. The relative replacement rates are calculated with respect to the seed frequencies at the beginning of the experiment (harvest 0).

By adding 28 or 100 g $CaCO_3$ per pot, pH values of 6·2 and 6·7 were obtained. At all pH values a high or low NPK-level was maintained by adding 20 cc 1 N KNO_3, 5 cc 1 N $MgSO_4$ and 5 cc 0·5 M $NH_4H_2PO_4$ (high) or a quarter of these quantities (low) after each 4-weekly cut.

To show the effect of the treatments on seedling establishment in Fig. 3 the relative seed frequencies ('harvest 0') instead of the relative yields of an arbitrary harvest are used to compare the yields of the successive harvests. The slope of the line from seeding to the first harvest increases considerably with increasing pH. This means that at the lowest pH the weight percentage of *A. tenuis* in the mixture is much higher than at the higher pH values. This is mainly caused by poor tillering of the seedlings of *D. glomerata* at the low pH.

From the first cut onwards, however, the lines representing the same NPK-treatment but different pH values show a high degree of parallelism. Hence the pH has no influence on the result of competition between full-grown plants. On the other hand the effect of the NPK-level after the first cut is considerable: at high NPK *D. glomerata* replaced *A. tenuis* very rapidly, whereas at low NPK *A. tenuis* replaced *D. glomerata* to a small extent.

VARIOUS ION RATIOS IN THE SOIL

It has been observed by de Vries and Dirven (1967) that *Alopecurus pratensis* L. (meadow fox-tail), *Poa pratensis* L. (meadow-grass) and *Festuca rubra* L. (red fescue) always have a much lower sodium content than *Lolium perenne* L. (perennial rye-grass), *Holcus lanatus* L. (Yorkshire fog) or *Anthoxanthum odoratum* L. (sweet vernal grass), even when growing closely

TABLE I

Minerals (in cm^3 solution) given per pot after the 2nd cut

	KNO_3 1N	$NaNO_3$ 1N	$MgSO_4$ 1N	$Ca(H_2PO_4)$ (mg)
Series I	0	40	10	250
II	14	26	10	250
III	26	14	10	250
IV	40	0	10	250

associated in the field. To examine whether this difference has implications in terms of their competitive abilities, an experiment in pots was carried out with *L. perenne*, *Alopecurus pratensis* and *Anthoxanthum odoratum* at four K/Na-ratios of the nutrient solution added to the soil (Table 1). Cutting treatments were imposed at monthly intervals.

In Fig. 4 the 3rd and 4th cut of *L. perenne* and *Alopecurus pratensis* are presented in replacement diagrams with the 2nd cut as a reference harvest.

The dry-matter production of the monocultures were not affected as long as potassium was supplied; without potassium (Series I) the production of the monoculture of *L. perenne* at the 4th harvest dropped to half the production of the other treatments and *Alopecurus pratensis* to a third. Nevertheless in competition *A. pratensis* replaced *L. perenne* when only sodium was applied and *L. perenne* replaced *A. pratensis* at an increasing rate with an increasing K/Na-ratio (Fig. 5).

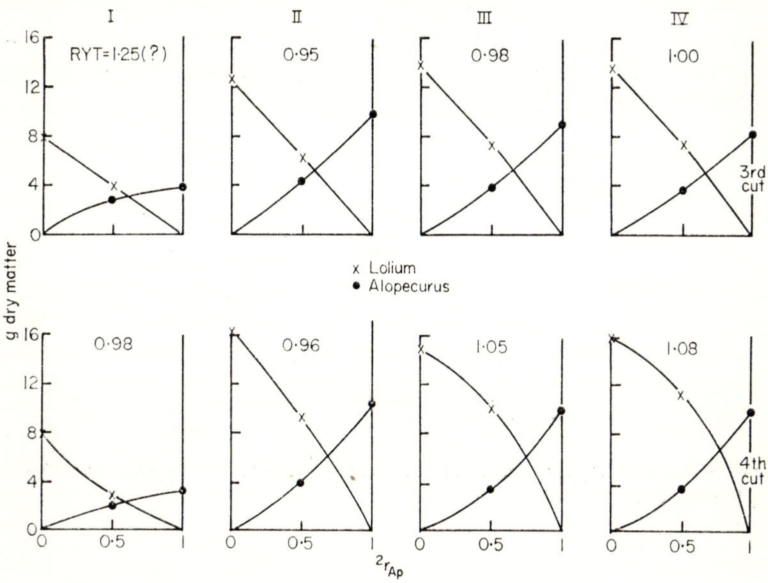

FIG. 4. Replacement diagrams of 2 successive harvests of a competition experiment with *Lolium* (L) and *Alopecurus pratensis* (Ap) at 4 K/Na-ratios (see Table 1). The observations concern averages of 4 pots.

The chemical composition of the species in the mixture at the 3rd cut is presented in Fig. 6. Whereas in *L. perenne* potassium is readily replaced by sodium with a decreasing K/Na-ratio of the nutrient solution, in *A. pratensis* the uptake of sodium is negligible as long as potassium is applied. When only sodium is applied, the sodium content of *L. perenne* is about 5 times as high as the sodium content of *A. pratensis*.

In the experiment with *L. perenne* and *Anthoxanthum odoratum* the result of competition was not affected by the K/Na-ratio, in spite of the specific responses of the monocultures to the low K/Na-ratio (Table 2): the growth

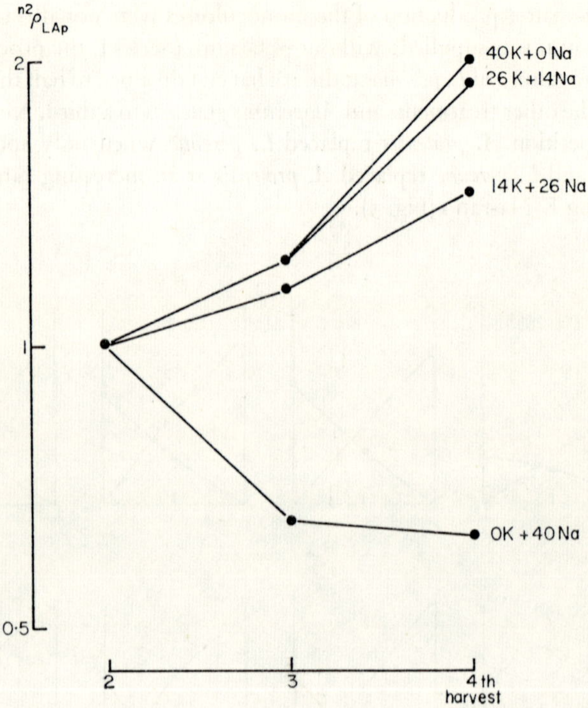

FIG. 5. Relative replacement rates of *Lolium* (L) with respect to *Alopecurus* (Ap) at 4 K/Na-ratios.

TABLE 2

Dry-matter yields of the monocultures of *Lolium* (*L*) and *Anthoxanthum* (*A*) at the 3rd and 4th cut with the (0 K+40 Na)-treatment expressed in percentages of the yields with the (40 K+0 Na)-treatment

Cut	3		4	
Species	L	A	L	A
0 K+40 Na	77	63	56	44

of *A. odoratum* being more depressed than *L. perenne*. Figure 7 shows that both species accumulated large amounts of sodium in the absence of potassium in the nutrient solution.

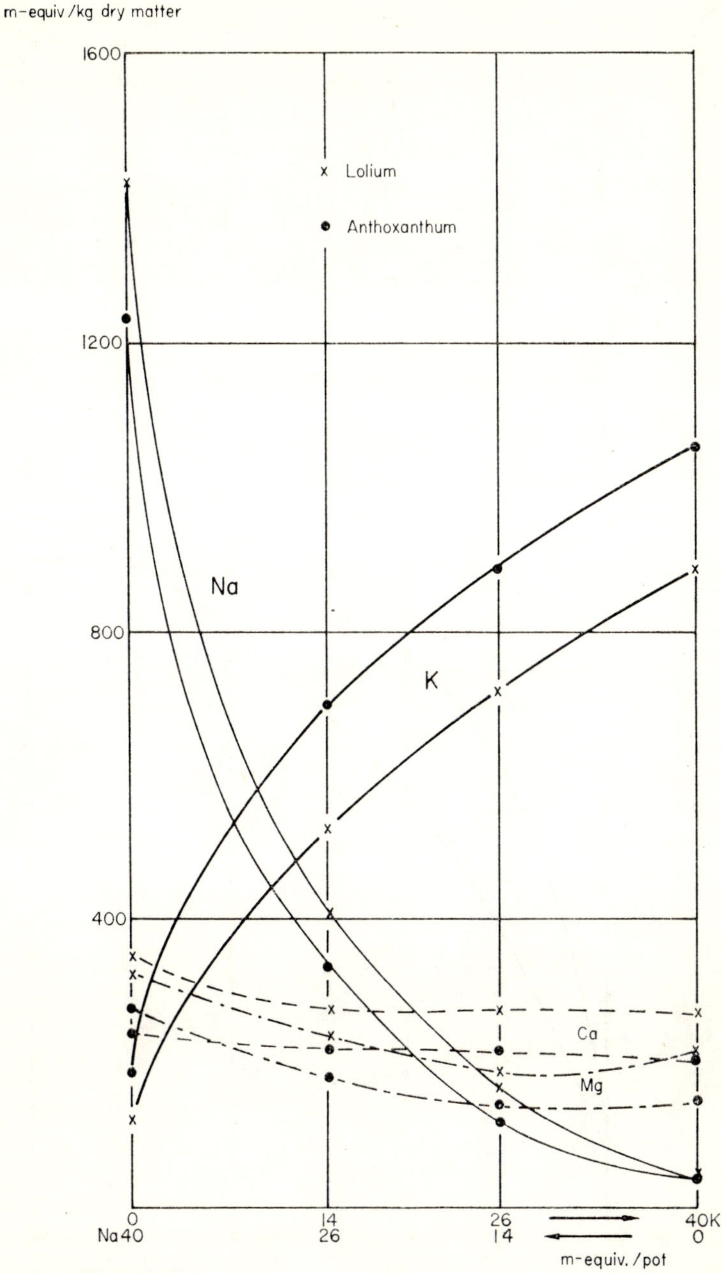

FIG. 6. Cation contents in *Lolium* and *Alopecurus* of the 3rd cut of the mixtures at a K-Na replacement series.

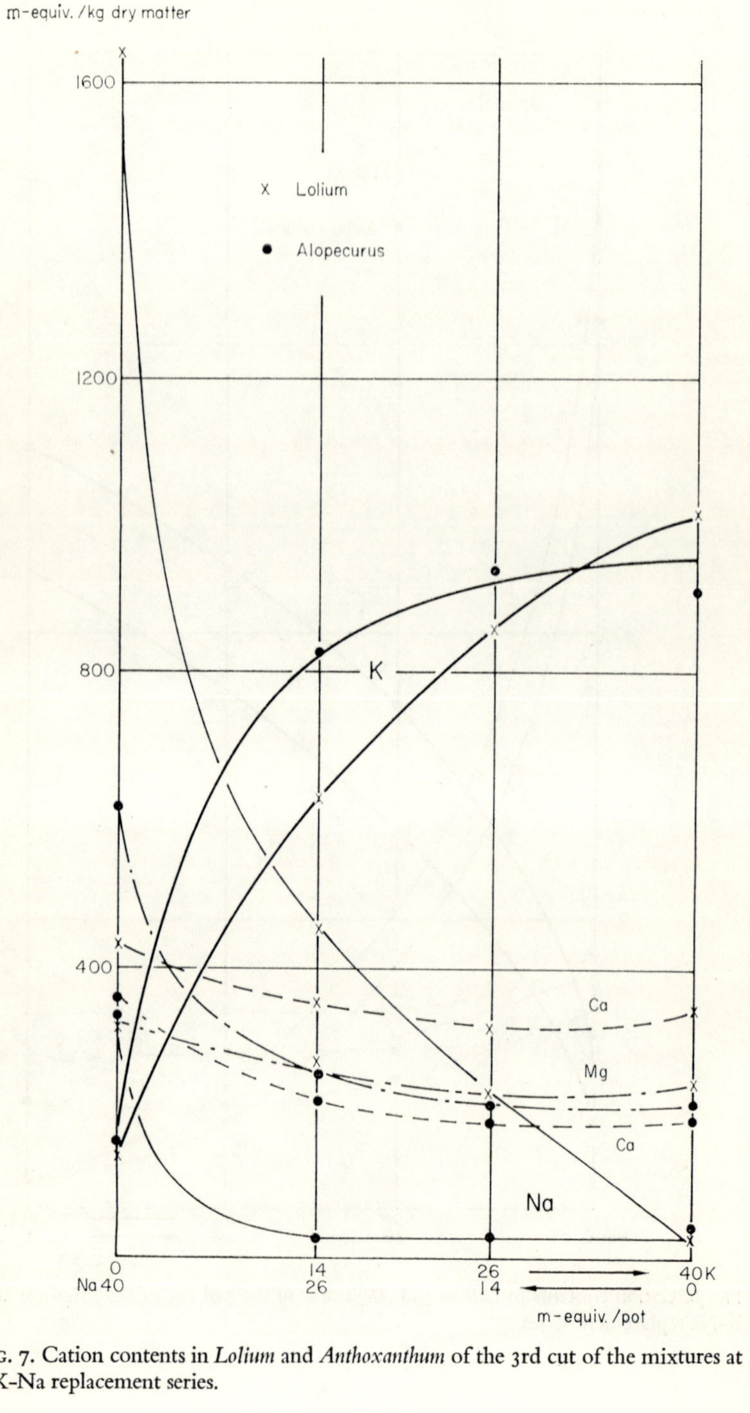

FIG. 7. Cation contents in *Lolium* and *Anthoxanthum* of the 3rd cut of the mixtures at a K-Na replacement series.

DISCUSSION

The great effect of the pH during seedling establishment and the absence of a later effect on the result of competition, was also observed in an experiment on plots ($1\cdot5 \times 0\cdot8$ m^2) carried out by de Vries (1940). In this experiment 11 species were sown in monoculture and in different mixtures on an acid, sandy soil (pH 4·6) of various pH values obtained by liming. The crop was harvested 3 times a year for 3 years.

In the first cut of the first year the composition of the mixture of *L. perenne* and *Agrostis tenuis* varied from 30 W (weight) % *A. tenuis* and 50 W% *L. perenne* at pH 4·6 to 1 W% of *A. tenuis* and 90 W% *L. perenne* at pH 6·8, but after this period of establishment *L. perenne* was replaced by *A. tenuis* at all pH values at the same rate (van den Bergh 1968).

Comparing the effect of the pH on the composition of the other mixtures, it is found that this effect is greatest with the mixture *A. tenuis–L. perenne* and decreases, if *A. tenuis* is subsequently associated with *Poa pratensis, Festuca rubra, Holcus lanatus* and *Anthoxanthum odoratum*. According to Kruijne and de Vries (1963) these species indicate lower pH values in the following order: *L. perenne, P. pratensis, F. rubra, H. lanatus, Anthoxanthum odoratum* and *Agrostis tenuis*. The perfect agreement of these sequences suggests that the interaction between the pH and the botanical composition in the field arises from the effect of the pH on the propagation by seed. This is also in agreement with the experience that liming permanent grasslands shows little effect on the botanical composition, whereas liming before seeding may improve the botanical composition of a ley considerably.

The opposite effect of the K/Na-ratio of the added nutrient solution on the monocultures of *L. perenne* and *Alopecurus pratensis* and on the result of competition may be explained by assuming that in the mixture the sodium excess is removed by ryegrass in favour of *A. pratensis*. In that case, however, the sodium content of *A. pratensis* in monoculture should be higher than that in the mixture. Since these contents appeared to be equal, this supposition is not plausible.

On the other hand, it is more often observed (van den Bergh 1968) that a high yielding species is less competitive in a mixture with a low yielding species under sub-optimal conditions. Looking for an explanation of this phenomenon, two factors seem to be of main importance.

Firstly, under optimal conditions after each cut, a closed crop surface is formed within a short time, so that light will limit photosynthesis in the lower layers of the canopy. In competition the growth of the low yielding species with its shorter tillers is relatively more retarded than the high yielding species with a relatively greater part of its leaf surface in the upper layers of the canopy. Under sub-optimal conditions the total leaf surface will be reduced, which results in an increased penetration of the light into the lower layers of the canopy. This will be more profitable to the low yielding species than to the high yielding one. This phenomenon is recognized also by Antonovics, Lovett and Bradshaw (1967) when discussing response curves of the same pattern but with differences in yield at optimum. They stated that for example in exposed conditions selection for small plants may occur.

Secondly, according to Curtis and Clark (1950) and Brouwer (1966) an increase in light intensity is likely to decrease the shoot/root ratio. In view of the foregoing, under sub-optimal conditions the light intensity will increase in the lower layers of the canopy. Hence, in competition for nutrients or water this will be again more advantageous to the low yielding species than to the high yielding one.

In the experiment with *L. perenne* and *A. pratensis* the low yielding *A. pratensis* had the advantage over the high yielding *L. perenne* when competing under sub-optimal conditions (low K/Na-ratio). This dominating influence of the production capacity at optimal conditions on the difference between the ρ-values at optimal and sub-optimal conditions makes it impossible to determine the specific influence of a growth limiting factor on competition.

In the experiment with *L. perenne* and *Anthoxanthum odoratum* the yields of the monocultures with the (40 K+0 Na)-treatment were about the same. In spite of specific responses to the low K/Na-ratio, this treatment did not affect the ρ-value. This illustrates again that the specific response to a growth limiting factor is of minor importance compared with the effect resulting from different production capacities of the species at optimal conditions.

The low yielding *Agrostis tenuis* replaced the high yielding *Dactylis glomerata* at low NPK as well as the high yielding *L. perenne* (in the experiment of de Vries (1940) the NPK-supply was rather low). Moreover, with the (0 K+40 Na)-treatment the low yielding *Alopecurus pratensis* replaced the high yielding *L. perenne*. Hence, under sub-optimal conditions the Montgomery effect is exhibited. This is quite comprehensible when it is realized that in going from optimal to sub-optimal conditions, the importance of above-ground competition will decrease in favour of below-ground competition. Assuming that the shoot/root ratio of low yielding species will decrease, at sub-optimal conditions this gives an advantage to the low yielding species.

REFERENCES

ANTONOVICS J., LOVETT J. and BRADSHAW A.D. (1967). The evolution of adaptation to nutritional factors in populations of herbage plants. *Isotopes in plant nutrition and physiology*. International Atomic Energy Agency, Vienna, pp. 549–567.

BAKHUIS J.A. and KLETER H.J. (1965). Some effects of associated growth on grass and clover under field conditions. *Neth. J. Agric. Sci.* **13**, 280–310.

VAN DEN BERGH J.P. (1968). An analysis of yields of grasses in mixed and pure stands. *Agric. Res. Rep. Wageningen*, **714**, 1–71.

VAN DEN BERGH J.P. and ELBERSE W.TH. (1962). Competition between *Lolium perenne* L. and *Anthoxanthum odoratum* L. at two levels of phosphate and potash. *J. Ecol.* **50**, 87–95.

BRADSHAW A.D., CHADWICK M.J., JOWETT D. and SNAYDON R.W. (1964). Experimental investigations into the mineral nutrition of several grass species. IV. Nitrogen level. *J. Ecol.* **52**, 665–676.

BROUWER R. (1966). Root growth of grasses and cereals. *The growth of cereals and grasses*. (Ed. F. L. Milthorpe and J.D. Ivins,) pp. 153–166. London.

CURTIS O.F. and CLARK D.G. (1950). *An introduction to Plant Physiology*. New York.

GUSTAFSSON A. (1951). Mutations, environment and evolution. *Cold Spr. Harbour Symp. on Quant. Biol.* **XVI**, 263–280.

KRUIJNE A.A. and DE VRIES D.M. (1963). Data concerning important herbage plants. *Meded. I.B.S. Wageningen*, **225**, 1–45.

MONTGOMERY E.G. (1912). Competition in cereals. *Bull. Nebr. Agric. Exp. Sta.* **XXVI**, art. V, 1–22.

DE VRIES D.M. (1940). Een oriënterende proef over den invloed van bekalking van zuren zandgrond op opbrengstvermogen en concurrentieverloop van eenige grassoorten en witte klaver. *Landbouwk. Tijdschr.* **52**, 8–17.

DE VRIES D.M. and DIRVEN J.G.P. (1967). Welke samenstelling van de zode is voor blijvend grasland wenselijk? *Kali* **71**, 3–8.

DE WIT C.T. (1960). On competition. *Versl. landbouwk. Onderz. The Hague* **66**(8), 1–82.

DE WIT C.T. and VAN DEN BERGH J.P. (1965). Competition between herbage plants. *Neth. J. Agric. Sci.* **13**, 212–221.

DE WIT C.T., TOW G.P. and ENNIK G.C. (1966). Competition between legumes and grasses. *Agric. Res. Rep. Wageningen*, **687**, 1–30.

INFLUENCE OF MINERAL NUTRITION ON THE ZONATION OF FLOWERING PLANTS IN COASTAL SALT-MARSHES

C. D. Pigott

Department of Biological Sciences, University of Lancaster

Salt-marshes develop along sheltered coasts on land which still lies within the tidal limits of the sea and their vegetation is composed entirely of plants which can grow on moderately saline soils and withstand immersion of their shoots in sea-water. Such plants are known as halophytes and as a group they differ from most plants of non-saline habitats which can usually neither germinate nor grow on saline soils and are injured or killed by immersion in sea-water.

It has been suggested by Walter (1961) that the special features of halophytes are that when grown on saline soils they can develop high osmotic pressures in their tissues and this is possible because they tolerate high concentrations of ions, including sodium and chloride, in their tissues. Similar concentrations of sodium in non-halophytic plants are apparently toxic and there is some evidence that an important effect of excess sodium is that it depresses uptake of potassium and causes severe potassium deficiency. This aspect of the nutrition of halophytes is discussed by Professor Epstein in a later paper in this symposium.

Moreover, although a high concentration of sodium chloride is not essential for the healthy growth of halophytes, and many and possibly all can be grown on non-saline soils, some species such as *Salicornia herbacea* and *Suaeda maritima* grow larger and clearly benefit from addition of sodium chloride in quantities far in excess of the minute amounts of sodium and chloride probably essential for the growth of all plants (Broyer, Carlton, Johnson and Stout 1954; Brownell 1965). This response to sodium is, of course, well known for *Beta vulgaris* (Wallace 1961) and is perhaps particularly characteristic of members of the Chenopodiaceae.

Although these general aspects of the mineral nutrition of halophytes have been studied experimentally, this approach to the problem of distribution of particular species within salt-marshes has been largely neglected. Within salt-marshes there is a conspicuous vertical zonation of species which is closely related to the levels of the sea during the tidal cycle. The upper limit of true salt-marsh vegetation coincides with the upper limit of high water of spring tides so that the whole marsh is then flooded. The lower limit of the marsh at least in the absence of *Spartina*, is a little above the upper limit of high water of neap tides so that the whole marsh is then exposed even at high water for one or two days in each fortnight.

The composition and structure of the vegetation in the zones of British marshes is greatly influenced by grazing, particularly by sheep. Heavily grazed marshes in the estuary of the Dovey have been described by Yapp, Johns and Jones (1917) and lightly grazed or ungrazed marshes on Scolt Head Island in Norfolk have been described by Chapman (1938, 1939). The principal species in the corresponding zones are shown in Table 1.

TABLE 1

Zonation of principal species in British salt-marshes

Level	Not grazed	Grazed
High water springs	*Juncus maritimus* and *Festuca rubra*	*Juncus maritimus* and *Festuca rubra*
	Armeria maritima and *Limonium vulgare*	*Festuca rubra*, *Armeria maritima* and *Glaux maritima*
	Halimione portulacoides and *Aster tripolium*	*Puccinellia maritima*
High water neaps	*Salicornia herbacea* agg.	*Salicornia herbacea* agg.

Most attempts to discover the underlying causes of zonation have been essentially analytical and because each species has been found to occupy a characteristic vertical range in relation to sea level, an explanation in terms of sensitivity to the period for which the species are submerged or exposed is attractive (Hinde 1954; Adams 1963). Except *Zostera*, most flowering-plants of salt-marsh have thick, often succulent photosynthetic organs with numerous stomata so that they are clearly dependent on gaseous carbon dioxide for photosynthesis, and the rate of diffusion to the leaves will be severely reduced during submergence. No measurements of the actual reduction in growth as a result of submergence during daylight appear to have been made, and it is not clear why species should differ in their sensitivity to this effect unless, for example, succulent species like *Salicornia* and *Aster* were capable of acid-metabolism allowing them to fix CO_2 while exposed and continue photosynthesis when submerged in shallow water.

During periods of exposure salt-marshes are affected both by evaporation and rain. The upper marsh is exposed at all times except the relatively short periods of high water of spring tides, but the lower marsh is only exposed for any period longer than a few hours at neap tides and even then the water-table remains high. This produces a variation in the water regime which is probably of great importance in determining the vertical distribution of species, but there is still insufficient experimental evidence. Measurements at Scolt Head during dry sunny weather at neap tides have demonstrated that a steep gradient in water-stress may develop down the marsh, and the leaves of *Limonium vulgare*, for example, were wilted and beginning to wither at the top of the marsh and were turgid and undamaged at lower levels.

This pattern of variation in water content of the soil produces gradients in aeration and salinity. Measurements made at Scolt Head show that impedance of aeration, when the soil pores remain water-filled, allows oxidation-reduction potentials below 300 mV to persist quite close to the surface of the mud in the lower marsh but the upper marsh values are normally above 400 mV. Evidence that this sort of difference may affect plant distribution has been obtained by Dr M.A. Zahran at Lancaster in an experimental study of *Festuca rubra* and *Puccinellia maritima*. Both species respond similarly to variation in salinity and nitrogen supply, but when grown in water-logged soils in pots and in the field *Puccinellia* continues to grow healthily, while *Festuca* produces stunted roots which turn dark and the leaves fail to expand and become chlorotic. These are symptoms characteristic of poisoning by ferrous iron but this has not been demonstrated.

Variation in salinity of the soil arises from the same interplay of tidal flooding, evaporation and rainfall. The open sea has an osmotic pressure of 24 atmospheres and a salt content of 33–38 g/L consisting of approximately 11 g/L (480 m.e./L) of sodium, 20 g/L (560 m.e./L) of chloride, 1·3 g/L (107 m.e./L) of magnesium and 2·8 g/L (63 m.e./L) of sulphate (Harvey 1963). The concentrations of soluble inorganic nitrogen compounds and phosphate are both very low during summer ($<0{\cdot}1$ μg-atoms P and $<1{\cdot}0$ μg-atoms N per litre). In coastal waters and estuaries some dilution occurs but nitrogen and phosphorus may be increased by pollution. The salinity of water pressed out of fresh samples of the top 5 cm of soil from different levels in the salt marsh, generally show concentrations of sodium less than half those in sea water (D.W. Jeffrey, unpublished). This shows that the soil is not in equilibrium with sea-water and implies that diffusion of salts or mass flow of sea-water at high water into the horizons where most roots occur is inadequate to mask the influence of rain-water draining down the marsh. Although the gradient of salinity is usually from high salinities in the lower marsh to low salinities in the upper, the upper marsh is subject to great variability. The effect of salinity on salt-marsh plants has been studied experimentally by Montfort and Brandrup (1927) and Chapman (1960) and it has been found that most species will not germinate in sea-water and require considerable dilution. Annual species of *Salicornia* are exceptional and will germinate not only in sea-water but even in more concentrated solutions of sodium chloride. There is some

TABLE 2

Germination of *Triglochin palustre* (P) and *T. maritimum* (M) in distilled water and dilutions of sea-water. Germination expressed as percentage of germination in distilled water after 20 days

		Percentage of sea-water						
Days	Species	0	5	10	25	50	75	100
5	P	53	53	42	31	8	0	0
	M	13	6	19	0	0	0	0
15	P	100	97	83	75	86	75	33
	M	100	88	63	50	6	13	0
20	P	100	97	83	75	86	75	44
	M	100	88	63	56	6	13	19

evidence to suggest that zonation is related to differences in salinity tolerance but even species which are restricted to the upper marsh are very tolerant and may indeed be subject to high salinities during warm dry weather. This has been demonstrated by A.R. Woolhouse in a comparative study of *Triglochin maritimum* and *T. palustre* which on salt-marshes is confined to the upper parts and probably to situations influenced by drainage from the land. These species have a similar sensitivity to salinity at germination (Table 2) but differ during subsequent growth (Table 3).

Although the special problem of salinity tolerance has attracted attention, other aspects of the mineral nutrition of halophytes have been largely neglected but may also be important in relation to distribution within salt marshes. It is well known that the seeds of *Suaeda maritima* and annual species of *Salicornia* are widely dispersed over the whole marsh during the winter, so that seedlings appear in abundance at all levels during the summer. However, the subsequent growth of the seedlings varies greatly. On the bare mud in the lower marsh and along the edges of creeks these seedlings grow rapidly and by August the plants are bushy, 15–20 cm high and green. Meanwhile the seedlings in the upper marsh have grown very little producing unbranched plants which usually do not exceed 2–3 cm in height. As well as being stunted the lower leaves of *Suaeda* and stem segments of *Salicornia* become shrunken and withered while the upper parts turn yellow or are tinged with red.

TABLE 3

Absolute and relative growth rates of *Triglochin palustre* (P) and *T. maritimum* (M) in sand culture with additions of sodium chloride

NaCl (g/L)	Species	Absolute growth rate (mg/plant/day)	Relative growth rate (mg/mg/day)
0	P	0·63	0·13
	M	0·29	0·10
13·5	P	0·01	0·02
	M	0·08	0·07
27·0	P	−0·01	−0·05
	M	0·01	0·01

These symptoms are suggestive of nitrogen and phosphorus deficiency and this possibility has been tested experimentally. At the end of May cores of soil containing seedlings of *S. europaea* and *Suaeda maritima* were collected from the salt-marsh at Holme-next-the-Sea in Norfolk (described by Marsh 1915). The cores were cut to fit into plastic plant pots. Some pots had the bottom sealed and others were left open. On the first occasion they were watered with sea-water and subsequently loss of water was made good with distilled water. Nutrients were added at weekly intervals and the plants were harvested on July 22nd after 7 weeks of growth in a well ventilated greenhouse. The results are presented in Table 4 and the size and form of representative specimens is shown in Fig. 1. The absence of any effect of drainage is almost certainly because the pores in the fine-grained sediment are small and remained water-filled in both treatments. It is clear that growth of both species on the upper marsh soil can be increased by addition of phosphate alone, and phosphate and nitrogen combined, while addition of phosphate to the mud from the lower marsh has no significant effect suggesting the natural supply is adequate. With additional nitrogen and phosphate growth of *S. maritima* on the upper marsh soil approaches that on the lower mud suggesting supplies of other nutrients are not limiting. Although growth is reduced by nutrient-deficiency, plants in the greenhouse do not become red.

When the cores were collected sodium phosphate and ammonium nitrate were also added to plots on the upper marsh. Although only one addition was made, *Salicornia europaea* had by late summer grown larger

TABLE 4

Dry weight of seedlings of *Salicornia europaea* and *Suaeda martima* grown on cores of soil for seven weeks with additions of sodium phosphate and ammonium nitrate (means and standard errors)

	No addition	+Phosphate	+Phosphate and nitrogen
Salicornia europaea			
Upper marsh (free drainage)	8·8 ± 1·5	20.1 ± 3·5	47.7 ± 5.5
Upper marsh (water-logged)	9·2 ± 1·8	15·9 ± 2·0	42·7 ± 7.4
Lower marsh (water-logged)	112·3 ± 10.4	134·6 ± 14·5	—
Suaeda maritima			
Upper marsh (water-logged)	23·1 ± 3·7	30·2 ± 5·1	108·0 ± 10·2
Lower marsh (water-logged)	122·0 ± 11·4	119·6 ± 10·8	—

and greener than plants of comparable age on the untreated control plots. There was, however, a general increase in vigour of the sward on the treated plots and this appeared to be suppressing further growth of *Salicornia.*

This experiment has been repeated with *S. dolichostachya* grown from seed on cores of silty mud from the mud flats and middle marsh at Sunderland Point near Lancaster. This experiment was designed to separate the effects of nitrogen and phosphorus and shows that in a greenhouse the main response is to nitrogen. On the cores from the middle marsh, the seedlings of *Salicornia* scarcely grow without addition of nitrogen, while on the bare mud growth is rapid, with or without addition. There is a significant but smaller response to addition of phosphate alone. In all experiments using cores of mud, the plants already present were not removed but their shoots were cut back to prevent shading of the seedlings.

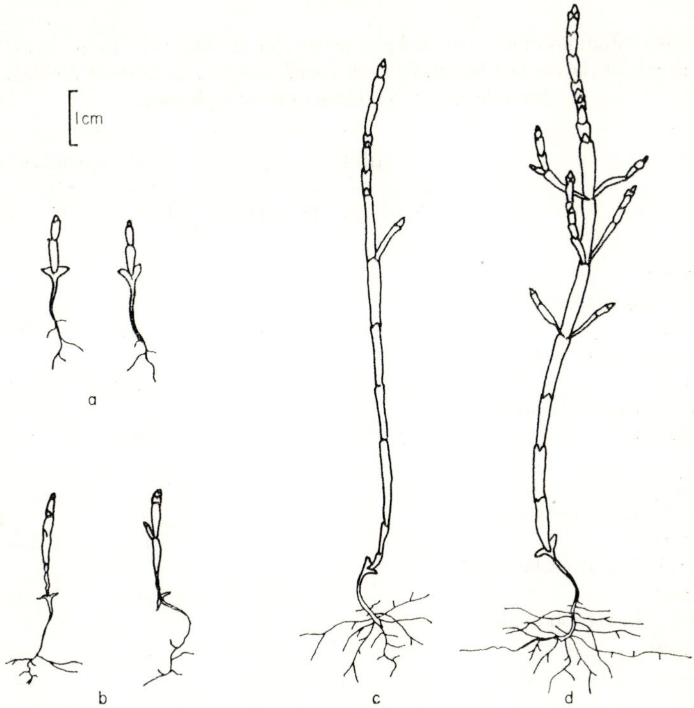

Fig. 1. Plants of *Salicornia europaea*. Seedlings at the start of the experiment (a) and after growth for seven weeks on cores of soil from the upper marsh without addition (b), and with addition of phosphate (c) and nitrogen and phosphate (d).

Analysis of plants of *S. dolichostachya* and *Suaeda maritima* collected from natural habitats shows that the normal concentration in the stems and leaves is high, at least when expressed on a dry weight basis (Table 5). Comparison of the concentrations of nitrogen and phosphorus in large and stunted plants shows consistently lower concentrations in the latter and this suggests deficiency of these nutrients may be restricting growth. As these species are annuals which do not germinate until May, it is clear that rapid growth can only be maintained if there is a plentiful supply of mineral nutrients. Although sea-water contains large amounts of potassium, calcium and magnesium, the concentration of phosphate and nitrogen is extremely low particularly during summer, and the supply of these elements to plant roots must be from sources in the soil.

TABLE 5

Concentrations of nitrogen and phosphorus (in mg/100 g dry tissue) in the shoots of *Salicornia dolichostachya* and *Suaeda maritima* from three different levels in salt-marsh (mean of three replicates)

	14th July 1965		16th September 1965	
	Nitrogen	Phosphorus	Nitrogen	Phosphorus
Salicornia				
Small plants from *Festuca* turf	1280	205	996	181
Small plants from *Puccinellia* turf	—	—	1212	144
Large plants from mud-banks	3503	252	2647	265
Suaeda				
Small plants from *Festuca* turf	—	—	1080	119
Small plants from *Puccinellia* turf	—	—	1176	63
Large plants from mud-banks	—	—	2205	164

The distribution of phosphorus and nitrogen in salt-marshes has recently been studied by D.W. Jeffrey at Sunderland Point and Bank End near Lancaster and at Scolt Head in Norfolk (Table 6). The total amount of both elements in the top 5 cm of the soils is low in these marshes, but there is a well-marked gradient in nitrogen which increases up the marsh. The gradient in phosphorus content is in the same direction at Scolt Head but reversed at Bank End which is heavily grazed.

Both elements presumably enter sediment as algal and animal detritus but fractionation of phosphorus in the soil (Chang and Jackson 1957) shows that in these marshes organic phosphorus is low and the principal form is an apatite mineral which could partly be formed *in situ* and partly be brought in as sediment by the flood tide. Monthly measurements of

TABLE 6

Nitrogen and phosphorus content of top 5 cm of soils from Cockle Bight, Scolt Head, Norfolk (mean of three replicates at A, B and D; one sample only at C)

Site	Total nitrogen (mg/g dry wt.)	Total phosphorus (mg/g dry wt.)	Fractionation of phosphorus (% of total phosphorus)					
			NH_4Cl soluble	calcium phosphates	aluminium phosphate	iron phosphates	reductant soluble	residual (including organic)
A	3136	528	0·4	66·1	1·7	2·6	1·9	27·3
B	577	171	4·5	74·4	2·0	2·7	1·3	15·1
C	528	122	2·2	65·7	3·0	3·6	2·0	23·5
D	376	152	3·8	63·9	4·1	6·8	2·6	18·8

A *Aster tripolium*, *Puccinellia maritima*, *Salicornia europaea* and *Spartina townsendii* on dark sticky mud.
B *Salicornia europaea* on sandy mud.
C *Zostera nana* on gravelly mud.
D Bare sandy mud with Chlorophyceae.

phosphate and nitrogen dissolved in the water pressed out of fresh soil samples show great variability and no significant pattern of differences between zones emerges.

The general similarity in the concentration of soluble nitrogen compounds and phosphate at all levels in salt-marshes suggests that the significant difference between the bare mud and the upper parts of the marsh is the extent to which the sediment is exploited by the roots of the perennial species already present. When nitrogen and phosphate are added, as in the experiments, the supply becomes adequate for both the existing plants and the annuals.

Some perennial plants are also apparently dependent for vigorous growth on the unexploited nutrients in fresh sediment. For example the common large-leaved race (var. *latifolia*) of *Halimione portulacoides* is generally confined to parts of the marsh where deposition of sediment is rapid and particularly to the levees of creeks. As the bases of the shoots become buried by fresh sediment numerous adventitious roots develop. If seedlings or small cuttings of *Halimione* are planted on old sediment from the middle or upper parts of a marsh they grow very slowly and the leaves turn yellow and are shed prematurely. Addition of phosphate produces a slight improvement but when nitrogen or nitrogen and phosphate are added growth is healthy and vigorous.

ACKNOWLEDGEMENTS

I wish to thank Dr D.W. Jeffrey for commenting on the draft of this paper and for his permission to present the analyses in Table 6, Mr A.R. Woolhouse for permission to publish the results of his experiments in Table 2 and 3, and Mr J.R. Wilkinson for technical assistance.

REFERENCES

Adams D.A. (1963). Factors influencing vascular plant zonation in North Carolina salt marshes. *Ecology*, **44**, 445–456.

Brownell P.F. (1965). Sodium as an essential micronutrient element for a higher plant (*Atriplex vesicaria*). *Pl. Physiol.* **40**, 460–468.

Broyer T.C., Carlton A.B., Johnson C.M. and Stout P.R. (1954). Chlorine a micronutrient element for higher plants. *Pl. Physiol.* **29**, 526-532

Chang S.C. and Jackson M.L. (1957). Fractionation of soil phosphorus. *Soil Sci.* **84**, 133–144.

Chapman V.J. (1938). Studies in salt-marsh ecology I–III. *J. Ecol.* **26**, 144–179.

Chapman V.J. (1939). Studies in salt-marsh ecology IV–V. *J. Ecol.* **27**, 160–201.

Chapman V.J. (1960). *Salt marshes and salt deserts of the world.* London.

Harvey H.W. (1963). *The chemistry and fertility of sea waters.* Cambridge.

Hinde H.P. (1954). The vertical distribution of salt marsh phanerogams in relation to tide levels. *Ecol. Monogr.* **24**, 209–225.

Marsh A.S. (1915). The maritime ecology of Holme-next-the-Sea, Norfolk. *J. Ecol.* **3**, 65–93.

Montfort G. and Brandrup W. (1927). Physiologische und pflanzengeographische Seesalzwirkungen 2. Ökologische Studien uber die Keimung und erste Entwicklung bei Halophyten. *Jb. wiss. Bot.* **66**, 902–946.

Wallace T. (1961). *The diagnosis of mineral deficiencies in plants by visual symptoms.* London, H.M.S.O.

Walter H. (1961). The adaptation of plants to saline soils. *Salinity problems in the arid zones* (Arid Zone Research 14). UNESCO, Paris.

Yapp R.H., Johns D. and Jones O.T. (1917). The salt marshes of the Dovey estuary. *J. Ecol.* **5**, 65–103.

THE APPLICATION OF ORDINATION TECHNIQUES

R. GITTINS

Institute of Statistics, North Carolina State University, Raleigh

INTRODUCTION

Study of the causal factors determining the distribution of plants and vegetation is a prime objective of ecology. The number of factors affecting plants, however, is very large and an initial problem in many investigations is to identify those factors which are most likely to be of overriding importance in determining the occurrence of particular species and kinds of vegetation. Ordination is a means of analyzing field observations for the purpose of recognizing such factors. The ultimate object of studies of this kind is to explain ecological behaviour in terms of the physiology of the individual plant. The term ordination was introduced and defined by Goodall (1954a) as, 'an arrangement of units in a uni- or multi-dimensional order'. In the context of vegetation, the units involved are generally either samples of vegetation or plant species. Information characterizing these entities is obtained by observation and measurement in the field. The resulting data form the starting point for ordination. These data usually consist of estimates of the composition of a number of vegetation samples in terms of the abundances of the species represented. Ordination involves associating these entities—either the stands *or* species—in some way such that the arrangement arrived at provides insight into the ecological processes which may have generated the observations. The assumption is made at the outset that a latent structure or organization of some kind exists within the observation set.

Ordination is an exploratory technique of greatest value in the intitial stages of a study. Its use is particularly appropriate where the connections between species' distribution and environmental variation are difficult to discern and crucial experiments difficult to conceive. It is the large number of variables involved and the complexity of their inter-relations which are responsible for these difficulties. In situations of this kind, ordination has proved valuable as a means of creating concepts or hypotheses bearing on the nature and effects of the major controls or sources of variation at work. The aim of analysis is to reveal precisely those relationships between plants and environment which are most likely to repay closer investigation.

Once these relationships have been provisionally identified, their effects in particular cases can be examined by controlled experiment. In this way the measurement foundations for a given ecological situation can gradually be established.

Recent developments in this field stem largely from the work of Goodall (1954b), Curtis (e.g. Bray and Curtis 1957; Curtis, 1959) and Dagnelie (1960, 1965a, b). The name ordination embraces a group of methods rather than a single technique. A number of these methods have been in existence for almost 40 years. It was not until the appearance of the studies mentioned, together with those of Greig-Smith (1957, 1964), however, that a general awareness of the possibilities of the approach spread to ecologists. The essential features of ordination will be illustrated here by reference to principal components analysis.

PRINCIPAL COMPONENTS ANALYSIS

Several excellent accounts of principal components analysis in ecological and more general biological contexts are available (see, for example, Kendall 1957; Reyment 1961, 1963; Seal 1964; van Groenewoud 1965; Orloci 1966, 1967). These, however, all assume at least some familiarity with the procedures of multivariate statistical analysis. An attempt is therefore made here to convey the essential ideas of components analysis in a relatively non-technical way. Recourse will be made to a geometric treatment wherever possible.

(1) *Geometric properties of field observations*

The initial observations in studies of vegetation can frequently be expressed in the form:

		Individuals (stands)				
		1	2	3	...	N
Attributes (species)	1	X_{11}	X_{12}	X_{13}	...	X_{1N}
	2	X_{21}	X_{22}	X_{23}	...	X_{2N}
	3	X_{31}	X_{32}	X_{33}	...	X_{3N}
	.	.	.	.	...	.
	n	X_{n1}	X_{n2}	X_{n3}	...	X_{nN}

where the X_{ji} represent the contributions of n species or other variates ($j=1, 2, \dots n$) to N samples (stands) of vegetation ($i=1, 2, \dots N$). An array of this kind is referred to as a data matrix. Any particular entry X_{ji} constitutes an *observed value*, the subscripts j and i identifying respectively the row and column in which the observation is located. In general, the X_{ji} consist of estimates of the representation of species j in the ith stand, the estimates being in terms of any suitable measure (cover, density, etc.). The *rows* of the matrix, therefore, indicate variation in the representation of the jth species over the N stands examined, while the columns indicate the composition of the ith stand with respect to the n species observed. Attributes† other than species, such, for example, as environmental characteristics of the sites may on occasion also be included in the data matrix.

The principal objectives in the analysis of vegetation are frequently: (i) to describe its composition and structure, indicating perhaps how it might be classified, and (ii) to identify the causes or processes which underlie and determine the behavior of the individual species which collectively form the vegetation. It is with the second of these that we are concerned here. Since the data matrix contains essentially all the information available for analysis, we must endeavor to operate on this in such a way that this objective is realized. We are faced, therefore, with the task of obtaining insight into the nature of the processes underlying differences in species' behavior, by means of suitable treatment or manipulation of the matrix of estimates of their abundance. Before turning to consider this problem, it will be helpful to consider some geometric properties of the data matrix. Although the observed values may be, and frequently are, analysed in their original form, it will be convenient for our purpose to consider them in standard form. This involves replacing each observed value X_{ji} by a new value z_{ji} where $z_{ji}=(X_{ji}-\bar{X}_j)/s_j$, $\bar{X}_j$ being the sample mean and s_j the sample standard deviation for species j. The set of all values z_{ji} ($i=1, 2, \dots N$) is called an attribute j in standard form. This transformation results in the origin of measurement being placed at the mean value of each attribute, and in the original unit or units of measurement being replaced by a new unit, the standard deviation. The end result of analysis is not affected by this transformation.

A useful concept where many variates are considered simultaneously is that of the sample space. By means of this, the behavior or variation of an attribute over a number of stands can be represented by a point or by a

† The terms attribute and variate are used interchangeably.

vector in higher dimensional space. This mode of representation is especially valuable where the attributes turn out to be merely different functions of a small number of independent variables. Where a number of attributes have been measured over N individuals, we may set up a Euclidean space of N dimensions, one dimension for each of the N individuals (stands). The entries of each row z_j: $(z_{j1}, z_{j2}, \ldots z_{jN})$ of the data matrix in standard form can then be regarded as the coordinates of a point representing the jth

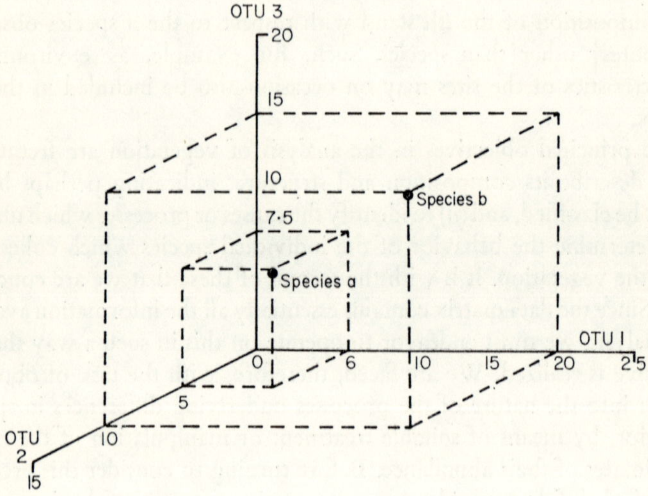

FIG. 1. Point representation of two species in sample space (N=3). The coordinate axes correspond to stands, the scales being in terms of any suitable measure of abundance, e.g. % cover. The position of points is uniquely determined by the abundance of the species in each stand. For OTU read stand. (Modified from van Groenewoud, 1965, by courtesy of *Ber. geobot. Inst. ETH, Stiftg. Rubel, Zürich*).

attribute in the N space. The position of each attribute will be uniquely determined by its particular combination of observed values. Species or attributes with similar sets of values will therefore occupy adjacent positions in the sample space. It follows that the positions and spatial relationships of the points are related to the behavior of the attributes corresponding to them over the vegetation samples examined. The closer the points, the greater the ecological similarity of the attributes and *vice versa*. If each point is joined to the origin of the coordinate system by a line, the resulting configuration is called the vector representation of the variables. In this way, the entire observation set can be regarded as determining a set of points or vectors corresponding to variates in a space, the dimension of which equals the number of stands examined. Examples of the point and vector representation of hypothetical sets of observations are shown diagrammatically in Figs. 1 and 2. Although an attempt is made to show only

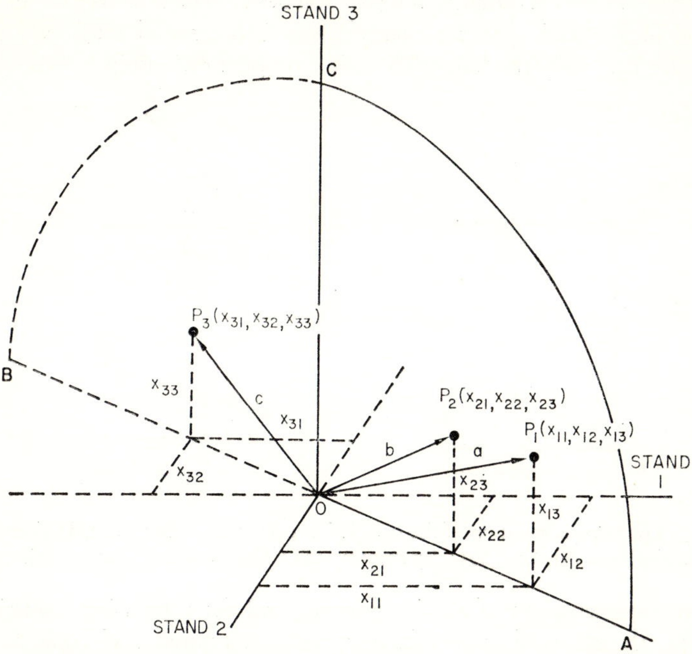

FIG. 2. Vector representation of three species (P_1, P_2, P_3) in sample space ($N=3$). The position of points is determined by species' abundance in the stands, e.g. P_1 by X_{11}, X_{12}, X_{13}. The vectors may then be obtained by drawing a line from the origin to each of the points. The vectors shown all lie in the plane ABC, i.e. within a sub-space of 3-dimensional sample space. (From van Groenewoud, 1965, by courtesy of *Ber. geobot. Inst. ETH, Stiftg. Rübel, Zürich.*)

three dimensions here (which correspond to three stands), no difficulty arises in practice in dealing with many dimensions (stands), at least if electronic computational facilities are available.†

† An alternative geometric representation of the data matrix is possible. A space of n dimensions may be postulated where each dimension corresponds to an attribute (species). The column entries z_i: (z_{1i}, z_{2i}, . . . z_{ni}), $i=1, 2, \ldots N$ can then be used as coordinates locating the ith *stand* in n dimensional attribute-space. (See Goodall, 1963; van Groenewoud, 1965; Williams and Dale, 1965). In the original model the points generally correspond to species, while in the second, they correspond to stands.

(2) *Geometric structure of a correlation matrix*

Interest in the point or vector representation of the field observations centers on the spatial relationships of the points. These relations may be measured and analyzed in terms either of the distance between pairs of points or in terms of the angular separation between pairs of vectors linking points to the origin (see, for example, van Groenewoud 1965). We will consider here only the latter. There is a simple relationship between the

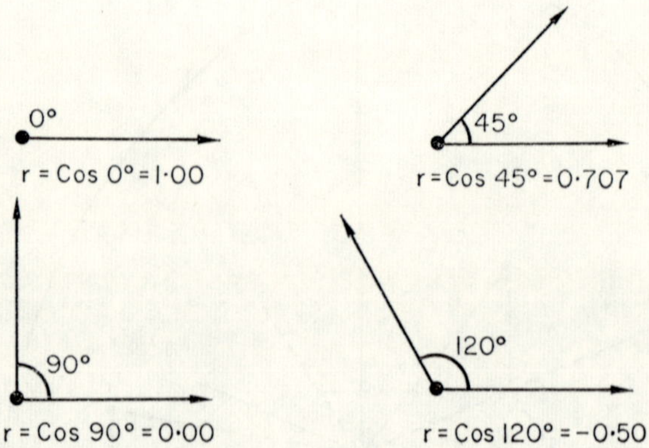

FIG. 3. Vector representation of the correlation coefficient, *r*, for selected values of *r* (see text). (From Fruchter, 1954, by courtesy of D. Van Nostrand Company, Inc.)

vector representation of a pair of species and the correlation coefficient relating the observed values of the species over a number of stands. If the variates are expressed in standard form, the cosine of the angle between the vectors in *N* space is identical to the value of the correlation coefficient between the variates. This may be expressed

$$\cos \phi_{jk} = r_{jk} \qquad (j, k = 1, 2, \dots n), \qquad (1)$$

where ϕ_{jk} is the angle of separation between vectors *j* and *k* and r_{jk} is the correlation coefficient between the representation of variates *j* and *k* over the *N* stands. The correlation coefficient, *r*, can therefore be used for analyzing the relations between attributes conceived of as points in sample space at the tips of vectors extending from the origin. The relationship between the angular separation of vectors and the value of the corresponding coefficient is an inverse one. The smaller the angle between vectors, the larger the correlation and *vice versa.* Consider, for example, two attributes occupying the same position in the sample space so that the angular separation of the vectors linking them to the origin is 0°. This would indicate perfect correlation between their observed values over the samples, since the cosine of 0° is +1, which, by equation (1), corresponds to a correlation coefficient of +1 also. Similarly, two points separated by an angle of 90° represent uncorrelated attributes, the cosine of 90° being zero, corresponding by (1) to a correlation coefficient of the same magnitude. In Fig. 3 the vector representation of these and a number of other values of the correlation coefficient are shown.

ATTRIBUTE	1	2	3	4
1	1·00	0·80	0·96	0·60
2	0·80	1·00	0.60	0·00
3	0·96	0·60	1·00	0·80
4	0·60	0·00	0·80	1·00

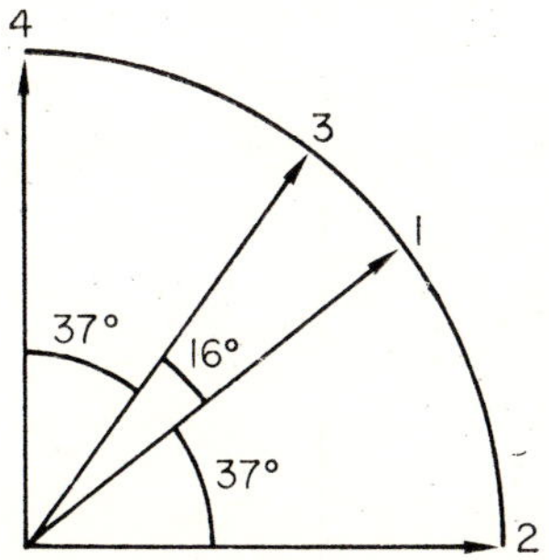

FIG. 4. Vector representation of a correlation matrix. Values of the correlation coefficient between variates considered in all possible pairs are contained in the table (matrix); the corresponding vector representation of the matrix is shown below. (Cos 16° = 0·96, cos 37° = 0·80, cos 53° = 0·60, cos 90° = 0·00). (From Fruchter, 1954, by courtesy of D. Van Nostrand Company, Inc.)

The expression of the resemblance between two variates by a pair of vectors or by a single correlation coefficient can easily be extended to cover the inter-relations between many variates. A matrix of correlation coefficients calculated between the observed values of the variates considered in all possible pairs, will, in fact, contain much of the information embodied in the sample space model. Construction of the vector model of the observation set is not itself required. In fact, there is an exact correspondence between the vector representation of a data matrix and the matrix of inter-correlations between the variates. A simple, hypothetical example of this involving four variates is illustrated in Fig. 4.

In most applications involving real data, correlations are frequently found between certain variates. Such correlations reflect the tendency for species to vary together in representation and hence for stands to resemble one another in composition to a greater or lesser extent. In terms of the

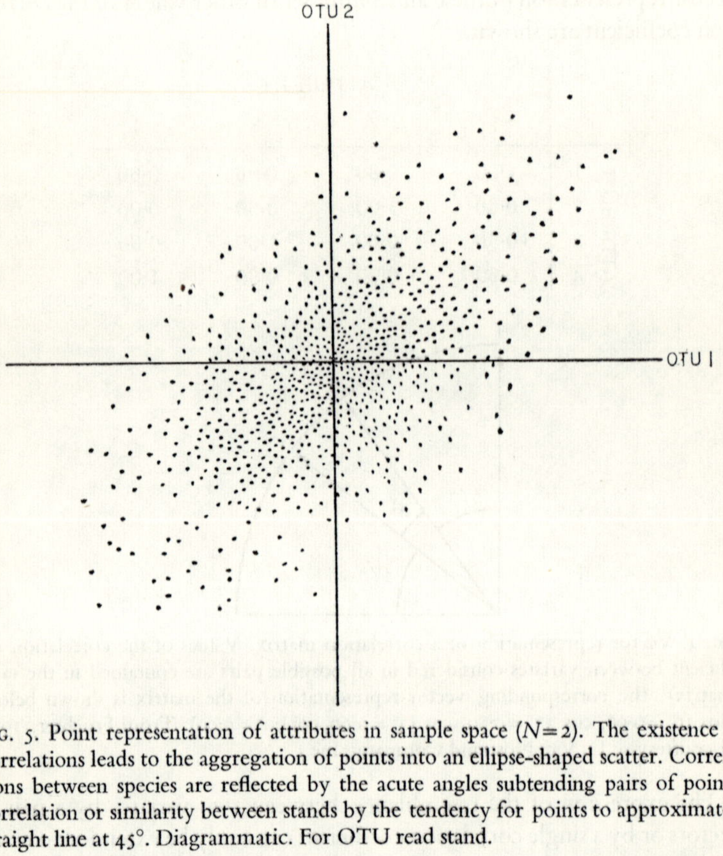

FIG. 5. Point representation of attributes in sample space ($N=2$). The existence of correlations leads to the aggregation of points into an ellipse-shaped scatter. Correlations between species are reflected by the acute angles subtending pairs of points; correlation or similarity between stands by the tendency for points to approximate a straight line at 45°. Diagrammatic. For OTU read stand.

point representation of attributes, this means that the points tend to lie in a more or less well defined ellipsoidal cluster and are not distributed throughout N space. In terms of vectors, it means that the vectors tend to be associated in groups or into 'cones' having rather small generating angles situated at the origin. In Figs. 5 and 6 the existence of inter-correlations between a number of attributes in N space are shown diagrammatically in point and vector form for the special cases where $N=2$ and $N=3$. The existence of inter-correlations of this kind has considerable practical significance. In the following section we will be concerned with a means of exploiting the existence of such correlations.

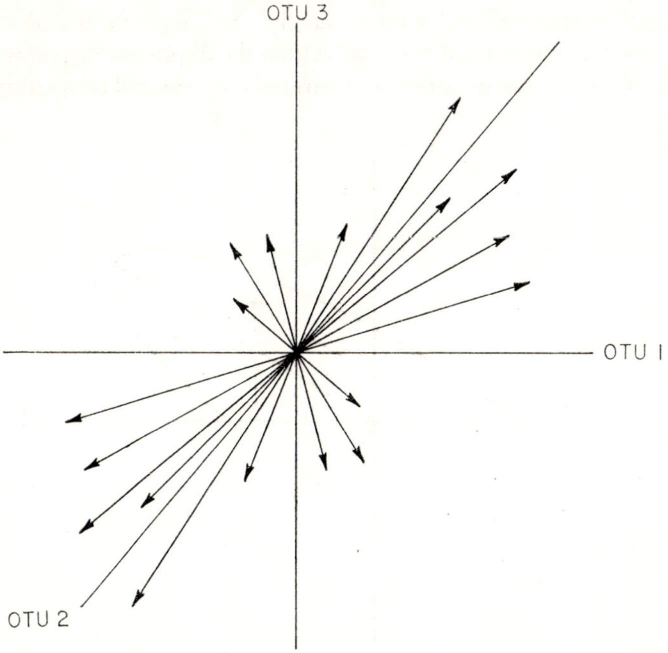

FIG. 6. Vector representation of attributes in sample space ($N=3$). The existence of strong correlation between certain variates is indicated by the acute angles between vectors. Diagrammatic. For OTU read stand.

(3) *Principal axes transformation*

Computation of the correlation matrix between variates normally forms the first step in ordination.† The existence of more or less strong correlations in the matrix implies that the *n* points (or vectors) are contained within

† The covariance matrix calculated between species may sometimes be employed in place of the correlation matrix (van Groenewoud 1965; Yarranton 1967.) For a discussion of the suitability of covariances and correlations in the present context, reference may be made to Lawley and Maxwell (1963), Seal (1964), Pearce (1965), and Morrison (1967).

Ordination may also proceed from a matrix calculated between stands (i.e. between *columns* of the data matrix) rather than between species (rows), provided an appropriate similarity coefficient is used. (See van Groenewoud 1965; Gower 1966, 1967; Orloci 1966, 1967.)

a restricted region of the *N* space. This enables new coordinate axes to be strategically placed through the major dimensions of the sub-region in which the points actually lie. The points can subsequently be described more efficiently with respect to the new axes than the original ones. Principal components analysis is essentially a technique for introducing a new coordinate system of this kind within the dispersion of points. The new coordinate axes are referred to variously as principal components or

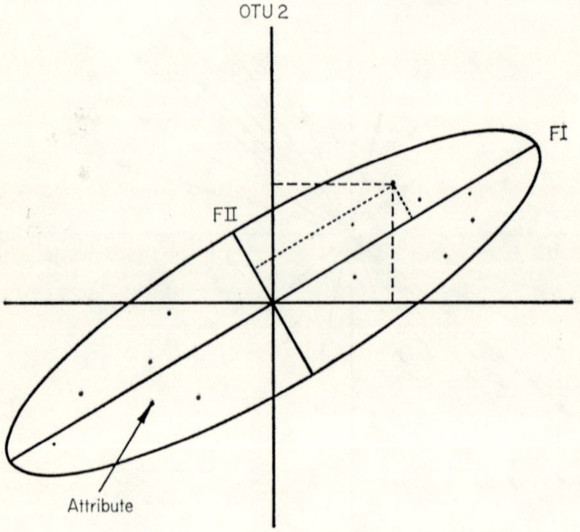

FIG. 7. Principal components. The positions of points are determined by reference to the coordinate axes corresponding to stands (marked OTU 1 and OTU 2). The principal axes or components of the scatter (F I, F II) provide an alternative and more convenient reference frame. The coordinates (loadings) of one species with respect to the original axes and the principal components are indicated.

principal axes. This form of analysis probably represents the most successful ordination technique currently employed in studies of vegetation.

As noted above, the correlations found in practice imply that the points in the sample spaces involved tend to form ellipsoidal clusters. From Fig. 7 the effectiveness of choosing as new reference axes those coinciding with the major and minor axes of an ellipse might be apparent. This is partly because the axes of an ellipse (or, in higher dimensional space, hyperellipsoid) can be ranked such that they become progressively shorter and therefore less important for descriptive purposes. For the particular case illustrated ($N=2$), the difference in lengths of the major and minor axes is obvious. The ordinates of points expressed with respect to the first (major) axis, provide a description of the distribution of points which for many purposes might be an acceptable approximation of their exact relations. Although information would be lost in ignoring the positions of points relative to the shorter axis, the ease of description which would result might be considered to more than outweigh this. Just how much information was lost would depend on the difference in length of the axes. This disparity in turn would itself be dependent on the strength of the correlations between variates. The higher the inter-correlations, the more elongated the point scatter and the greater the difference in the lengths of its axes. For example, if no correlation existed (i.e. $r_{jk}=0$ for all j, k), the scatter of points would be circular (a sphere or hypersphere where $N \geqslant 3$) and all reference axes would be of equal length and importance. In the limiting case where all correlation is supposed perfect (i.e. $r_{jk}=\pm 1$ for all j, k) the points would fall on the line (plane or hyperplane where $N \geqslant 3$) corresponding to the principal axis. The difference in length is then at its maximum, the second (and subsequent) axis vanishing entirely. In practice the degree of correlation found is always between these extremes. This renders certain axes or dimensions of less descriptive importance than others, and, provided a suitable reference frame is chosen, can simplify the tasks of description and interpretation. A suitable reference frame is one coinciding with the principal axes of the hyperellipsoid formed by the points. Another useful property of this set of axes is that those variates which are correlated with respect to the original set of axes are un-correlated in the new system; that is, the co-ordinates of these variates are transformed to independence by the particular change of axes involved. This may render subsequent ecological interpretation of the results more straightforward than would otherwise be the case.

Principal components analysis leads to the establishment of a reference system of exactly the kind described. Reference to Fig. 7 shows that this involves simply a rotation of the original reference axes about the origin. However, it can also be appreciated from the figure that the original axes might conceivably be rotated to any position in the plane shown. Some criterion is therefore required which will lead to a uniquely determined set of axes. The criterion adopted in component analysis relates to the proportion of the total variance attributable to each new axis or component. The requirement is that the first component be positioned in the point scatter such that it accounts for as much of the total variance as possible (i.e. it is located in such a way that the sum of squares of the perpendicular projections of points onto it is the maximum possible); the second component is then located such that it accounts for as much of the remaining variance as possible—which leads to its being at right angles to the first;

and so on for subsequent components until all the variance is accounted for. Geometrically, the resulting axes are those which coincide with the principal axes of the postulated hyperellipsoid.† The lengths of successive components (axes) are proportional to that part of the total variance accounted for by each. It may be that the first two or three components account for almost the whole of the variation, for example for as much as 80 or 90% of the total. The variation can then be represented approximately by the first two or three reference axes and in certain circumstances it may be permissible to neglect the remainder. In such cases this has the effect of reducing the dimensionality of the system from an *N* dimensional system to one of more manageable number. It is the achievement of precisely this economy which is a major object of component analysis.

The ordination is complete when the coordinates of points are specified with respect to the principal components. It is usual in addition to indicate the variance, i.e. the length, or importance, of successive components. This is related to the 'latent root' or 'eigenvalue' associated with each component. It may be worth emphasizing that, while the coordinates of points differ with respect to the original and rotated axes, the configuration of points itself is unaffected by the transformation. In particular, the distances and angular relations between points are invariant, i.e. preserved, under this transformation. The same information occurs in each case, it is merely more conveniently displayed by the principal components.

(4) *Stand ordination*

In many cases analysis is carried a stage further leading to an ordination of stands. To accomplish this, the standardized observed values of species in a given stand are multiplied by their coordinates on a particular component Summation of the products over the species then yields the coordinate of the stand in relation to that component. This process repeated for all stands and all components results in an ordination of the stands with respect to the components extracted from the between-species correlation matrix. The species' coordinates—or loadings—on the components, function here as *a posteriori* weighting coefficients in describing relations among stands.

† Recently, analyses of ecological data in which the principal components were eventually rotated to other positions in sample or attribute space have appeared (Dagnelie, 1965b; Ivimey-Cook and Proctor, 1967; Webb *et al.*, 1967). Webb *et al.* found the use of *oblique* axes helpful in their analysis. The results of these analyses indicate that rotations of these kinds may well become more important in ordination studies.

Thus, starting from a point or vector representation of species in sample space, we have arrived at an ordination of stands (in component space) comparable in some ways to that referred to in the footnote on p. 41 above. Stands and species can therefore each be arranged or ordered in the component space. Ordinations of both kinds are of ecological interest.

Although the account above has been based largely on illustrations in two- or three-dimensional space, generalization to higher dimensional spaces is straightforward. In practice, of course, ordination is not performed graphically in the way outlined, the equivalent algebraic operations normally being executed by digital computer. Algebraically components analysis may be expressed as follows:

$$F_j = a_{1j}z_1 + a_{2j}z_2 + a_{3j}z_3 + \ldots + a_{nj}z_n \qquad (j = 1, 2, \ldots, n), \qquad (2)$$

where F_j is the jth principal component of the observed variates $z_1, z_2, \ldots, z_n$, and the $a_{1j}\ a_{2j}, \ldots, a_{nj}$ are coefficients weighting the relative importance of each species or other variate z_j in the derived component. From equations (2) it is clear that the components F_j are weighted sums of the original variates. Moore (1965) provides a clear account of the nature of the weighting coefficients and of their possible interpretation in biological terms. Equations (2) can be rewritten in the form:

$$z_j = a_{j1}F_1 + a_{j2}F_2 + a_{j3}F_3 + \ldots + a_{jn}F_n \qquad (j = 1, 2, \ldots, n), \qquad (3)$$

where the n species z_j are represented as weighted sums of the components, $F_1, F_2, \ldots, F_n$. Using this relation, the observed values of any species j over the i stands are given by the expression:

$$z_{ji} = a_{j1}F_{1i} + a_{j2}F_{2i} + a_{j3}F_{3i} + \ldots + a_{jn}F_{ni} \qquad (j = 1, 2, \ldots, n;\ i = 1, 2, \ldots, N)$$

All n components F_j are required to reproduce the observations z_{ji} exactly. The components, however, are always calculated in order of decreasing magnitude or importance. This enables the observed values z_{ji} to be approximated ($z_{ji}^\star$) by the first m, say, components, F_j:

$$z_{ji}^\star = a_{j1}F_{1i} + a_{j2}F_{2i} + a_{j3}F_{3i} + \ldots + a_{jm}F_{mi}. \qquad (4)$$

Usually m is much smaller than n, the number of observed variates.

It is the numerical values of the F's and a's in equations (2) and (4) respectively which are of practical importance. The calculated values for the F's are the coordinates of stands in relation to the principal axes and which thus provide a stand ordination. Similarly, the a's, when suitably scaled, may be employed as coordinates of species on the principal axes to yield a species ordination. The basic problem is to obtain estimates for the coefficients a_{ji}. The required estimates are the elements of the latent vectors of the correlation matrix between variates (see, for example, Orloci, 1966). The vectors, appropriately scaled, yield a species ordination, and, together with the initial observations z_{ji} lead to the determination of the F's in equations (2).† Thus, while equations (2) are solved explicitly, there is rarely any need in practice to complete the solutions to equations (4). In equations (4), which attempt to describe or represent the structure of the observations z_{ji} mathematically, using a small number of derived 'factors' or components F_j, it is the possible significance of the components F_j ecologically which attracts attention. Further details of the algebraic and computational aspects of components analysis are beyond the scope of this account. Reference may be made to the works cited on p. 38 for additional information bearing on these aspects. Other useful accounts of components analysis and related methods are provided by Cooley and Lohnes (1962), Horst (1965) and Harman (1967).

(5) *The interpretation of results*

Before considering some ecological results of ordination, it is of interest to examine the results of a components analysis. The results usually take the form of a table of the kind shown in Table 1. This analysis relates to an area of grassland vegetation. The field observations consisted of estimates of the abundance of thirty species and the expression of three soil variables in 45 stands (i.e. $n=33$, $N=45$). The main body of the table consists of numbers which, when read by columns, relate to the principal components when read by rows, correspond to the variates. Since the smaller components are generally of little interest, only the first five or ten components are usually involved in such tables. The table is related to equation (4), the row entries comprising the a coefficients of the relevant components (F's) of the equation. These entries are interpretable both as the coordinates (loadings) and as the correlations of the variates with the principal com-

† Orloci (1967) in a valuable paper points out that the required stand and species ordinations are sometimes more conveniently calculated by an alternative route. Here the latent vectors of the between *stand* similarity matrix are the basic requirement. The latent vectors themselves provide the stand ordination while their elements may be used as coefficients weighting the row entries of the data matrix to yield the components scores of species (i.e. the species ordination). This approach is especially valuable where the number of species exceeds the number of stands. Identical ordinations result from either route provided (1) the data matrix is suitably standardized, (2) appropriate similarity functions are used and (3) the latent vectors are suitably scaled.

ponents. The column labeled h^2 contains the sum of the squares of the entries of each row

$$\left(h_j^2 = \sum_{i=1}^{5} a_{ji}\right).$$

This value indicates the proportion of the total original variance of a variate which is preserved by the m components retained. When all n components are taken into account, the value of h^2 is always 1·00. Thus, in the table, the nearer a given h^2 value approaches unity, the more completely the variate in question is accounted for by the components shown. At the foot of the table, the sum of the squares of each column is shown. These values are the variances or latent roots of the principal components. They show clearly the principal of maximum contribution to the total variance of each successive component. Together the five components account for almost 73% of the variance of the observations. Since the latent roots are proportional to the lengths of the components, they enable the shape of the dispersion of points to be visualized. In this case they show that, while the dispersion may be considered to form a hyperellipsoid, this is not as elongated as is sometimes found.

The similarity of the table to a data matrix is worth noticing. This is not really surprising since all that has been done is to relate the points to a new set of coordinate axes. Apart from the smaller number of columns, the chief remaining difference concerns the nature of the columns. In a data matrix, the columns relate to actual stands (showing the abundance of each species present), and, in the sample space model of this, we set up a correspondence between the columns and the coordinate axes, letting each stand be represented by an axis. In the matrix of species' component loadings, the columns define the rotated coordinate axes of the sample space, the column entries indicating the location of species with respect to these. In sample space, therefore, the coordinate axes of the data matrix correspond to real, physical entities—the stands. It remains to be considered whether any similar correspondence can be shown between the rotated axes (principal components) and external reality. This topic is discussed in some detail below.

The results of a stand ordination are similar in appearance to those of a species analysis. In Table 2, the corresponding stand ordination of the grassland data is given. Stands are listed at the left and the entries of the table read by rows are the coordinates of each stand on the principal components. These coordinates are often referred to as the *component scores* of stands. Although much insight into species' behaviour and vegetation structure can be obtained from such tables, interpretation is often aided

TABLE 1

Principal components analysis of grassland vegetation, Anglesey, North Wales. Loadings of species and soil variates (attributes) on the first five components

Variate	a_1	a_2	a_3	a_4	a_5	Variance h^2
1. *Helianthemum chamaecistus*	0·775	−0·407	0·208	0·155	−0·123	0·849
2. *Thymus drucei*	0·869	−0·001	−0·310	−0·243	0·091	0·919
3. *Lotus corniculatus*	0·195	−0·654	−0·075	0·106	0·095	0·492
4. *Plantago lanceolata*	−0·374	−0·643	−0·095	0·115	−0·321	0·679
5. *Galium verum*	−0·064	0·559	−0·081	0·409	0·239	0·548
6. *Carex flacca*	0·023	−0·798	0·320	−0·103	−0·004	0·750
7. *C. caryophyllea*	0·499	−0·615	0·202	0·382	0·034	0·815
8. *Poterium sanguisorba*	−0·447	−0·687	−0·052	0·170	−0·131	0·721
9. *Hieracium pilosella*	0·404	−0·401	0·320	−0·114	0·180	0·472
10. *Luzula campestris*	−0·626	−0·568	0·184	0·282	−0·190	0·864
11. *Taraxacum laevigatum*	0·561	−0·061	−0·239	0·377	−0·063	0·522
12. *Viola riviniana*	−0·262	−0·356	0·230	−0·188	−0·320	0·386
13. *Briza media*	0·761	−0·364	0·009	0·266	−0·070	0·787
14. *Koeleria cristata*	0·660	0·037	0·269	0·357	0·003	0·637
15. *Anthoxanthum odoratum*	0·044	−0·365	0·523	−0·234	−0·181	0·496
16. *Sieglingia decumbens*	0·281	−0·368	0·251	−0·449	0·420	0·655
17. *Dactylis glomerata*	−0·774	0·508	0·044	−0·102	0·098	0·879
18. *Agrostis tenuis*	−0·818	−0·391	0·040	−0·085	0·231	0·884
19. *Helictotrichon pubescens*	−0·902	−0·023	0·253	0·132	−0·058	0·899
20. *Phleum bertolonii*	−0·421	0·802	0·168	−0·123	0·127	0·880
21. *Holcus lanatus*	−0·679	−0·296	−0·440	−0·371	−0·385	0·909
22. *Centaurea nigra*	−0·721	−0·378	−0·279	−0·118	−0·059	0·759
23. *Trifolium pratense*	−0·918	−0·069	0·004	0·228	0·019	0·900

24. *T. repens*	−0·718	−0·277	−0·429	−0·101	−0·320	0·889
25. *Anthyllis vulneraria*	−0·041	0·002	−0·438	0·377	0·120	0·350
26. *Cladonia impexa*	0·796	−0·084	−0·291	−0·225	−0·234	0·831
27. *Pseudoscleropodium purum*	0·919	−0·141	−0·227	0·030	−0·067	0·921
28. *Dicranum scoparium*	0·905	−0·057	−0·066	−0·127	−0·184	8·877
29. *Rhytidiadelphus squarrosus*	−0·106	−0·134	−0·141	0·612	0·370	0·561
30. *Hypochaeris radicata*	−0·662	−0·359	−0·269	0·087	−0·123	0·662
31. Soil depth	−0·800	−0·176	0·152	−0·167	0·340	0·838
32. Soil phosphate	−0·495	0·493	0·342	0·328	−0·308	0·807
33. Soil potassium	−0·011	0·377	0·486	0·292	−0·389	0·615
Variance (latent root)	12·257	5·784	2·315	2·143	1·557	24·056
Percent of total variance	37·14	17·53	7·01	6·49	4·72	72·90
Accumulated percent	37·14	54·67	61·68	68·18	72·90	—

Total variance (trace): 33·00

by plotting the positions of species or stands in relation to pairs of components. Figures 8 and 9 are, respectively, ordinations of species and stands plotted with reference to the first two components of Tables 1 and 2. Interest in scatter diagrams of this kind centers on the spatial relations (distances) between points, and on their position in relation to the origin. The closer the points, the greater the ecological similarity of the entities they represent, and *vice versa*. Strong negative correlation is implied by points which are diagonally opposite; such stands or species may well be

TABLE 2

Principal components analysis of grassland vegetation, Anglesey, North Wales. Component scores of stands (individuals) on the first five components

Individuals (Stands)	F_1	F_2	F_3	F_4	F_5
1	4·096	0·989	1·956	−0·678	−0·455
2	−3·224	7·897	−0·499	1·292	0·937
3	3·702	0·201	1·415	0·014	−0·217
4	1·312	−0·292	−2·854	0·457	2·032
5	−5·513	−3·445	1·756	−1·267	−0·453
6	3·803	0·153	1·602	−0·716	−0·851
7	−5·292	−1·275	1·198	0·042	−1·990
8	0·877	−1·709	−1·972	0·183	1·331
9	−3·677	8·197	−0·631	0·991	0·201
10	−5·435	−2·734	1·886	−0·986	0·359
11	2·672	0·933	−0·514	−1·089	−0·659
12	2·361	−0·269	−0·381	−1·145	−0·879
13	−3·909	−1·634	−0·529	−0·448	−2·444
14	2·095	−0·212	−0·629	−1·324	1·376
15	2·775	−2·001	0·319	0·690	−2·350
16	1·961	−0·924	−0·199	−0·075	−1·053
17	0·512	−1·818	−2·091	0·419	1·281
18	−1·371	0·257	−2·435	−3·777	0·268
19	−4·707	2·917	−0·679	−0·754	−1·515
20	−7·412	−3·386	2·354	−0·360	3·491
21	2·839	−0·698	1·717	0·613	−2·198
22	3·789	2·303	1·748	−0·313	0·576
23	−4·148	−0·095	−0·642	−0·488	−0·368
24	1·025	−0·102	−0·099	−2·422	−1·626
25	−5·184	−0·255	2·523	0·112	1·414
26	3·981	1·124	1·860	1·449	0·606
27	2·230	0·149	0·328	−2·572	−1·313
28	−3·193	3·688	−1·555	−0·195	−0·676
29	−2·077	−2·519	−1·156	2·658	−0·947

TABLE 2—*continued*

Individuals (Stands)	F_1	F_2	F_3	F_4	F_5
30	0·366	−2·147	−1·521	2·488	−0·515
31	2·978	−1·870	−0·907	2·089	−0·332
32	1·643	−1·646	−1·426	1·430	1·609
33	1·993	1·036	−0·473	−2·054	0·027
34	3·088	2·929	0·788	−0·829	0·509
35	−6·375	1·143	2·970	1·452	−0·033
36	2·708	0·575	−0·008	1·067	0·916
37	3·436	0·770	1·118	1·444	0·453
38	3·441	−0·168	−0·540	0·454	2·036
39	1·186	−0·982	−0·600	−1·067	1·105
40	−2·585	−2·665	−1·785	2·675	−0·011
41	−2·017	−0·502	−1·758	−2·530	0·571
42	3·739	0·936	2·152	−0·855	0·509
43	4·314	−0·803	1·059	1·716	0·295
44	−0·199	−2·000	−1·290	1·579	−0·231
45	−3·234	2·023	1·573	−0·896	−0·786

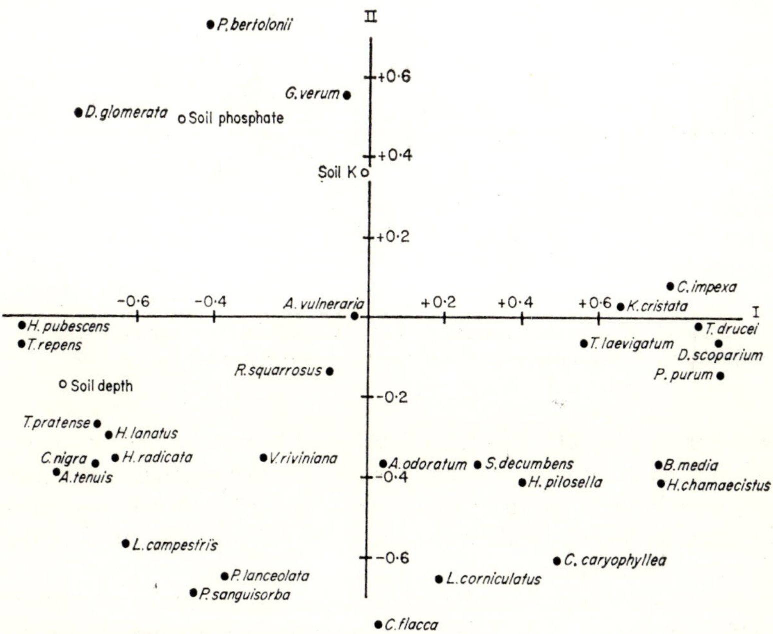

FIG. 8. Grassland vegetation, Anglesey, North Wales. Principal components analysis; projection of species and three soil variables on components I and II. (The variates represented are listed in Table 1.)

ecologically alike to the extent that they are responding to a common influencing factor of some kind (e.g. soil base status, perhaps, or soil moisture content). They differ, however, in reacting to this influence in opposite senses. The distance of a point from the origin is given by h, which is the standard deviation of the variable. It follows from the interpretation of h^2 given above, that, the further a point is from the origin, the more the entity in question is accounted for by, or related to, the components shown.

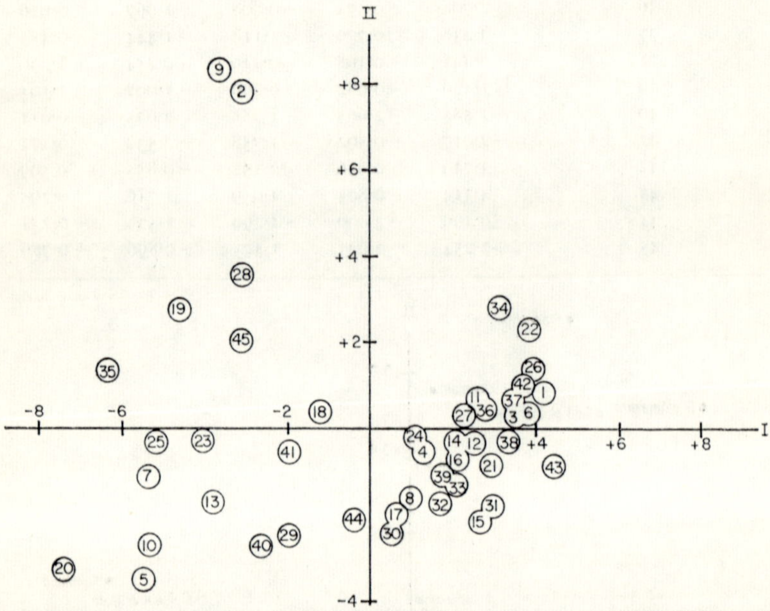

FIG. 9. Grassland vegetation, Anglesey, North Wales. Principal components analysis; projections (component-scores) of stands on components I and II.

The construction of three-dimensional models provides an even more effective aid to the interpretation of results than the preparation of scatter diagrams. In these, species or stands, represented by small balls supported by rods, are located by their coordinates in the 3-space spanned by the first three components.

These various devices are all attempts to portray the sample or attribute spatial conceptions of the field observations. Some distortion of relationships between points arises in these representations from the failure to use all the information available. As a result, however, the salient features of the data stand out more clearly, which in turn may simplify their ecological assessment.

The possible ecological significance of the new reference frame is also of interest. Although the principal components strictly speaking are purely mathematical constructions, it is frequently found that they correspond also to ecological features of the analysis. Indeed, only where the reference axes can be considered to reflect ecological factors or processes of some kind, is the ordination likely to be regarded as completely satisfying. Identification of the ecological factors involved is a matter for interpretation, and, as such, it is dependent in part on the judgement of the investigator. Usually it involves an attempt to evaluate the results of the ordination against information external to the analysis.

In species' ordinations where there are at least some species whose general ecological requirements are known, the interpretation of components is usually straightforward. An indication of the ecological control or influence expressed by different components may then be provided by species which have a high loading on one or other of them. As an example, consider the species ordination shown in Fig. 8. Here *Thymus drucei* and *Trifolium repens* are among the species with high loadings on the first axis. They differ between themselves, however, in occupying positions at opposite ends of the axis. The 'preference' of *T. drucei* for shallow, well-drained soils and of *T. repens* for deeper soils with more balanced water relationships is well known from general field experience. Using this knowledge, component I may, as an initial hypothesis, be regarded as an expression of the possible control exerted by soil depth and the other conditions which go with it, on species' behaviour. Since this axis accounts for more of the total variance of the data than any other (see Table 1) the factor-complex of soil depth may, with some justification, be regarded as the major ecological control operative in the community. The same process of interpretation can be extended to other axes or components although components after the first three or four usually prove difficult to relate to ecological factors. Since successive axes are perpendicular to one another, it is to be expected that they will correspond to habitat factors which operate largely independently of each other, as van Groenewoud (1965) has pointed out. Where environmental variates are included in species' ordinations, interpretation is simplified. In Fig. 8, for example, the positions of soil depth and exchangeable soil potassium in relation to axes I and II respectively, provide clear indications of the habitat influences which may possibly underlie these components. In cases where nothing is known of the ecology of the species involved, insight gained in this way of the nature of the chief ecological factors operative and of species' response to them, although of a provisional nature, is especially valuable. A wise choice of environmental variables for inclusion, however, requires the exercise of experience and care.†

In stand analyses, the possible ecological meaning of the reference frame may be explored by simple or multiple regression analysis. This involves regression of the intensity of an ecological factor over the stands, on the position of stands along a particular axis or axes. Familiarity with the vegetation is helpful in deciding which environmental variables may most profitably be tested in this way with a given axis. In some instances, correlation might be considered to provide a more suitable test than regression. As a preliminary step, it may be worthwhile to plot the levels of the factor against stand positions on a component. An example of this, taken from van Groenewoud (1965), is shown in Fig. 10(a). Here a measure of light conditions (lux hours/day × 1000) in fifteen stands of Swiss forest vegetation, is plotted against stand positions along the first axis of an ordination. A well defined linear relationship is apparent, which is also statistically significant (van Groenewoud, 1965). This suggests that the axis may correspond to the effect of light on vegetation composition. Species' behavior can be examined in the same way. In Fig. 10 (b, c), the representation (% cover) of *Picea abies* and *Abies alba* in relation to stand positions on the same axis are shown. The behavior of the species is evidently linearly related to the component, although in opposite directions. From the nature of the component, it follows that the behavior of the species may be regarded provisionally as controlled largely by variation in light. There is a second approach to the identification of components from a stand ordination. This is to plot the distribution of stand component-scores at the appropriate stand positions on a base-map of the area involved. Particular environmental

† There is a difference of opinion concerning the inclusion of variates measured on different scales in the same analysis. This arises partly from the fact that the components are not invariant under changes of scale in the variates. Thus, for example, it may be held undesirable for, say, species' abundance estimated as percentage cover, soil cations expressed as m.e./100g dry soil and soil acidity expressed in pH units, to be analyzed together, even when the original units of measurement are first standardized. See Anderson (1958, p. 279); Lawley and Maxwell (1963, p. 47); Seal (1964, p. 117); Pearce (1965, p. 195); Williams and Dale (1965, p. 56); Cattell (1966, p. 221); Morrison (1967, p. 223). Nevertheless, certain analyses based on such data have yielded apparently consistent results (Ferrari *et al.*, 1957; Sokal *et al.*, 1961; Rayner, 1966; cf. also Fig. 8 above). Further work on this aspect of ordination is required. A quite different approach to the joint analysis of species and environmental variates of equal or perhaps of greater promise is canonical correlation analysis (see Kendall, 1957; Anderson, 1958; Cooley and Lohnes, 1962; Dagnelie, 1965b).

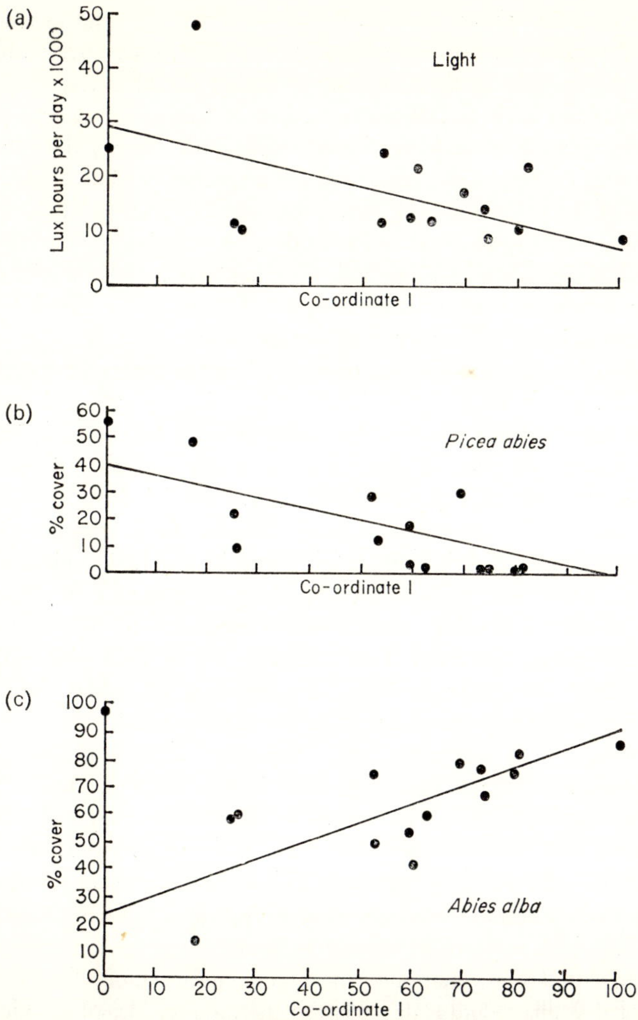

FIG. 10. Forest vegetation, Switzerland. (a) Interpretation of components from stand ordinations; relation of light conditions (lux hours/day) of stands to component-scores of stands on component I. (b, c) species' behavior (% cover) in relation to component I. (From van Groenewoud, 1965, by courtesy of *Ber. geobot. Inst. ETH, Stiftg. Rübel, Zürich*).

variables can be treated in the same way and the resulting distribution maps contoured and compared. The maps may then suggest the existence of connections between habitat factors and components, leading to identification of the latter (see Goodall 1954b). A refinement of this technique is provided by trend-surface analysis. By means of this, the total variance of a mapped variate can be sub-divided into portions, each attributable to separate causes. The method consists of fitting the observed values of mapped variates such as component-scores or habitat factors, to polynomial functions of the map coordinates of the stands examined. Functions of increasing degree, each accounting for a specific portion of the total variation of the variate, may be fitted. The surfaces corresponding to the functions are readily mapped and may then be examined and compared for their ecological implications. This approach may be most useful in the case of components whose nature has proved difficult to establish by the methods mentioned previously. (See Gittins 1968.)

Interpretation of the reference frame enables the nature of the major ecological controls to be appreciated, at least in a provisional way. It therefore turns out that the principal components, which were arrived at initially for the purpose of descriptive economy, are also of value as *explanatory* variables of ecological significance. Thus, it is the interpretation or explanation of relationships embodied in the observations, together with their efficient description, which comprise the basic objectives of ordination. Once these aims are accomplished, the ground is prepared for the use of experimental methods. The components of ordinations of stands and species extracted from the same set of field observations, generally lead to similar ecological conclusions. The chief difference concerning such components is that, in the stand ordination it is a highly loaded stand which exemplifies the component, whereas in the species ordination it is a highly loaded species (or other variate). Despite their similarities, both ordinations have intrinsic interest since they each contain information not present in the other. This, however, draws attention to an inherent weakness of the approach. This is that, within the framework of the spatial models of the observation set (pp. 39–41 above), there is no way of simultaneously manipulating individuals and attributes so as to arrive at a *unified* representation of the essential information of the system. This has led Williams and Dale (1965), to entertain the idea of abandoning spatial models altogether.

(6) *An investigation of factors influencing grassland yield*

As an example of the use of ordination, reference is made to a study by Ferrari, Pijl and Venekamp (1957), of the yield of 50 stands of grassland vegetation in the Netherlands. Appreciable differences in yield between stands were apparent, and the study was conducted in order to identify

TABLE 3

Principal components analysis (centroid approximation) of variates from grassland in the Guelderland Valley, The Netherlands. Loadings of variates on the first four principal components (modified from Ferrari, Pijl and Venekamp, 1957, by courtesy of *Netherlands J. agri. Sci.*)

Variate	a_1	a_2	a_3	a_4	h^2
1. pH (KCl)	0·20	0·24	0·41	−0·09	0·27
2. Content of organic matter	−0·38	0·75	0·02	−0·15	0·73
3. Content of clay particles	−0·10	0·80	0·04	−0·20	0·69
4. Content of fine sand	0·17	0·68	−0·13	0·13	0·53
5. U figure (specific surface)	−0·04	0·88	−0·14	0·13	0·81
6. P-citr (P status)	0·42	0·08	0·45	−0·20	0·43
7. K-HCl (K status)	0·62	−0·04	0·06	−0·04	0·39
8. MgO content (Mg status)	0·27	0·61	0·44	0·02	0·64
9. Thickness of humus layer	0·43	−0·11	−0·02	0·22	0·25
10. Distance of farm (management)	−0·34	0·05	−0·09	−0·34	0·24
11. Ground water level	0·52	−0·39	−0·09	0·01	0·42
12. Fluctuation ground water level	−0·29	0·59	−0·05	0·17	0·46
13. Nitrogenous fertilization	0·59	0·05	0·45	0·11	0·57
14. Phosphatic fertilization	0·37	0·01	0·84	0·00	0·84
15. Potassic fertilization	0·43	0·18	0·71	0·04	0·72
16. Grade of quality pasture	0·57	−0·09	−0·04	0·20	0·37
17. *N* content of grass	0·36	−0·02	0·00	0·63	0·53
18. P_2O; content of grass	0·47	−0·13	0·14	0·58	0·59
19. K_2O content of grass	0·68	−0·13	−0·08	0·30	0·58
20. MgO content of grass	−0·03	−0·05	0·22	0·15	0·07
21. Spring yield of % of annual yield	0·90	−0·01	−0·09	−0·31	0·91
22. Spring yield	0·98	0·03	−0·02	−0·11	0·97
23. Annual yield	0·75	0·07	0·03	−0·38	0·71
Variance (latent root)	5·66	3·48	2·12	1·14	12·72

the factors responsible for the differences. Twenty-three soil and vegetational attributes were measured, so that $n=23$ and $N=50$. The variates are shown in Table 3, from which it can be seen that the majority fall roughly into five categories: physical soil characteristics, soil base status, fertilizer application, mineral content of grass and dry weight yield. Yield refers to the overall or total yield of stands, the contributions of individual species to this not being determined. The data were analysed by an approximation of the principal components technique. The account which follows is based on results presented by Ferrari *et al.* (1957), but does not necessarily express their views.

The results of the ordination of attributes are summarized in Table 3 and Fig. 11. Table 3 contains the component loadings of the variates, while

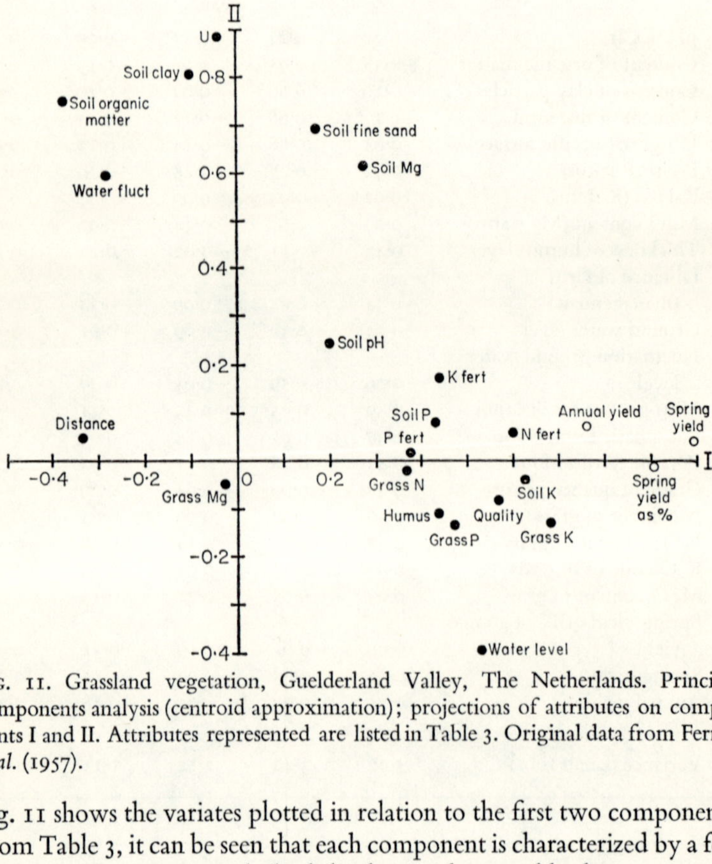

FIG. 11. Grassland vegetation, Guelderland Valley, The Netherlands. Principal components analysis (centroid approximation); projections of attributes on components I and II. Attributes represented are listed in Table 3. Original data from Ferrari *et al.* (1957).

Fig. 11 shows the variates plotted in relation to the first two components. From Table 3, it can be seen that each component is characterized by a few variates with conspicuously high loadings. These enable the components to be interpreted as follows: F_1: yield; F_2: soil physical properties; F_3: fertilizer application; F_4: mineral composition of grass. Variates with loadings of this kind on F_1, (21, 22, 23) and F_2, (2, 3, 4, 5) respectively, are uniquely accounted for by these components. This is apparent from the near-zero loadings of the variates on all other components. In particular, since the variates of each set have near-zero loadings on the component characterized by the other, the variates concerned are evidently independent. On the other hand, variates exemplifying F_3, (6, 13, 14, 15) and F_4, (17, 18, 19) are not uniquely associated with these components; they also have relatively high loadings on F_1. Similarly, the variances of ground water level and fluctuation (11, 12), are distributed over components F_2 and F_1. The ecological implication of these relationships is that, while yield might conceivably be dependent on variates correlated with F_1, it seems likely to be largely independent of variation in the soil physical properties examined. Other variates with high loadings on F_1, and therefore to which yield might be related, are soil potassium status (7) and distance of farm (10). The latter is a convenient index of farm management. Many of these points are brought out by Fig. 11. The most striking feature of this figure is the clear distinction between variates associated either with F_1 or F_2. Attention is also drawn to the relatively large distance of most of the variates from the origin, and to the very high loadings of the indices of yield on F_1 and their near-zero loadings on F_2. Of the variates which are associated with F_1 (see Fig. 11), experience suggests that soil potassium and phosphate status (7, 6), soil moisture relations (11, 12) and distance (10) are those most likely to influence yield. Soil potassium and phosphate, with positive loadings, are evidently related to yield in a direct sense. Distance (equivalent to poor management), on the other hand, with a negative loading, is inversely related to yield. The inverse relationship between distance and soil potassium and phosphate status is also worth noticing. The independence of all three variates from soil physical properties is clearly implied by their extremely small loadings on F_2. The behavior of the measures of soil moisture differs in an important respect from that of the variates just considered. Although ground water level and fluctuations have moderately high loadings (in opposite senses) on F_1, and are therefore correlated with yield, they are also related to F_2. Soil moisture relations are evidently dependent on the soil physical properties measured. Therefore it seems that, while the soil physical properties do not themselves influence yield, they are of significance in their partial control of soil moisture conditions. The contrast in the behavior of the two indices of soil moisture (Fig. 11) is a feature of some interest and can be explained in the following way. The opposite loadings of the variates on F_2 apparently reflects that drier soils are those which experience the greatest fluctuations in water level. The differing effects of these conditions on yield is shown by the contrast in the loadings of the variates on F_1. Dry soils with fluctuating water conditions are shown by the negative loading of ground water fluctuation (12) on F_1 to be inversely related to yield; moist soils with balanced water relations are shown by the positive loading of ground water level (11) on F_1 to be directly related to yield.

The relationships considered by no means exhaust all those provided. They are, however, sufficient to indicate the kind of information to be obtained from ordination. The salient features of the analysis may now be drawn together. The direct (i.e., immediate) controls of yield appear to be soil potassium and phosphate status and soil moisture. These factors can themselves be related to more remote influences. Soil potassium and phosphate are likely to be dependent on fertilizer applications, an aspect of management, and soil moisture on external water supply and soil physical properties. Yield may therefore be regarded ultimately as a function of management, water supply and soil texture, through the control exerted by them on soil potassium and phosphate status and soil moisture conditions. On the basis of these results, Ferrari *et al.* (1957) made a selection of five variates as the starting point for an analysis designed to explore the relationships exposed further. In retrospect, many of the relationships found may appear largely self-evident. This is partly because the ecology of temperate pastures is already reasonably well known, and partly a reflection of the nature of the data used. Overall yield, as a measure of vegetation, provides no information on the effects of differences in species composition on yield. Appreciable differences in yield are likely to exist between species with respect to the controls identified (cf. Fig. 8). In general, such differences between species with respect to the controls operative in a given community, and, indeed, the nature of the controls themselves, are likely to be poorly delineated. There seems little reason to doubt that ordination could do much to resolve them.

ACKNOWLEDGEMENTS

The greater part of this work was carried out at the Department of Entomology of the University of Kansas and this paper is contribution 1404 from that department. Financial aid was provided in part by Public Health Service Research Grant GM-11935 from the National Institute of General Medical Sciences to Robert R. Sokal. I should like to acknowledge my indebtedness to Dr F. James Rohlf, of the Entomology Department, The University of Kansas, for certain of the ideas expressed here. I am also much indebted to Dr R.R.Sokal, for making my visit to Kansas possible and for his continued support and encouragement.

The analyses of grassland data from North Wales were performed at the Computation Center of the University of Kansas (GE 625), using the FORTRAN IV Numerical Taxonomy SYStem of multivariate statistical programs (NT—SYS) by F. James Rohlf, J. Kishpaugh and R. Bartcher of the Entomology Department, The University of Kansas, Lawrence, Kansas.

REFERENCES

Anderson T.W. (1958). *An introduction to multivariate statistical analysis.* New York.

Bray J.R. and Curtis J.T. (1957). An ordination of the upland forest communities of southern Wisconsin. *Ecol. Monogr.* **27**, 325–349.

Cattell R.B. (1966). The meaning and strategic use of factor analysis. In *Handbook of multivariate experimental psychology* (ed. R.B. Cattell), pp. 174–243. Chicago.

Cooley W.W. and Lohnes P.R. (1962). *Multivariate procedures for the behavioral sciences.* New York.

Curtis J.T. (1959). *The vegetation of Wisconsin: an ordination of plant communities.* Madison, Wisconsin.

Dagnelie P. (1960). Contribution à l'étude des communautés végétales par l'analyse factorielle. (Engl. summ.). *Bull. Serv. Carte phytogéogr. Sér. B,* **5**, 7–71, 93–195.

Dagnelie P. (1965a). L'étude des communautés végétales par l'analyse statistique des liaisons entre les espèces et les variables écologiques: principes fondamentaux. (Engl. summ.). *Biometrics,* **21**, 345–361.

Dagnelie P. (1965b). L'étude des communautés végétales par l'analyse statistique des liaisons entre les espèces et les variables écologiques: un example. (Engl. summ.). *Biometrics,* **21**, 890–907.

Ferrari Th.J., Pijl H. and Venekamp J.T.N. (1957). Factor analysis in agricultural research. *Netherlands J. Agric. Sci.* **5**, 211–221.

Fruchter B. (1954). *Introduction to factor analysis.* New York.

Gittins R. (1968). Trend-surface analysis of ecological data. *J. Ecol.* **56**, 845–69.

Goodall D.W. (1954a). Vegetational classification and vegetational continua. *Angew. PflanzSoz.* **1**, 168–182.

Goodall D.W. (1954b). Objective methods for the classification of vegetation. III. An essay in the use of factor analysis. *Aust. J. Bot.* **2**, 304–324.

Goodall D.W. (1963). The continuum and the individualistic association. (French summ.). *Vegetatio,* **11**, 297–316.

Gower J.C. (1966). Some distance properties of latent root and vector methods used in multivariate analysis. *Biometrika,* **53**, 325–338.

Gower J.C. (1967). Multivariate analysis and multidimensional geometry. *The Statistician,* **17**, 13–28.

Greig-Smith P. (1957). *Quantitative plant ecology.* 1st edn. London.

Greig-Smith P. (1964). *Quantitative plant ecology.* 2nd edn. London.

Harman H.H. (1967). *Modern factor analysis.* 2nd edn. Chicago.

Horst P. (1965). *Factor analysis of data matrices.* New York.

Ivimey-Cook R.B. and Proctor M.C.F. (1967). Factor analysis of data from an east Devon heath: a comparison of principal component and rotated solutions. *J. Ecol.* **55**, 405–413.

Kendall M.G. (1957). *A course in multivariate analysis.* London.

Lawley D.N. and Maxwell A.E. (1963). *Factor analysis as a statistical method.* London.

Moore C.S. (1965). Inter-relations of growth and cropping in apple trees studied by the method of component analysis. *J. hort. Sci.* **40**, 133–149.

Morrison D.F. (1967). *Multivariate statistical methods.* New York.

Orloci L. (1966). Geometric models in ecology. I. The theory and application of some ordination methods. *J. Ecol.* **54**, 193–215.

Orloci L. (1967). Data centering: a review and evaluation with reference to component analysis. *Syst. Zool.* **16**, 208–212.

Pearce S.C. (1965). *Biological statistics: an introduction.* New York.

Rayner J.H. (1966). Classification of soils by numerical methods. *J. Soil Sci.* **17**, 79–92.

Reyment R.A. (1961). Quadrivariate principal component analysis of *Globigerina yeguaensis. Stockh. Contr. Geol.* **8**, 17–26.

Reyment R.A. (1963). Multivariate analytical treatment of quantitative species associations: an example from palaeoecology. *J. Animal Ecol.* **32**, 535–547.

Seal H. (1964). *Multivariate statistical analysis for biologists.* London.

Sokal R.R., Daly H.V. and Rohlf F.J. (1961). Factor analytical procedures in a biological model. *Univ. Kans. Sci. Bull.* **42**, 1099–1121.

van Groenewoud H. (1965). Ordination and classification of Swiss and Canadian coniferous forests by various biometric and other methods. *Ber. geobot. Inst. ETH, Stiftg. Rubel, Zürich 36*, 28–102.

Webb L.J., Tracey J.G., Williams W.T. and Lance G.N. (1967). Studies in the numerical analysis of complex rain-forest communities. I. A comparison of methods applicable to site/species data. *J. Ecol.* **55**, 171–191.

Williams W.T. and Dale M.B. (1965). Fundamental problems in numerical taxonomy. *Adv. bot. Res.* **2**, 35–68.

Yarranton G.A. (1967). Principal components analysis of data from saxicolous bryophyte vegetation at Steps Bridge, Devon. I. A quantitative assessment of variation in the vegetation. *Canad. J. Bot.* **45**, 93–115.

AN INVESTIGATION OF THE ECOLOGICAL SIGNIFICANCE OF LIME-CHLOROSIS BY MEANS OF LARGE-SCALE COMPARATIVE EXPERIMENTS

J. P. GRIME and J. G. HODGSON
Department of Botany, University of Sheffield

INTRODUCTION

In the study of the determination of plant distribution by environment there are two main objectives. The first is the solution of individual problems, and the second is the attempt to put together the results of separate investigations in such a way that the data achieves a predictive value with respect to other species and environments. Where the ecologist, however successfully, adopts an *ad hoc* approach to attain the first objective, he may contribute little towards the second. With this difficulty in mind, an attempt has been made in the introduction to this paper to evaluate three main approaches to the study of the mechanisms affecting plant distribution. The remainder of the paper consists of an illustration of a method of study designed to increase the possibility of extrapolation.

The Direct Approach

In the British Isles where a flora, including many species which are ubiquitous on a geographical scale coincides with an intimate and complex mosaic of vegetation types, it is likely that plant distribution is primarily an expression of tolerance of environment rather than effectiveness of dispersal. Under these circumstances it may be considered legitimate to confine the investigation of plant distribution to the attempt to recognize the reasons for failure in the environments from which a plant is excluded. As such, the study may be regarded as a form of plant pathology and seems to represent the most direct approach conceivable to the study of plant distribution.

The objective in this approach is to observe and analyse the process of failure as it occurs in the field. Investigations of this kind may clearly only be carried out at sites where natural transgression of vegetation frontiers takes place frequently and where the circumstances of such invasions are sufficiently defined and are amenable to study. Sharp boundaries between plant communities are particularly suitable for this purpose (Grime 1963b). Localized studies of this type are most effective where the vegetation pattern is imposed by rapid and catastrophic effects of environment. More typically however, fatalities are due to a complex of factors and in particular, the contribution of mineral nutritional factors remains obscure. Seedlings may persist for an indefinite period in a state of chronic nutrient deficiency and whilst it is often possible to recognize terminal phenomena, it is difficult to measure the extent to which plants may be predisposed to killing factors, by nutritional disorders.

In the most favourable situations therefore, it is often difficult to press the direct approach to a conclusion. Elsewhere it is difficult to apply the method with confidence either because natural invasions are infrequent and of low density or because the vegetation pattern is determined by occasional phenomena. Where the difficulty arises from a scarcity of invading propagules the inoculum may be increased artificially (Rorison 1960a; Davison 1964). Here it is probably essential that the simulated invasion should reproduce exactly the circumstances of natural invasion.

The direct approach is often extremely difficult to apply in older perennial communities where a plant distribution is the result of a long history of events arising from interactions between a plant population and a changing environment. In vegetation of some antiquity therefore it may be impossible to recognize mechanisms critical in the past, by examination of the contemporary situation.

Where the direct method is inappropriate, one may resort to deducing the critical mechanisms. Indirect approaches fall under two broad headings (a) Comparisons between environments, (b) Comparisons between plants.

Indirect approaches

(*a*) *Comparisons between environments*

In numerous investigations, the attempt has been made to recognize the mechanisms controlling plant distribution by comparing the environments in which a plant occurs with those from which it is excluded. In this approach, the amplitude (Curtis and McIntosh 1951; Goodall 1954) or scale (Greig-Smith 1964) of variation in plant density or vigour is compared with the amplitude or scale of environmental variation on one or more selected occasions. This method has been particularly successful in situations where the gross pattern of plant distribution has been imposed by land management in the present or past (e.g. Anderson 1961). With respect to the influence on plant distribution of more elusive factors such as soil nutrients, the approach often achieves little penetration. Correlations between plant frequency and soil features such as pH are readily obtained but these do not always lead to an understanding of mechanism. Where evidence of mechanism is drawn from spatial correlations between plant frequency and environmental variation, there are many difficulties both of procedure and of interpretation. These have been considered previously (Waring and Major 1964; Grime 1965) and may be restated as follows: (i) Practically, the number of environmental variables which can be measured and the frequency with which measurements can be carried out is limited. (ii) Some variables, e.g. mineral nutrient availability, are not easily measurable. (iii) In situations of particular interest, e.g. community boundaries, many environmental changes occur. All are correlated with a change in vegetation though few, if any, are causes of the change. (iv) Many correlations are due to the determination of environment by vegetation rather than the reciprocal relationship. (v) Frequently there is a time-lag between an event in the environment and the resulting change in the vegetation. (vi) There are great differences in tolerance between seedlings and mature plants. Where distribution is determined by an interaction of seedling and environment, correlations involving mature plants may be misleading.

(b) Comparisons between plants

Mechanisms excluding a species or ecotype from a habitat may be suggested on the basis of differences in requirements or in tolerance which distinguish the plant from those successful in the habitat concerned. To this end comparative measurements have been carried out in experiments varying in complexity from transplant studies in ecosystems to enzyme studies *in vitro*. Regardless of their complexity, such comparisons have in common the difficulty of establishing the importance in the field of the differences they reveal. The difficulties are of two kinds. First, interspecific differences are numerous and it is therefore difficult to determine which, if any, are of ecological significance. This may be resolved to a certain extent by confining attention to the more consistent differences between large numbers of species successful and unsuccessful in the habitat concerned. Better still, ecotypes where available, may be expected to differ with respect to a smaller number of more critical characteristics.

The second type of difficulty arises from the fact that there is no guarantee that the basis of the comparison (i.e. the experimental conditions and the plant features chosen for examination) is one which will expose the differences of importance in the field. However, comparisons in which the rooting medium and climate are simplified have the advantage that it is easier to ensure that differences noted between plants are occurring with respect to the same known variable or group of variables. Experiments in which field conditions are more closely simulated (e.g. by use of natural soils) may fractionally increase the probability of finding the critical differences but they transfer to the laboratory much of the environmental complexity which has defied analysis in the field.

There is no ideal approach to the study of plant distribution and it seems likely that each ecologist will continue to adopt an amalgam of direct and indirect methods peculiar to each problem. Nevertheless it is desirable that methods are selected with some regard to the basis on which information from different investigations may be integrated.

There are at least two possible bases for a classification of the mechanisms affecting plant distribution.

(1) a classification based on the environments in which they operate.
(2) a classification based on the plant components affected.

At Sheffield, we are working with an eye to the second possibility. In practice, groups of plant species or ecotypes drawn from particular habitats are examined under a variety of standardized conditions, some involving a particular environmental stress. In this way a list of attributes may be compiled for each species or ecotype and it is possible to recognize unusual characteristics common to plants from similar habitats. These characteristics, in certain cases, provide an indication of the plant components which are (or have been) the subject of natural selection in the habitat concerned; as such they may provide a clue to mechanisms which exclude other plants from the same habitat. A further possibility lies in recognition of components of the flowering plant which have a limited capacity for phenotypic adjustment. Where such components occur in fundamental plant processes, it seems likely that they have provided a recurrent focus for conflicting selection pressures during the evolution of the flowering plant. The comparative approach gives the opportunity to analyse both the evolution of the flowering plant and the distribution of contemporary genotypes in terms of internal constraints which limit the potentiality of the flowering plant. The constraints exercised by more obviously phenotypically-inflexible plant characters, e.g. quantity of seed reserve (Grime and Jeffrey 1965) are immediately evident. However, the existence of additional and less obvious 'components of limited potentiality' is suggested first by the apparent relationships in flowering plants between success in one particular type of environment and failure in another and, secondly, by the widespread coincidence of particular adaptive characters and physiologically related features disadvantageous in other environments (Grime 1965, 1967).

In this paper an attempt is made to define one such 'component of limited potentiality' in the mineral nutrition of the flowering plant.

CHOICE OF SPECIES FOR COMPARATIVE STUDY

The comparative experiments described in this paper are concerned with susceptibility to two nutritional disorders, lime chlorosis, which occurs on calcareous substrata, and, aluminium toxicity, prevalent on acid soils. In order to examine the range of sensitivity to these disorders it was clearly desirable to use species known to differ in their affinity for acid and calcareous substrata. Using data collected from the extremely varied grasslands of the Sheffield region by Dale *et al.* (1965–68) it has been possible to classify the majority of the common species with respect to frequency of occurrence on various geological substrata and on soils of different pH. For the present work species were selected on the basis of the relative frequency of occurrence over the range in surface (0–3 cm) pH provided by 340 random samples from grasslands on six geological substrata (Toadstone, Millstone Grit, Coal Measures, Carboniferous Limestone, Magnesian Limestone and Bunter Sandstone). A detailed argument supporting the use of the relationship between frequency of occurrence and surface pH as an index of edaphic tolerance has been given elsewhere (Grime 1963a). It is clear that pH : frequency distributions such as those illustrated in Fig. 1 are compound effects of soil, climate and biotic factors; they nevertheless describe the range of soil types experienced by the species in the field. The pH-frequency diagrams obtained confirm that it would be misleading to attempt a separation of the species into two distinct groups, calcicoles and calcifuges. Instead there is a continuous array of distribution containing at least seven types: (1) strongly calcifuge, e.g. *Deschampsia flexuosa*, (2) strongly calcicole, e.g. *Scabiosa columbaria*, (3) catholic, e.g. *Festuca ovina*, (4) widely-ranging with a peak on acid soils, e.g. *Rumex acetosa*, (5) widely-ranging with a peak on neutral soils, e.g. *Teucrium scorodonia*, (6) widely-ranging with a peak in the pH range 4–6, e.g. *Agrostis tenuis* and (7) restricted to soils in the pH range 4–6, e.g. *Potentilla erecta*.

In some of the experiments additional species have been used, including several which occur in vegetation other than grassland. These have been characterized in Fig. 1, by reference to the illustrated pH : frequency distribution which most closely resembles that of the species concerned. It is noteworthy that grasses are prominent among the species which are widely ranging with respect to soil pH. However, *F. ovina* is the only species of frequent occurrence on both highly acidic and highly calcareous soils. In view of the known existence (Bradshaw 1952, Snaydon 1962) of

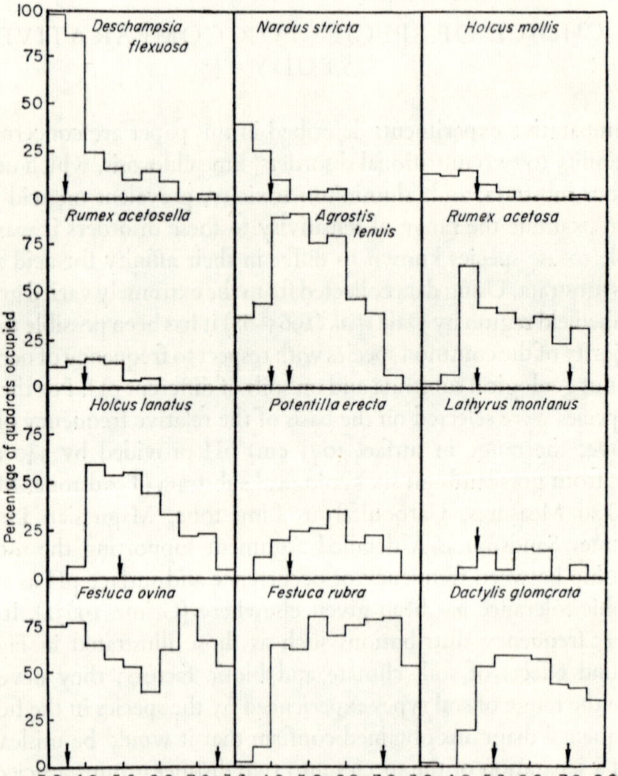

FIG. 1. Diagrams to show the frequency of occurrence of twenty-four species over the range in surface pH (0–3 cm) provided by 340 random m^2 quadrats in established grassland. Data was collected from 41 sites in the extremely varied grasslands which occur in the vicinity of Sheffield. In the histograms, the minimum number of samples in a half-unit pH class is twelve. The arrows indicate the pH of the surface soil at the sites of seed collections used in experimental work.

The pH : frequency distributions of additional species mentioned in this paper may be roughly classified as follows:

(1) Strongly calcifuge (*Deschampsia flexuosa* type)—*Calluna vulgaris, Erica tetralix, Eriophorum angustifolium, E. vaginatum, Galium saxatile, Rumex acetosella.*

edaphic ecotypes at least among wide-ranging species, it was considered necessary to specify the soil type of the areas from which seeds were collected for the experiments. The surface pH at each seed source is indicated on the pH frequency diagrams in Figs. 1 and 2 and elsewhere as a suffix to the specific name of the plant, e.g. *D. flexuosa* (3·5).

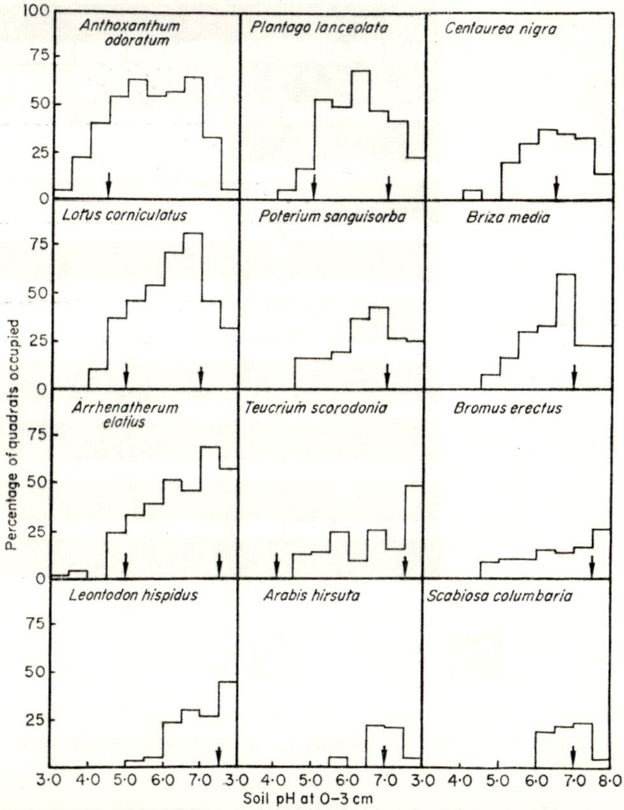

(2) Strongly calcicole (*Scabiosa columbaria* type)—*Arabidopsis thaliana, Carlina vulgaris, Cirsium acaule, Draba muralis, Helianthemum chamaecistus, Hypericum hirsutum, H. montanum, Koeleria cristata, Plantago media, Sesleria caerulea, Thymus drucei.*

(3) Catholic (*Festuca ovina* type)—none.

(4) Widely ranging with peak on acid soils (*Rumex acetosa* type)—*Molinea caerulea.*

(5) Widely ranging with peak on neutral soils (*Teucrium scorodonia* type)—*Trisetum flavescens, Urtica dioica.*

(6) Widely ranging with peak in pH 4–6 (*Agrostis tenuis* type)—*Galium verum, Juncus effusus, Lathyrus pratensis, Serratula tinctoria, Ulex europaeus.*

(7) Restricted to soils in pH range 4–6 (*Potentilla erecta* type)—*Betonica officinalis, Digitalis purpurea, Genista tinctoria, Hypericum pulchrum, Viola lutea.*

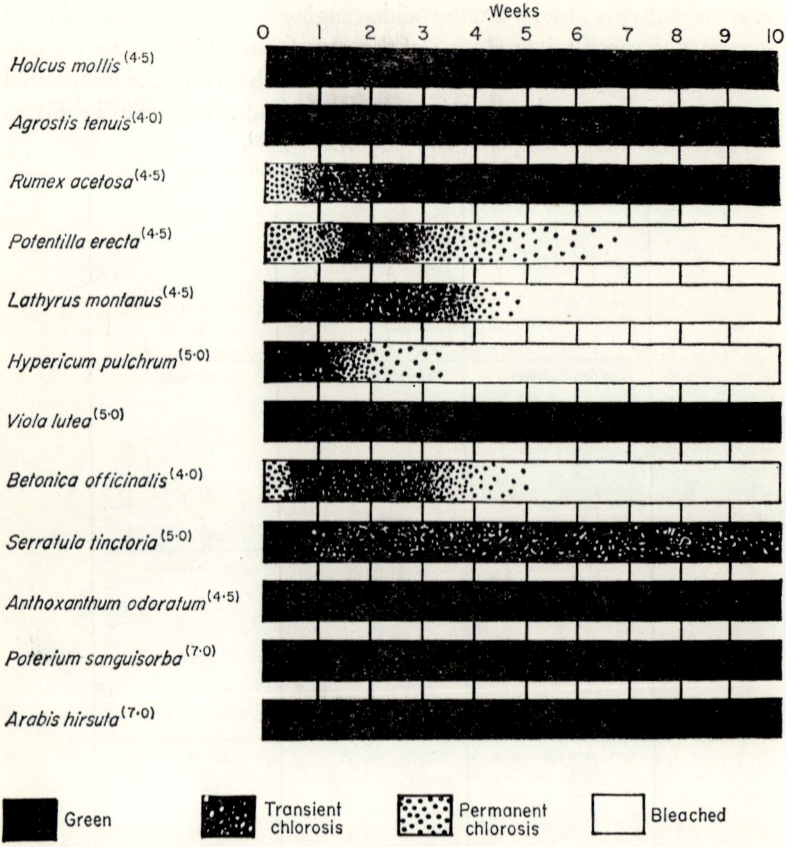

FIG. 2. The course of leaf chlorosis in seedlings of a range of grassland plants germinated and grown for 10 weeks on a Coombsdale rendzina of moderate chlorosis potential.

COMPARATIVE EXPERIMENTS ON SUSCEPTIBILITY TO LIME-CHLOROSIS

A number of inter-specific comparisons have been made with respect to susceptibility to lime-chlorosis and the results of these are summarized in Table 1. There is no certainty that the chloroses recorded in different investigations are due to the same causes, although there is considerable evidence in the majority of cases that iron status of the plant is implicated. Only a few of these experiments have involved native plants and in consequence it is difficult to generalize from the results with regard to the

PLATE 1. Transient lime-chlorosis in the terminal leaves of a seedling of *Viola lutea*, 16 weeks after germination on a Coombsdale rendzina of moderate chlorosis potential.

PLATE 2

PLATE 3

PLATE 4. Seedlings of *Betonica officinalis* about 30 weeks after germination.

1. Grown on a brown earth.

2–6. Chlorotic plants grown on a rendzina of moderate chlorosis potential.

2. Leaves expanded but showing interveinal chlorosis.

3–6. Young leaves bleached and stunted.

5. Many unexpanded laminae at shoot apex.

PLATE 2. A seedling of *Potentilla erecta* at the five-leaf stage, growing on a Coombsdale rendzina of moderate chlorosis potential. Numbers 1–5 are the foliate leaves in order of appearance. Chlorosis may be observed in the youngest expanded leaf.

PLATE 3. The seedling illustrated in Plate 2 at the eight-leaf stage. Numbers 1–8 are the foliage leaves in order of appearance. A progressive chlorosis is apparent.

PLATE 5. Asymmetrical response in leaf colour and leaf expansion in the shoot of a chlorotic transplant of *Betonica officinalis* grown on a calcareous soil but provided with ferrous sulphate through a single root (arrowed).

TABLE I

Recorded differences between flowering plants with respect to susceptibility to lime-chlorosis

(a) INTERSPECIFIC DIFFERENCES

SPECIES		AUTHORITY
Susceptible to chlorosis	*Resistant to chlorosis*	
Galium saxatile	*G. pumilum*	Tansley (1917)
Deschampsia flexuosa	*Cannabis sativa* *Secale cereale* *Sinapis alba*	Olsen (1958)
Galium saxatile *Holcus mollis*	*Asperula cynanchica* *Scabiosa columbaria*	Rorison (1960a,b)
Paspalum dilatum	*Chloris gayana*	Mokady (1961)
Eucalyptus dalrympliana *E. gomphocephala* *E. gunnii*	17 *Eucalyptus* species	Lacaze (1963)
Lathyrus montanus	*L. pratensis*	Grime (1965)
Deschampsia flexuosa	*Arrhenatherum elatius* *Holcus mollis* *Koeleria cristata* (as *K. gracilis*)	Woolhouse (1966)
E. baxteri	*E. diversifolia* *E. incrassata*	Parsons and Specht (1967)

(b) INTRASPECIFIC DIFFERENCES

SPECIES	AUTHORITY
Glycine max (soybean)	Weiss (1943) Brown and Holmes (1955)
Trifolium repens	Snaydon (1962)
Teucrium scorodonia	Hutchinson (1966)
Plantago lanceolata	Payne (1966)
Zea mays	Brown (1967)

relative sensitivity of species from calcareous and acid soils. Moreover, since the experiments involve different soils and species, it is not possible to estimate the range in susceptibility represented in these investigations.

Nevertheless, the results suggest that calcifuge plants are unusually susceptible to lime-chlorosis. This conclusion is supported by the demonstration chlorosis-sensitive ecotypes from acid soils in *Trifolium repens*

(Snaydon 1962), *Teucrium scorodonia* (Hutchinson 1967c) and *Plantago lanceolata* (Payne 1966).

In the course of exploratory experiments (Grime 1958) it became clear that lime-chlorosis was readily induced in calcifuge species by growing these plants on immature soils from the Carboniferous limestone of Derbyshire. With the development of a rapid method of measuring leaf chlorosis by colour matching (Grime 1961) it was possible to compare the chlorosis potential of a wide range of soil types (Grime 1964), to carry out large scale comparisons of chlorosis susceptibility on the seedlings of Derbyshire plants (Tables 2 and 3) and to measure the changes in the intensity of chloroses which often occur during the seedling phase (Fig. 2).

Comparative experiments showed clearly that lime-chlorosis is peculiar to strong calcifuges and to plants which are to some extent restricted to acid substrata. Although all the calcicolous species examined were resistant, not all calcifuges were sensitive to lime-chlorosis. With the exception of *D. flexuosa*, grasses from acidic soils, e.g. *Nardus stricta*, *Holcus mollis* and *Agrostis tenuis*,were surprisingly resistant to chlorosis. Moreover, susceptible species differed in the severity and course of development of symptoms during the seedling phase. For example, in the same experiment (Fig. 2) transient chlorosis occurred in *Rumex acetosa* during the first three weeks after germination, and appeared after 14 weeks in seedlings of *Viola lutea* (Plate 1), whilst in *Potentilla erecta* (Plates 2 and 3) there was a rapid, severe and progressive chlorosis.

On soils of moderate chlorosis potential (Table 3) there was no evidence of ecotypic variation with respect to chlorosis and species of common occurrence on calcareous and mildly acidic soils resembled calcicoles in being resistant to chlorosis.

If data from three comparative experiments are combined with those obtained by Hutchinson (1967a) who carried out an extensive comparison of seedling chlorosis on a Derbyshire rendzina, we can classify many of the commoner species of grassland, heathland and bog with respect to chlorosis sensitivity (Fig. 3).

Although the main purpose of these investigations was to examine the distribution of chlorosis sensitivity among plants of different habitats, two additional facts emerge. Firstly, appearance of chlorosis coincides with other symptoms including stunting of the youngest leaves (Plate 4) and inhibition of root growth, an effect which often anticipates the appearance of symptoms in the shoot. A connection between chlorosis and root stunting is suggested by the failure to detect root inhibition in species resistant to chlorosis (Table 4). A second feature of these experiments is the evidence

TABLE 2

Comparison between seedlings of six species with respect to leaf colour and the concentration of iron, manganese and aluminium in the shoot (Peaslee and Grime, unpublished data).

Seedlings of calcicolous species, e.g. *Scabiosa columbaria*, failed rapidly on the acid soil. To circumvent this difficulty and to obtain sufficient material for spectographic analysis the following procedure was adopted. The seedlings were grown for 7 weeks on the calcareous soil, at which time the seedlings in each container were separated into two matched sets. Measurements of leaf colour index (Grime 1964) and of Fe, Mn and Al concentration in the shoot were carried out on the first set. The seedlings of the second set were washed carefully in distilled water and transplanted onto the acid soil. After a further 2 weeks, comparable measurements were carried out on the second set.

Leaf colour measurements: *transplanting to acid soil associated with a significant ($P<0{\cdot}05$) change (yellow→green) in leaf colour.

Plant analysis: Values above the line refer to concentration of element in the shoot after 7 weeks growth on the calcareous soil. Values below the line refer to concentration of element after 7 weeks growth on the calcareous soil and 2 weeks after transplanting out to the acid soil.

	LEAF COLOUR INDEX			mg % dry weight of shoot		
SPECIES	Rendzina pH 7·3	Brown earth pH 4·0	Difference	Iron	Manganese	Aluminium
Deschampsia flexuosa	15·85	28·95	13·10*	0·015	0·002	0·015
(3·5)				0·077	0·118	0·282
Lathyrus montanus	13·65	26·00	12·35*	0·015	0·025	0·008
(4·5)				0·049	0·085	0·087
Digitalis purpurea	24·80	30·10	5·30*	0·012	0·006	0·009
(4·5)				0·080	0·109	0·217
L. pratensis	21·94	28·54	6·60*	0·022	0·009	0·010
(5·0)				0·058	0·200	0·102
Briza media	27·45	30·04	2·59	0·008	0·007	0·001
(7·0)				0·027	0·037	0·053
Scabiosa columbaria	30·05	32·00	1·95	0·013	0·003	0·013
(7·0)				0·081	0·097	0·195

TABLE 3

A comparison of susceptibility to lime-chlorosis in twenty-two species of flowering plants.

Ten seedlings were germinated and grown on each of two fertile neutral soils, a Derbyshire mull-rendzina of moderate chlorosis potential, the other a garden soil of low chlorosis potential. Measurements of leaf colour index were carried out on the youngest expanded leaves of each seedling at weekly intervals and chlorosis (indicated ★) was recognized as a significant ($P < 0{\cdot}05$) depression in L.C.I. of the seedlings on the rendzina in comparison with those on the garden soil.

SPECIES		DIFFERENCE IN LEAF COLOUR INDEX			
		7 days	14 days	21 days	28 days
Deschampsia flexuosa	(3·5)	−6·6★	−1·7★	−9·5★	−13·5★
Nardus stricta	(4·0)	−0·1	2·5	0·5	−2·4
Holcus mollis	(4·5)	−1·0	−1·3	−1·7	−0·3
Digitalis purpurea	(4·5)	−0·4	0·1	−0·2	0·5
Agrostis tenuis	(4·5)	0·2	−0·9	−0·3	0·2
Ulex europaeus	(4·5)	−0·4	−5·2★	−13.2★	−10·4★
Lathyrus montanus	(4·5)	1·2	−5·1★	10·8★	−10·3★
Betonica officinalis	(4·0)	−2·0★	−3·8★	−6·7★	−14·7★
Lathyrus pratensis	(6·0)	−1·0	−1·8	0·5	−0·2
Urtica dioica	(6·5)	0·4	0·4	−0·3	−0·6
Poterium sanguisorba	(7·0)	0·0	−0·4	0·0	0·2
Briza media	(7·0)	−1·0	0·4	0·4	0·4
Bromus erectus	(7·5)	−0·2	−0·2	−0·6	−0·2
Hypericum hirsutum	(7·5)	1·3	0·7	0·4	−1·5
Arabis hirsutum	(7·0)	1·1	0·5	0·7	0·2
Arabidopsis thaliana	(7·5)	1·0	0·4	−0·6	0·2
Rumex acetosa	(4·5)	−2·3★	−1·2★	−0·1	0·2
R. acetosa	(7·0)	−3·5★	−0·9★	0·1	0·0
Festuca ovina	(3·5)	−0·8	−0·2	−2·1	−0·7
F. ovina	(7·5)	−0·4	0·0	−0·1	0·4
Dactylis glomerata	(5·0)	0·0	0·8	1·0	1·6
D. glomerata	(7·0)	−1·0	0·7	0·0	0·7
Plantago lanceolata	(5·0)	0·0	−0·2	0·7	0·9
P. lanceolata	(7·0)	0·0	0·9	0·6	0·4
Lotus corniculatus	(5·0)	−1·2	−0·2	0·1	0·6
L. corniculatus	(7·0)	−0·6	−2·4	−1·4	0·1
Arrhenatherum elatius	(5·0)	0·0	−0·2	0·7	0·9
A. elatius	(7·5)	−0·8	0·5	0·7	1·8

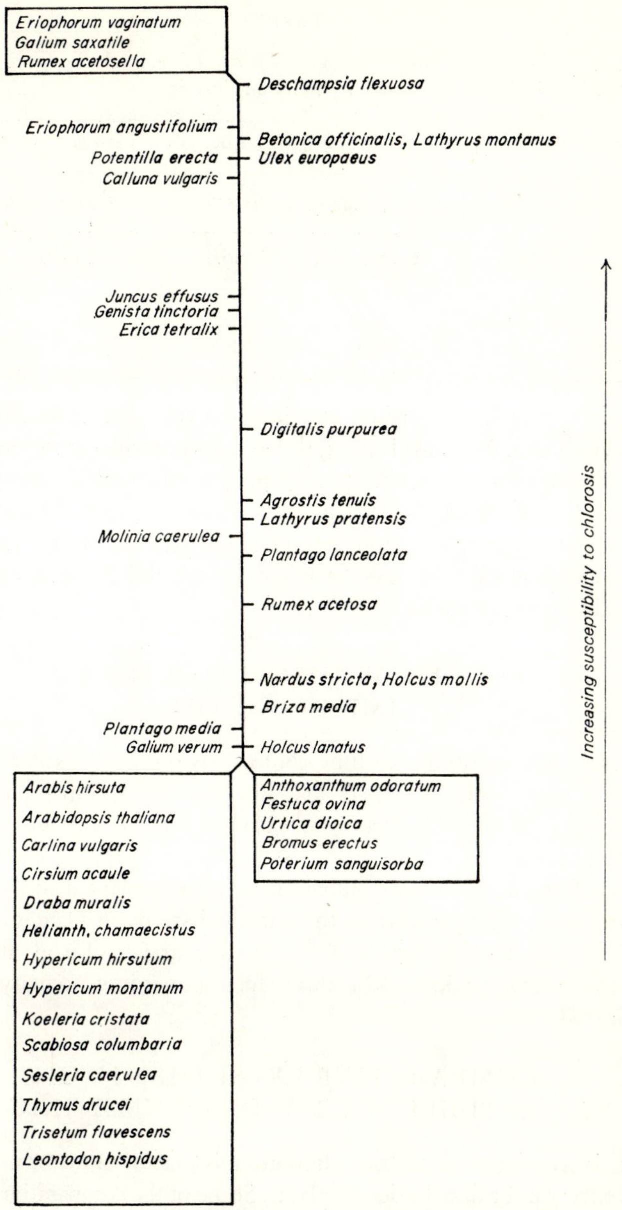

FIG. 3. Scheme illustrating the susceptibility to lime-chlorosis in a range of flowering plants from habitats in grassland, heath and bog in the vicinity of Sheffield. Estimates of chlorosis susceptibility are based on data from comparative experiments such as those referred to in Tables 2 and 3 and Fig. 2 and from Hutchinson (1967a).

TABLE 4

Comparison of root growth on a brown earth and on a calcareous soil in two species which differ in sensitivity to lime-chlorosis.

Measurements were carried out 80 days after germination at which time chlorosis was apparent in *Betonica officinalis* but not in *Genista tinctoria*.

	Betonica officinalis (4·0)		*Genista tinctoria* (5·0)	
	Brown earth	Rendzina	Brown earth	Rendzina
Mean dry wt. of root (mg.)	20·2 ±4·9	4·7 ±2·6	7·1 ±1·2	7·2 ±2·9

suggesting that iron nutrition is causally involved in lime-chlorosis on Derbyshire soil. Analysis (Table 2) shows that the level of iron (and possibly also of manganese) in the shoots of plants grown on calcareous soils is low irrespective of whether the plant is chlorotic or green. Sensitive plants respond in dry weight to applications of chelated iron to the soil (Grime 1964) and leaf chlorosis can be cured by supplying inorganic iron to individual roots (Plate 5).

ORIGIN OF SUSCEPTIBILITY TO LIME CHLOROSIS

It is clear that sensitivity to lime-chlorosis is peculiar to calcifuges. It is therefore reasonable to suspect that susceptibility to lime-chlorosis is a consequence of natural selection for survival on acid soils and that lime-chlorosis arises in calcifuges from the continued operation on calcareous soils of features of mineral nutrition which are adapted to the acid soil environment. Comparisons of the various hazards to plant survival on acid soils (e.g. Hewitt 1952) indicate the importance of iron, aluminium and manganese toxicities. In this paper aluminium toxicity will be considered.

COMPARATIVE EXPERIMENTS ON SUSCEPTIBILITY TO ALUMINIUM TOXICITY

There is growing evidence that aluminium is a factor of major importance in its effect on the distribution of plants. Some of the comparative work on aluminium sensitivity is summarized in Table 5.

The results of different workers are not of course directly comparable and the tolerance estimate obtained is dependent on which species are used in the comparison. This is why, for example, *Lolium perenne* is variously defined as sensitive and resistant (Hackett 1965, 1967).

TABLE 5
Recorded differences between flowering plants with respect to susceptibility to aluminium toxicity.

(a) INTERSPECIFIC DIFFERENCES

SPECIES			AUTHORITY
Very susceptible	*Susceptible*	*Resistant*	
Barley Beets Lettuce Timothy	Cabbage Oats Radishes Rye Sorghum	Corn Redtop Turnips	McLean and Gilbert (1926)
Medicago sativa *Onobrychis viciifolia* (as *O. sativa*)		*Lupinus luteus* *Ornithopus sativus*	Rorison (1958)
Asperula cynanchica *Scabiosa columbaria*		*Galium saxatile* *Holcus mollis*	Rorison (1960a,b)
Barley	Pea	Mangold	Jones (1961)
Carex lepidocarpa		*C. demissa*	Clymo (1962)
Lettuce	Turnip Radish	Cucumber Maize Rice (variety) Squash	Aimi and Murakami (1964)
Mustard	Barley Buckwheat Bushbean Corn Hemp	Oats	Foy and Brown (1964)
Medicago sativa		*Trifolium subterraneum*	Munns (1965)
Alopecurus pratensis *Festuca pratensis*		*Deschampsia flexuosa* *Lolium perenne*	Hackett (1965)
Agrostis stolonifera	*A. canina* *A. tenuis*	*A. setacea*	Clarkson (1966)
Lolium perenne		*Deschampsia flexuosa*	Hackett (1967)
Schoenus nigricans	*Briza media* *Chenopodium bonus-henricus* *Silene dioica* *Urtica dioica*	*Molinea caerulea*	Sparling (1967)

TABLE 5—*continued*

(b) INTRASPECIFIC DIFFERENCES

SPECIES	AUTHORITY
Barley	Foy, Armiger, Briggle *et al.* (1965); MacLean and Cheason (1966); MacLeod and Jackson (1967); Aimi and Murakami (1964)
Wheat	Foy, Armiger, Briggle *et al.* (1965)

The effect of aluminium on root growth was determined for seedlings of the species which had previously been used in the chlorosis screening. Following the technique of Sparling (1967) the aluminium concentration required to cause 50% inhibition or root growth was measured. The results

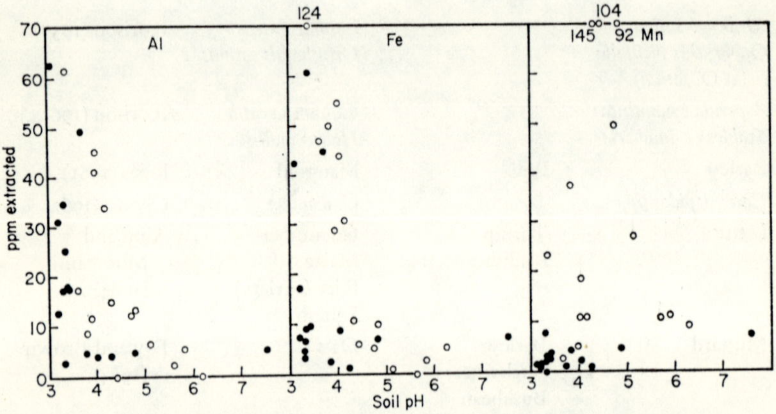

FIG. 4. Concentrations of iron, manganese and aluminium obtained in a 'saturation extraction' from organic (●) and mineral (○) horizons of a wide range of grassland soils in Derbyshire and Northumberland. (Welsh and Davison, unpublished data.)

indicate a close correlation between the field distribution and the sensitivity of the seedling to aluminium toxicity. *D. flexuosa*, *N. stricta* and *H. mollis*, all species of common occurrence on very acid soils, were very resistant, *D. flexuosa* especially so. *Poterium sanguisorba*, *Centaurea nigra*, *Briza media* and *Leontodon hispidus*, species mainly restricted to calcareous soils, were very sensitive to aluminium.

The degree of aluminium sensitivity is inversely related to the excursion of the species in the field into soils below pH 4·5. Several species, notably *R. acetosa* and *H. lanatus*, show a catastrophic decline in % occurrence below

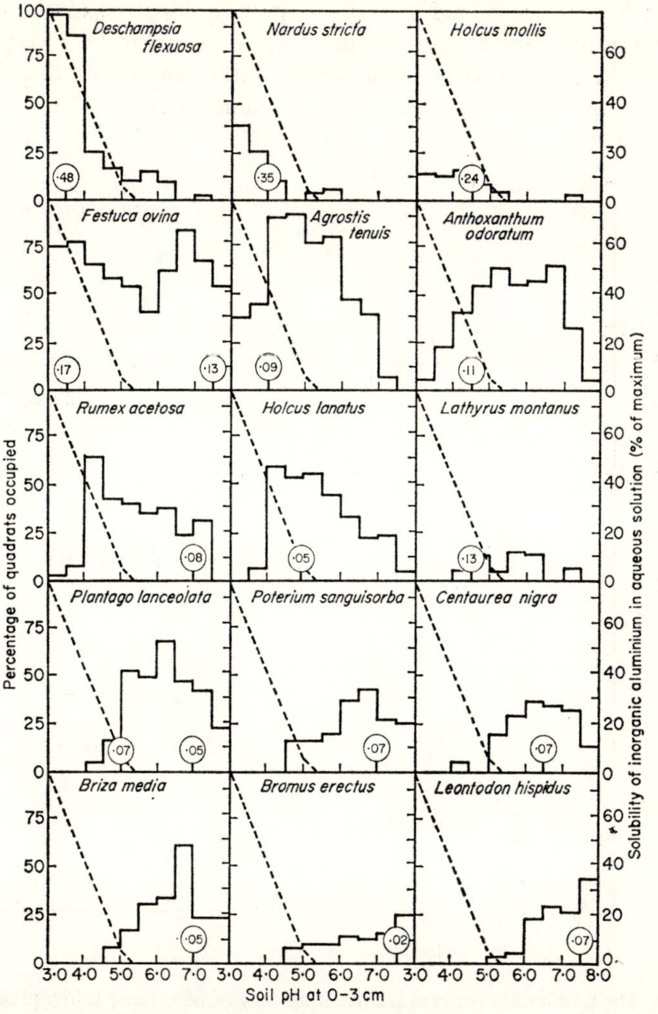

FIG. 5. Diagrams showing the relationship between the frequency of the species over the range in soil pH 3–8, the solubility of inorganic aluminium in aqueous solutions over the same pH range (Magistad 1925). Encircled values refer to mMoles of aluminium required to inhibit root growth in 1-week-old seedlings of the species. Test of aluminium sensitivity carried out using the techniques of Sparling (1967). The position of the circle indicates the soil pH of the surface soil at the seed source.

this value which corresponds with the marked effect of pH on the solubility of aluminium in inorganic and soil solution (Magistad 1925, Figs. 4 and 5).

The importance of pH 4·5 with respect to resistance to aluminium toxicity is apparent in Fig. 6 where there is shown to be a very close correlation between *relative frequency* of occurrence on soils of pH<4·5 and the senstitivity of the species to aluminium toxicity.

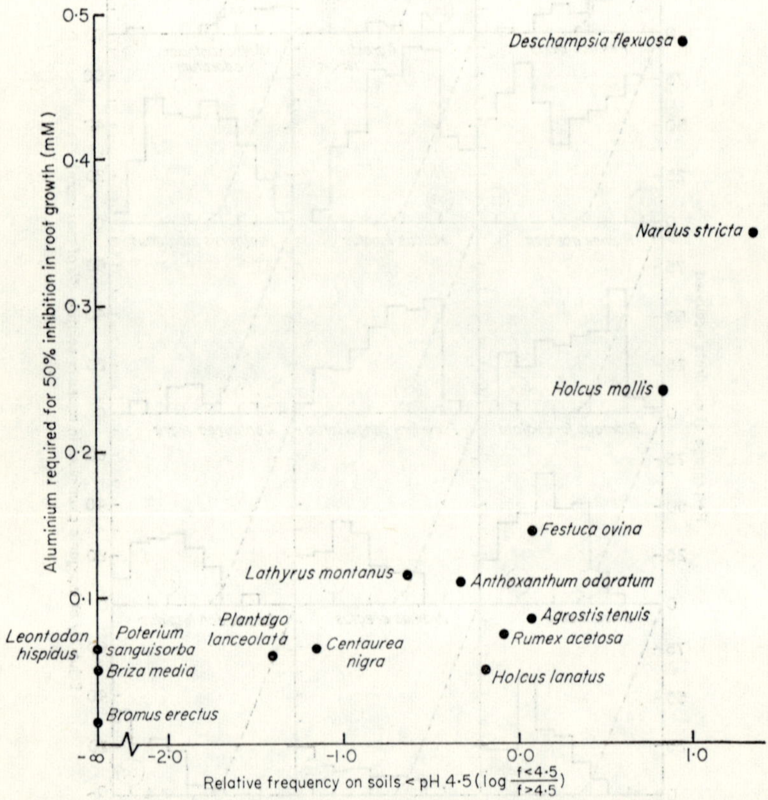

FIG. 6. The relationship between relative frequency of occurrence below pH 4·5 and resistance of the seedling root to aluminium toxicity (measured using techniques of Sparling (1967) in a range of grassland species.)

It was interesting to find low resistance to aluminium in wide-ranging species such as *R. acetosa*, *H. lanatus* and *Plantago lanceolata*. This is to be expected since these plants do not occur to any extent below pH 4·5 (Fig. 5). The very wide-ranging *Festuca ovina*, however, showed considerable resistance in both the acid and calcareous races tested.

RESISTANCE TO ALUMINIUM TOXICITY IN RELATION TO LIME-CHLOROSIS AND THE IRON NUTRITION OF THE ROOT

The inverse correlation between resistance to aluminium toxicity and resistance to lime-chlorosis

All strongly calcifuge species so far examined appear to be resistant to aluminium toxicity and many are susceptible to lime-chlorosis. It is worth considering therefore whether the two attributes arise from a common mechanism. Before considering the physiological and biochemical evidence for such a connection, two possible objections to this suggestion deserve consideration.

First, it seems likely that resistance to aluminium toxicity is only one among many possible adaptations in the mineral nutrition of the calcifuge. Aluminium toxicity is however one of the most widespread and drastic of the effects reducing the fertility of acid soils.

A second objection arises from the fact that certain calcifuge grasses, e.g. *N. stricta*, are resistant to both aluminium toxicity and lime-chlorosis. In this connection it may be of importance to recognize that chlorosis in the shoot is a secondary symptom of a nutritional disorder originating in the root. As such chlorosis is subject to the modifying effects of the growth rate of the plant, the length and efficiency of the translocation path to the leaf and the distribution and affinity for iron of sinks in the root and shoot.

In view of the possibility that chlorosis is a fallible index of net iron stress in the plant, the attempt has been made to study directly the relationship between resistance to aluminium toxicity and the iron nutrition of the root.

Positive responses to aluminium

To date there has been no convincing demonstration that aluminium is an essential micronutrient for the flowering plant. It is interesting therefore to find in the literature on this subject a number of reports of plant responses in root length, root weight or shoot weight on the addition of small quantities of aluminium to water cultures (Neger 1923; Sommer 1926; Lipman 1938; Ho and Nakamura 1954; Clymo 1962; Clarkson 1966; Hackett 1962, 1964, 1965). A response in dry weight to aluminium supply has also been recorded for *Chlorella vulgaris* by Bertrand and De Wolfe (1966). Responses have been noted in both calcicoles and calcifuges (Hackett 1964; Clarkson 1966) and in one experiment carried out by

Clarkson (1966) in the calcifuge species *Agrostis setacea* and *A. tenuis* responses were sustained up to concentrations of aluminium inhibitory to the calcicole, *A. stolonifera* (Table 6).

TABLE 6

Comparison of the effect of aluminium concentration on root extension in three species of *Agrostis*. (After Clarkson 1967.)

Root lengths were measured after a week's growth and are expressed as a percentage of lengths attained in the absence of aluminium.

Aluminium concentration (mM)	*Agrostis stolonifera* (calcicolous)	*Agrostis tenuis* (calcifuge)	*Agrostis setacea* (strongly calcifuge)
0·0	100	100	100
0·1	72	175	137
0·2	48	134	155
0·4	46	104	106
0·8	19	77	98
1·2	9	49	91
1·6	9	8	63

One possible explanation for the stimulative effect of aluminium may be suggested. In water cultures from which aluminium is excluded, it is conceivable that a binding mechanism in the root, which normally operates to detoxify this element, is effective on one or more nutrient cations to an extent that a deficiency condition arises within the root. In this circumstance the addition of aluminium would be expected to result in at least partial engagement of the detoxification mechanism by aluminium with a commensurate reduction in the fixation of nutrient ions.

Response of iron-deficient roots to aluminium

In the preceding section it was suggested that the stimulative effects of aluminium may be due to alleviation of nutrient stress. This implies that some water cultures provide an inadequate supply of one or more nutrient ions, at least with respect to calcifuge plants.

In experiments carried out by Hackett (1965) a response to aluminium by the calcifuge *D. flexuosa* occurred on a nutrient solution which had been shown experimentally to contain adequate concentrations of the major nutrients for this species. However in this work, as in other experiments in which aluminium response has been recorded, doubts must be raised with regard to the adequacy of the supply of iron (and also possibly that of micronutrients such an manganese). In particular, where iron is not supplied in the form of a chelate, the levels of iron designed for solution culture may be suboptimal for calcifuges (Olsen 1958). In addition it remains to be shown that levels of iron sufficient to avoid an obvious chlorosis are invariably adequate for maximal root growth.

A possible explanation for the stimulative effects of aluminium is that roots contain a binding mechanism which (a) is relatively unselective between aluminium and iron (and perhaps also manganese) and (b) which differs in capacity between calcicoles and calcifuges. In order to test this hypothesis measurement was made of the effect of aluminium on root growth in seedlings of species from acid and calcareous soils, growing in water cultures under conditions of iron stress.

Precipitation of aluminium phosphate was avoided by using the technique of Stoklasa (1911). Aluminium was supplied to the seedling in a single salt solution (+Al) in periods alternating with those in which the root was immersed in a complete nutrient solution (−Al) containing a suboptimal concentration of iron as ferrous sulphate. Details of the solutions and experimental procedure are given in Fig. 7.

Some difficulty was experienced in inducing iron stress in seedlings of certain species. The onset of deficiencies was affected by the size and source of seed and by the growth rate of the seedling. Evidence was also obtained of differences between species with respect to the external concentration of iron sufficient for root growth. The extreme calcicole *S. columbaria* showed optimal root growth at concentrations below 0·036 mM, a level insufficient for maximum root growth in a majority of the other species tested.

In all experiments in which conditions of iron deficiency were successfully induced evidence was obtained of a response in root growth in seedlings treated with low concentrations of aluminium sulphate. The concentration range in which stimulation occurred varied considerably according to species and curves such as those illustrated in Fig. 9 were obtained only after preliminary experimentation. In Fig. 8 there is drawn a generalized version of the curves obtained in the experiments and an attempt is made to explain the response in terms of the effects of aluminium concentration on growth during the +Al and −Al phases of the experiment. With exposure to increasing concentrations of aluminium four responses in root growth may be distinguished ($a-b$, $b-c$, $c-d$ and $d+$).

In ($a-b$) over a range of low aluminium concentrations, root growth falls rapidly to a minimum. This appears to represent progressive inhibition of growth in the +Al phase due to the toxic effect of aluminium. From the work of Clarkson (this volume) it seems likely that this will persist for some time into the −Al phases, but here at least growth will also be limited by iron-deficiency. At (*b*) therefore there is little or no growth in both phases. In (*b*−*c*) with a further increase in aluminium concentration there is a stimulation of root growth and at (*c*) the rate of growth may

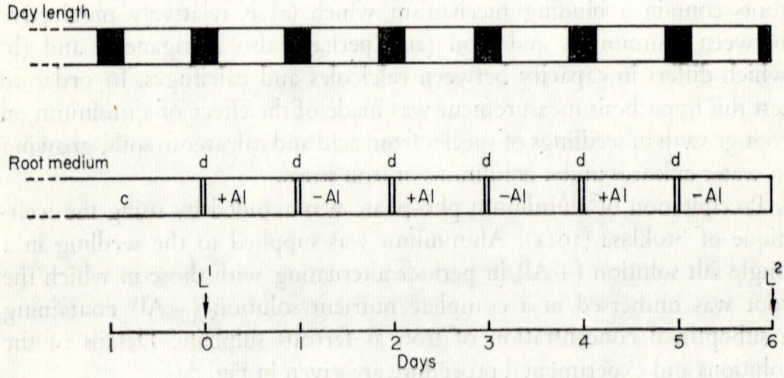

FIG. 7. Scheme describing the experimental procedure and the solutions used to investigate the response of iron-deficient roots to aluminium treatment.

Day length: shaded areas represent dark periods.

Root medium:

−Al solution—A complete nutrient solution designed by Rorison (1958). The pH was adjusted to 4·0. Iron at 2 ppm. (0·5 ppm for *Scabiosa columbaria*) as ferrous sulphate was dissolved in the solution immediately before use and the solution was discarded after 24 hours.

+Al solution—A solution of aluminium sulphate adjusted to pH 4·0.

c solution—as −Al but with Fe E.D.T.A. as iron source.

d—distilled water washing between +Al and −Al treatments.

L^1—initial measurement of total root length of seedling taken approximately 1 week after germination.

L^2—final measurement of total root length.

exceed that of plants at (*a*) which are immersed in distilled water during the +Al phase (e.g. *Briza media*, Fig. 9). We conclude that this stimulated root growth occurs in the −Al phase.

In (*c*−*d*) further additions of aluminium cause a decline to zero in root growth and probably corresponds to the extension of the inhibitory effect of aluminium into the −Al phase.

(*d*+) represents a range of concentrations over which root growth is completely inhibited in both phases by the toxic effect of aluminium.

In examining aluminium response curves of particular species such as those in Fig. 9 it is clear that the height of the response at (*c*) is determined by the inherent rate of root extension of the species and the degree of iron stress induced and as such it is of secondary importance. The curves are of particular interest in relation to (1) differences between and within species with respect to the concentration ranges (*a*–*b*), (*b*–*c*) and (*c*–*d*) and (2) the nature of the stimulatory effect of aluminium in (*b*–*c*).

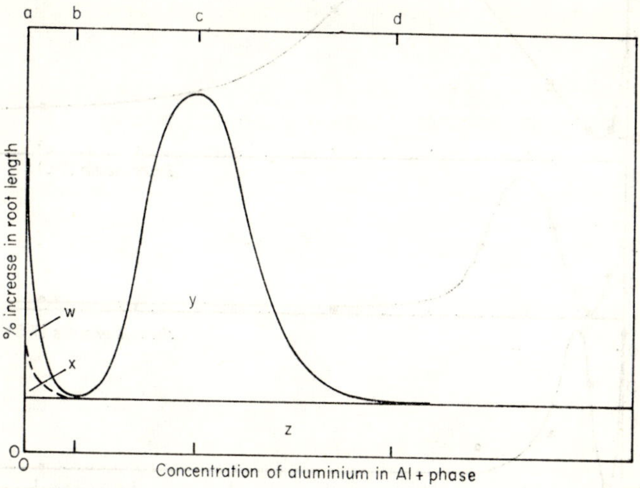

FIG. 8. Generalized version of the response of an iron deficient root system to aluminium treatment of the type described in Fig. 7 and a suggested partitioning of the root growth observed.

w and *y*—growth during –Al phases.
x—growth during +Al phases.
z—growth occurring between initial measurement and inhibition of growth by first aluminium treatment.
a, *b*, *c*, and *d*—see text.

(*a*) *Differences in response to aluminium*

(*i*) *Between species.* Of the five species in Fig. 9 two, *N. stricta* and *Ulex europaeus* are calcifuges, known to be resistant to aluminium toxicity. The former is resistant to lime-chlorosis; the latter susceptible. *N. stricta* shows considerable resistance to the inhibitory effect of aluminium in the (*a*–*b*) section of the curve and both species show some root growth at levels of aluminium above 0·15 mM. In the calcicolous species *B. media* and *Bromus erectus* the response curves are of the same form as for calcifuges but they are contracted, complete inhibition of root growth occurring at about

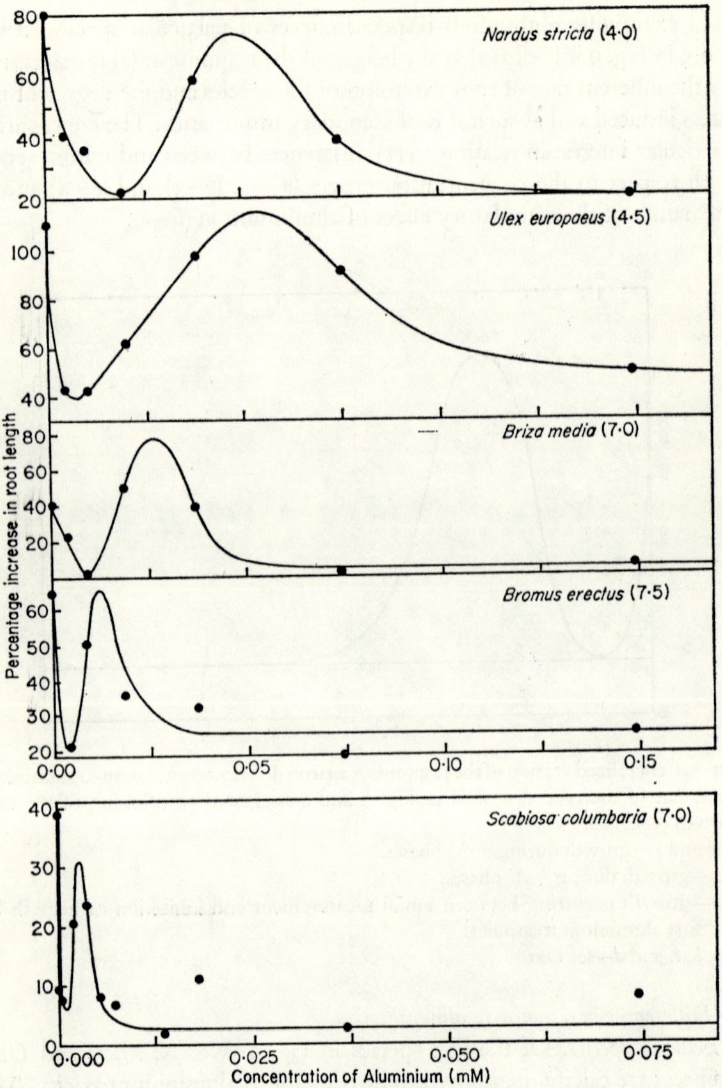

FIG. 9. Response in root length to aluminium treatment of iron deficient root systems in seedlings of five species.

0·1 mM aluminium. In the extreme calcicole *S. columbaria*, (*d*) occurs at 0·007 mM aluminium. There is however a distinct response to aluminium below this concentration.

(ii) *Between ecotypes*. Hutchinson (1967, 1968) has described ecotypes of

Teucrium scorodonia which differ in chlorosis susceptibility. Using seed kindly provided by Dr Hutchinson, an experiment was carried out to compare the response to aluminium of plants from two populations differing in susceptibility to chlorosis. Preliminary experiments with seedlings confirmed that the two populations differed both with respect to lime-chlorosis on a calcareous soil (Fig. 10 (a)) and with respect to iron-deficiency chlorosis when cultured on an acid nutrient solution (Fig. 10 (b)). The seedlings were then transplanted onto a garden loam and grown for one year at which time cuttings were taken at random from plants of the two populations. Using the technique described in Fig. 7 the response to aluminium was measured in the acid soil ecotype one week after the appearance of roots on the cuttings (Fig. 10c) and again after two weeks (Fig. 10d). The results showed that one week after rooting the plants were extremely sensitive to aluminium. After two weeks however considerable root growth occurred up to 0·1 mM aluminium. A comparison of aluminium response between cuttings of the two ecotypes was commenced after 3 weeks of root growth had occurred. The results (Fig. 10e) indicate a marked difference between the ecotypes in response to increasing aluminium concentration. The peak in response, (*c*), occurred at 0·012 mM in the calcareous soil (chlorosis sensitive) ecotype and at 0·040 mM in the acid soil (chlorosis resistant) ecotype.

(*b*) *The nature of the stimulatory effect of aluminium on root extension*

In these experiments the response to aluminium was recorded under conditions in which root growth was limited by iron supply and it is suggested that this stimulation is due to an increase in the utilization of absorbed iron following aluminium treatment. Evidence supporting this hypothesis was obtained in an experiment comparing the aluminium response in *S. columbaria* at two sub-optimal levels of iron supply (Fig. 11). With zero applied iron (which was therefore available only as a contaminant) the response to aluminium is truncated. The effect of raising the external iron supply by addition of 0·009 mM iron as ferrous sulphate was to double the rate of root growth at both (*a*) and (*c*). This demonstration of the necessity of an iron supply for the aluminium response suggests that the stimulatory effect of aluminium is to occupy sites at which iron might otherwise become bound. Further evidence of competition between iron and aluminium for the same binding mechanism was obtained in this experiment. Raising the external iron supply caused the ($c-d$) section of the curve to be deflected to the left,† i.e. with an additional supply of iron

† In view of the variability in the data the positions of the curves in Fig. 11 are tentative and require confirmation.

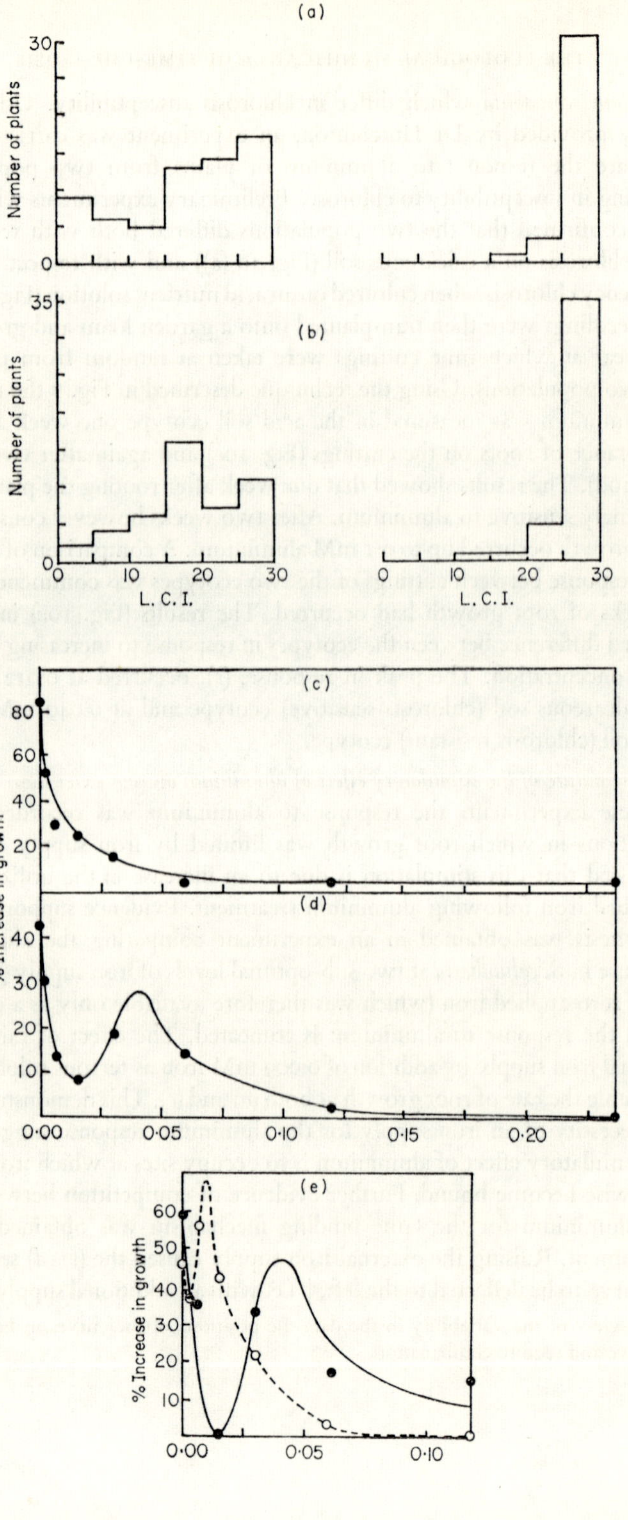
(a)
30
Number of plants
0
(b)
35
Number of plants
0
0
10
20
30
L. C. I.
0
10
20
30
L. C. I.
(c)
80
60
40
20
% Increase in growth
(d)
40
30
20
10
0·00
0·05
0·10
0·15
0·20
(e)
60
50
40
30
20
10
% Increase in growth
0·00
0·05
0·10

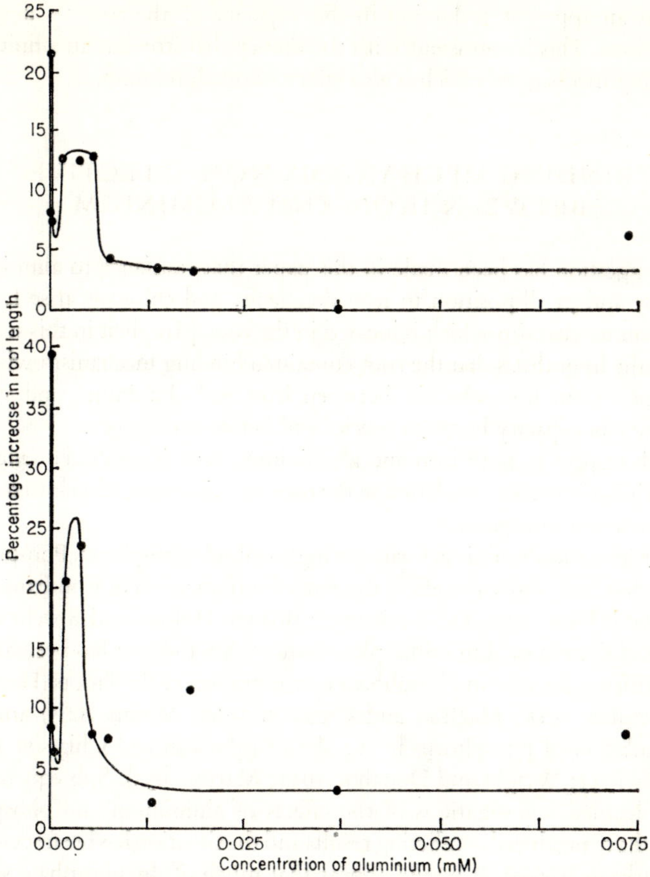

FIG. 11. Response in root growth to aluminium treatment in seedlings of *Scabiosa columbaria* provided with no additional iron which was therefore available only as a contaminant (above) and iron at 0·009 mM (below).

FIG. 10. (a) Comparison of an acid soil ecotype (left) and a calcareous soil ecotype (right) of *Teucrium scorodonia* with respect to lime-chlorosis in the youngest expanded leaves of seedlings growing on a calcareous soil. Values of leaf-colour index (L.C.I) <25 indicate some degree of chlorosis.

(b) Comparison of the same ecotypes with respect to chlorosis in an iron-deficient nutrient solution, pH 4·0.

(c) Response to aluminium of roots treated one week after appearance on cuttings of the acid soil ecotype of *T. scorodonia*.

(d) Response to aluminium of roots treated two weeks after appearance on cuttings of the acid soil ecotype.

(e) Response to aluminium of roots treated three weeks after appearance on cuttings of the acid soil ecotype (●) and the calcareous soil ecotype (○).

there is an apparent reduction in the capacity of the root to detoxify aluminium. This is consistent with the theory that iron has an affinity for a binding mechanism which is also effective on aluminium.

BINDING MECHANISMS NON-SELECTIVE BETWEEN IRON AND ALUMINIUM

The suggestion has been made in this paper that resistance to aluminium toxicity and predisposition to iron deficiency and chlorosis arise from a common mechansim which is located in the root.† Implicit in this suggestion is the hypothesis that the root contains a binding mechanism which is to some extent non-selective between iron and aluminium and which may vary in capacity between species and between ecotypes.

With respect to both iron and aluminium, there is published evidence of precipitation or accumulation in the root or of incorporation into mobile but unreactive compounds.

Both iron and aluminium form sparingly-soluble phosphates. Phosphorus supply has been shown to affect the rate of mobilization of iron from root to shoot (Olsen 1935; De Kock 1955; Brown, Holmes and Specht 1955) and precipitation of aluminium phosphate within the root has been associated with a reduction in phosphorus concentration in the shoots (Hartwell and Pember 1918; Maclean and Chiasson 1966; Munns 1965) and the accumulation of phosphorus in the shoots (Maclean and Chiasson 1966; Wright 1943; Wright and Donahue 1953; Martin 1955). Subsequent and more detailed investigations of the effects of aluminium and phosphate supply have produced conflicting resúlts and the situation has been reviewed by Clarkson (1966a). It would appear that much of the phosphate which accumulates in the root is fixed at the cell-surface by an adsorption-precipitation reaction (Rorison 1955; Clarkson 1966a) and that the depletion of phosphorus in the shoot arises primarily as a consequence of the reduction in absorbing surface resulting from the inhibition of root growth by aluminium toxicity. Perhaps the most serious objection to phosphate as a binding site for aluminium is to be found in the work of Hackett (1967) using the aluminium tolerant species *D. flexusoa*. In these experiments, *D. flexuosa* showed considerable resistance to aluminium but no evidence was found of a major effect of aluminium supply on the concentration of phosphorus in roots or shoots.

† This is not to preclude the possibility of a similar mechanism operating in the shoot.

The cell walls of the root have a measurable capacity for adsorption of cations. Clarkson (1966a) has reported that after a short period of uptake, 85–95% of the aluminium present in the root is adsorbed on to the cell wall and can be removed by washing. In his paper, Dr Turner (this volume) has described a range of organisms, including the flowering plant *A. tenuis* in which resistance to a specific heavy metal is due to 'irreversible' binding mechanisms associated with the cell wall. However such binding mechanisms appear to be extraordinarily specific (Turner, this volume) and it seems most unlikely that iron and aluminium would compete for attachment to the same sites on the cell wall. Moreover, there is evidence that in calcifuges aluminium and iron may be translocated to the shoot in bound form. There are numerous examples in which species resistant to aluminium toxicity have been found to accumulate high concentrations of aluminium in the shoot (Chenery 1948; Hutchinson 1943; Hackett 1967; Table 2). Similarly, chlorosis often coincides in susceptible plants with levels of iron in the shoot which are comparable with those in healthy specimens of chlorosis-resistant species (Table 2) suggesting that iron reaches the leaves in a form unavailable for chlorophyll synthesis.

A more promising explanation of a relatively unselective binding capacity in the root involves the organic acid component. Small (1946) has suggested that in 'acidophilous' plants organic acids occur at high concentrations and function as a strong buffer system. Support for this suggestion was obtained by Chenery (1948) who showed that species tolerant of high internal concentrations of aluminium had cell sap values below pH 5·3. Various organic acids, such as malic, citric, oxalic and tartaric acids which occur widely in flowering plants will complex with both iron and aluminium *in vitro*. In experiments using root mascerates from various species, Jones (1961) showed that the capacity of the cell sap to bind aluminium varied according to species and directly in relation to the resistance of the species to aluminium toxicity. Both Hutchinson (1943) and Jones (1961) suggest the occurrence of aluminium-organic acid complexes within the plant.

In the light of the research into aluminium tolerance, it is interesting to find that in recent studies of translocation within plants (Tiffin and Brown 1962; Brown and Tiffin 1965; Tiffin 1965) iron has been found in complexed form as ferric citrate, malonate and malate. It is of particular interest to find that the difference in chlorosis susceptibility between ecotypes of *Teucrium scorodonia* (such as those found to differ in resistance to aluminium in the present investigation) coincides with a qualitative difference in iron transport from root to shoot (Hutchinson 1968).

CONCLUSIONS

To date, lime chlorosis has been investigated primarily as a nutritional disorder arising on calcareous soils. This paper draws attention to an alternative line of enquiry, namely the consideration of lime chlorosis as the expression of a physiology adapted to acid soils.

The calcifuges studied in this investigation are resistant to aluminium toxicity and it appears that this resistance is due to a chelating mechanism which also has an affinity for iron. Such a mechanism provides a possible explanation for the susceptibility of calcifuges to lime chlorosis and is iconsistent with the results of recent research into the mechanism of both aluminium tolerance and lime chlorosis. In both of these fields recent work indicates the importance of metal-organic acid complexes.

This investigation has been exploratory and concerned exclusively with iron and aluminium. It remains therefore to (1) characterize the chelation mechanisms of various plant species and ecotypes (2) examine their specificity and (3) estimate the relative importance of particular elements in the evolution of the difference in chelating capacity which appears to distinguish calcicoles from calcifuges.

REFERENCES

AIMI R. and MURAKAMI T. (1964). Cell-physiological studies on the effect of aluminium on growth of crop plants. (1) Toxicity of aluminium to plant cells. (2) Effect of aluminium on plant growth. (3) Effect of aluminium on tissue elongation. (4) Simple method for detecting susceptibility of crop plants to acid soil. *Bull. Nat. Inst. Agric. Sci. Tokyo* 11*D*, 331–396.

ANDERSON D. J. (1961). The structure of some upland plant communities in Caernarvonshire. II. The pattern shown by *Vaccinium myrtillus* and *Calluna vulgaris*. *J. Ecol.* **49**, 731–738.

BERTRAND D. and DE WOLF A. (1966). Aluminium, a dynamic minor element for higher plants. *C.r. herbd. Searc. Acad. Sci., Paris* 262*D*, 479–481.

BRADSHAW A.D. (1952). Populations of *Agrostis tenuis* resistant to lead and zinc poisoning. *Nature, London*, **169**, 1098.

BROWN J.C. (1967). Differential uptake of Fe and Ca by two corn genotypes. *Soil Sc.* **103**, 331–338.

BROWN J.C. and HOLMES R.S. (1955). Iron the limiting element in chlorosis. Part I. Availability and utilization of iron dependent upon nutrition and plant species. *Pl. Physiol.* **30**, 451–457.

BROWN J.C., HOLMES R.S. and SPECHT A.W. (1955). Iron the limiting element in a chlorosis. Part II. Copper phosphorus induced chlorosis dependent upon plant species and varieties. *Pl. Physiol.* **30** 457–462.

BROWN J.C., HOLMES R.S. and TIFFIN L.O. (1958). Iron chlorosis in soybeans as related to the genotype of rootstock. *Soil Sci.* **86**, 75–82.

BROWN J.C. and TIFFIN L.O. (1965). Iron and citrate in stem exudate. *P. Physiol.* **40**, 395–400.

CHENERY E.M. (1948). Aluminium in plants and its relation to plant pigments. *Ann. Bot.* **12**, 121–136.

CLARKSON D.T. (1966a). Aluminium tolerance in species within the genus *Agrostis. J. Ecol.* **54**, 167–178.

CLARKSON D.T. (1966b). Effect of aluminium on the uptake and metabolism of phosphorus by barley seedlings. *Pl. Physiol.* **41**, 165–172.

CLYMO R.S. (1962). An experimental approach to part of the calcicole problem. *J. Ecol.* **50**, 701–731.

CURTIS J.T. and MCINTOSH (1951). An upland forest continuum in the prairie-forest border region of Wisconsin. *Ecology* **32**, 476–496.

DAVISON A.W. (1964). Some factors affecting seedling establishment on calcareous soils. *Ph.D. Thesis, University of Sheffield.*

FOY C.D., ARMIGER W.H. and BRIGGLE L.W. *et al.* (1965). Differential aluminium tolerance of wheat and barley varieties in acid soils. *Agron. J.* **57**, 413–417.

FOY C.D. and BROWN J.C. (1964). Toxic factors in acid soils. II. Differential aluminium tolerance of plant species. *Soil Sci. Soc. Amer. Proc.* **28**, 27–32.

GOODALL D.W. (1954). Objective methods for the classification of vegetation. III An essay in the use of factor analysis. *Aust. J. Bot.* **2**, 304–324.

GREIG-SMITH P. (1964). *Quantitative plant ecology.* 2nd edn. London.

GRIME J.P. (1959). A study of the ecology of a group of Derbyshire plants with particular reference to their nutrient requirements. *Ph.D. thesis, University of Sheffield.*

GRIME J.P. (1961). Measurement of leaf colour. *Nature*, London. **191**, 614.

GRIME J.P. (1965a). Comparative experiments as a key to the ecology of flowering plants. *Ecology* **46**, 513, 515.

GRIME J.P. (1965b). The ecological significance of lime chlorosis (an experiment with two species of *Lathyrus*). *New Phytol.* **64**, 477–488.

GRIME J.P. and JEFFREY D.W. (1965). Seedling establishment in vertical gradients of sunlight. *J. Ecol.* **53**, 621–642.

HACKETT C. (1962). Stimulative effects of aluminium on plant growth. *Nature*, London, **195**, 471–472.

HACKETT C. (1965). Ecological aspects of the nutrition of *Deschampsia flexuosa* (L) Trin. II. The effect of Al, Ca, Fe, K, Mn, N, P and pH on the growth of seedlings and established plants. *J. Ecol.* **53**, 299–314.

HACKETT C. (1967). Ecological aspects of the nutrition of *Deschampsia flexuosa* (L) Trin. III. Investigation of phosphorus requirement and response to aluminium in water culture and a study of growth in soil. *J. Ecol.* **55**, 831–840.

HARTWELL B.L. and PEMBER F.R. (1918). The presence of aluminium as a reason for the difference in the effect of so-called acid soils on barley and rye. *Soil Sci.* **6**, 259–279.

HEWITT E.J. (1948). The resolution of the factors in soil acidity. The comparative effects of manganese and aluminium toxicities on some farm and market garden crops. *A.R. Long Ashton Agric. Hort. Res. Stat.* 58–65.

HEWITT E.J. (1952). A biological approach to the problems of soil acidity. *Trans. Int. Soc. Soil Sci. Jt. Meet. Dublin* **1**, 107–118.

HEWITT E.J. (1966). Sand and water culture methods used in the study of plant nutrition. *Comm. Ag. Bur. Tech. Comm. Bur. Hort.* **22** (Ed. 2).

HUTCHINSON G.E. (1943). The biogeochemistry of aluminium and certain related elements. *Quart. Rev. Biol.* **18**, 1–29, 129–153, 242–262, 331–363.

HUTCHINSON T.C. (1966). An investigation of populations of *Teucrium scorodonia* with particular reference to their susceptibility to lime-induced chlorosis. *Ph.D. thesis, University of Sheffield.*

HUTCHINSON T.C. (1967a). Lime-chlorosis as a factor in seedling establishment on calcareous soils. I. A comparative study of species from acid and calcareous soils in their susceptibility to lime-chlorosis. *New Phytol.* **66**, 697–705.

HUTCHINSON T.C. (1967b). Ecotypic differentiation in *Teucrium scorodonia* with respect to susceptibility to lime-induced chlorosis and to shade factors. *New Phytol.* **66**, 439.

HUTCHINSON T.C. (1967c). The occurrence of coralloid root systems in plants showing lime-chlorosis. *Nature, London*, **214**, 943.

HUTCHINSON T.C. (1968). A physiological study of *Teucrium scorodonia* ecotypes which differ in their susceptibility to lime-induced chlorosis. *Pl. Soil.* **28**, 81–105.

ITO S. and NAKAMURA T. (1954). Studies on the critical reaction of the protoplasm of crop plants. II. On the variation of rice, barley, and wheat varieties in relation to the resistance to the toxic action of iron and aluminium ions. *Proc. Crop. Sci. Soc. Japan*, **22**, (304) 115–116.

JONES L.H. (1961). Aluminium uptake and toxicity in plants. *Pl. Soil* **13**, 297–310.

LACAZE M. (1963). The resistance of eucalyptus to $CaCO_3$ in the soil: report of a preliminary test. *F.A.O. World Consutt. for Genet. Stockholm No. F.A.O./FORGEN* **63/–4/10**, pp. ii, 8.

LIPMAN C.B. (1938). Importance of silicon, aluminium and chlorine for higher plants. *Soil Sci.* **45**, 189–198.

MACLEAN A.A. and CHIASON T.C. (1966). Differential performance of two barley varieties to varying aluminium concentrations. *Can. J. soil Sci.* **46**, 147–153.

MACLEAN F.T. and GILBERT B.E. (1925). The relative aluminium tolerance of crop plants. *Soil Sci.* **20**, 163–175.

MACLEOD L.B. and JACKSON L.P. (1967). Aluminium tolerance of two barley varieties in nutrient solution, peat and soil culture. *Agron. J.* **59**, 359–363.

MAGISTAD O.C. (1925). The aluminium content of the soil solution and its relation to soil reaction and plant growth. *Soil Sci.* **20**, 181–226.

MARTIN J.B. (1955). Interrelationships of calcium and aluminium in turf species used for highways. *Diss. Abstr.* **25**, 5478.

MOKADY R. (1961). Soil aggregation and lime content as factors in lime-induced chlorosis. *Pl. Soil* **15**, 367–376.

MUNNS D.N. (1965). Soil acidity and growth of a legume. II. Reactions of aluminium and phosphate in solution and effects of aluminium, phosphate, calcium and pH on *Medicago sativa* L. and *Trifolium subterraneum* L. in solution culture. *Aust. J. agric. Res.* **16**, 743–755.

NEGER F.W. (1923). Neue Methoden und Ergebrisse der Mikrochemie der Pflanzen. *Flora, Jena*, **116**, 323–330.

OLSEN C. (1958). Iron uptake in different plant species as a function of pH value of the nutrient solution. *Physiol. Plant*, **11**, 889–905.

PAYNE J.A. (1966). Edaphic ecotypes in *Plantago lanceolata*. *B.Sc. thesis, University of Sheffield.*

PARSONS R.F. and SPECHT R.L. (1967). Lime chlorosis and other factors affecting the distribution of *Eucalyptus* on coastal sands in Southern Australia. *Aust. J. Bot.* **15**, 95–105.

RORISON I.H. (1958). The effect of aluminium on legume nutrition. *Nutrition of Legumes* (Ed. E.G. Hallsworth), pp. 43–61, London.

RORISON I.H. (1960a). Some experimental aspects of the calcicole-calcifuge problem. I. The effects of competition and mineral nutrition upon seedling growth in the field. *J. Ecol.* **48**, 585–599.

RORISON I.H. (1960b). The calcicole-calcifuge problem. II. The effects of mineral nutrition on seedling growth in solution culture. *J. Ecol.* **48**, 679–688.

RORISON I.H. (1965). The effect of aluminium on the uptake and incorporation of phosphate by excised sainfoin roots. *New Phytol.* **64**, 23–27.

SMALL J. (1946). *pH and Plants.* Baillière, Tindall and Cox, London.

SNAYDON R.W. (1962). The growth and competitive ability of contrasting natural population of *Trifolium repens* L. on calcareous and acid soils. *J. Ecol.* **50**, 439–447.

SOMMER A.L. (1926). Studies concerning the essential nature of aluminium and silicon for plant growth. *Univ. Calif. Publ. Agric. Sci.* **5**, 57–81.

SPARLING J.H. (1967). The occurrence of *Schoenus nigricans* L. in blanket bogs. II. Experiments on the growth of *Schoenus nigricans* L. under controlled conditions. *J. Ecol.* **55**, 15–31.

STOKLASA J. (1911). De l'importance physiologique du manganese et de L'aluminium dans la callule vegetale. *Compt. Rend. Acad. Sci. (Paris)*, **152**, 1340–1342.

TANSLEY A.G. (1917). On competition between *Galium saxatile* L. (*G. hercynicum* Weig) and *G. sylvestre* Poll. (*G. asperum* schreb.) on different types of soil. *J. Ecol.* **5**, 173–179.

TIFFIN L.O. (1965). Translocation of iron by citrate in plant exudates. *P. Physiol.* **40**, xii.

TIFFIN L.O. and BROWN J.C. (1962). Iron chelates in soybean exudate. *Science, N.Y.* **135**, 311–313.

WARING R.H. and MAJOR J. (1964). Some vegetation of the California Coastal Redwood region in relation to gradients of moisture, nutrients, light and temperature. *Ecol. Monogr.* **34**, 167–215.

WEISS M.G. (1943). Inheritance and physiology of efficiency in iron utilisation in soybeans. *Genetics*, **28**, 252–253.

WOOLHOUSE H.W. (1966). The effect of bicarbonate on the uptake of iron in four related grasses. *New Phytol.* **65**, 372–375.

WRIGHT K.E. (1943). Internal precipitation of phosphorus in relation to aluminium toxicity. *Pl. Physiol.* **18**, 708–712.

WRIGHT K.W. and DONAGHUE B.A. (1953). Inactivation of phosphorus in roots by aluminium. *Pl. Physiol.* **28**, 675–680.

DISCUSSION ON MINERAL NUTRITION AND PLANT DISTRIBUTION

Recorded by: Dr M. J. CHADWICK and Mr C. P. HARDING

Dr Rorison opened the discussion by questioning Mr van den Bergh about the constancy of pH levels in his pot cultures. Mr van den Bergh replied that these had been measured at the beginning and end of the experiment and there had been little change. Dr Glentworth pointed out that if a comparison of *Anthoxanthum* and *Lolium* had been made using different soil-types from the ones used in the experiment the results obtained would probably have been quite different. While agreeing, Mr van den Bergh laid stress upon the method as a measure of *relative* differences caused by an external factor. Mr Merton drew attention to the effect of cutting frequency upon botanical composition and asked Mr van den Bergh whether or not a different cutting regime might not have altered his results. Mr van den Bergh said that in experiments where cutting occurred less frequently and plants were allowed to flower the influence of pH was in the same direction although differing slightly in its extent.

Dr Woolhouse asked Professor Pigott if there were any estimates of atmospheric nitrogen fixation in the lower slopes of salt marshes. Professor Pigott said that it had been observed that surface-living blue-green algae were more abundant on the lower muds and drew attention to the fact that nitrogen builds up rapidly round our shores due to the influence of sewage. Dr Woolhouse pointed out that in Professor Pigott's experiments the *Salicornia* on the lower slopes had yielded over four times as much as that on the artificially fertilized upper slopes and suggested some other factor must be involved. Professor Pigott replied that the nutrient levels applied were rather low and higher levels would probably have given a greater response. Dr Clarkson asked about the seasonal periodicity of mineral nutrients, particularly phosphate, in sea water and was told that nitrogen, at least, showed a marked rise in summer due to the increasing amounts of animal excreta. Animal populations were large in the lower muds but less important in the upper regions. Dr Freijsen wondered whether plants of the upper marsh and transition zones, where sodium concentration and mineral nutrients generally are lower, would show different responses from the euhalophytes. Professor Pigott replied that he had done no work on this but commented that *Armeria maritima*, an upper marsh plant, shows virtually no response to nitrogen and phosphorus when grown on upper marsh soils that induce deficiency symptoms in *Salicornia*.

Replying to a query from Dr Hackett, Dr Gittins agreed that significance tests were usually lacking in vegetation work utilizing principal components analysis. He said, however, that as long as the data complied with the usual normality requirements there was no reason why components should not be tested for significance. Dr Gittings informed Dr Woolhouse that vegetational data would usually require transformation before tests of significance were applied. Dr Bannister pointed out that most species' responses are curvilinear whilst the method of principal components depends upon linear relationships between variables. Dr Gittings said that this was true, and care was needed in interpreting the results. In reply to Dr Snaydon, Dr Gittins said that the use of different quadrat sizes in the same community had remarkably little effect on the results obtained using principal components analysis. He also said that the method separated correlations between factors rather than reflecting correlations.

Dr Rorison asked Dr Hackett if, in his experience, the growth-enhancing effects of low levels of aluminium in some solution cultures could be due to the lowering of a supra-optimal level of phosphate by its precipitation with aluminium. There was, however, apparently no experimental evidence to support this view. Dr Clarkson expressed surprise that the response of *Scabiosa columbaria* to aluminium was more marked when iron was present in high quantities. Implicit in all that was said was that there must be competitive uptake or binding between aluminium and iron, yet by increasing the concentration of iron the system became more sensitive. Dr Grime replied that he was suggesting that the increased iron saturated the chelating sites and thus free aluminium remained, constituting a toxic level. Dr Grime said that leaf sprays had not been used as the main interest was in root phenomena. However, Dr Jennings pointed out that iron could be translocated from leaf to root. Dr Rorison emphasized that this would be in a very different form from the ionic iron and aluminium outside the plant which became adsorbed on to exchange sites on the root surface. Several speakers thought, however, that the form of aluminium and iron in the cell and its equilibrium with these elements in their bound forms was important from the point of view of plant response. Dr Wilkins expressed the hope that a soil could be found where aluminium and iron concentrations did not run concurrently in order that the tolerance mechanism to each might be elucidated separately in plants that had evolved tolerance to only one of the elements. Professor Bradshaw was doubtful about this approach. He thought it unlikely that aluminium and iron would be bound in a single system as suggested. All the heavy metal work suggested great specificity of bindings even for closely related elements.

SECTION TWO

MINERAL NUTRIENT SUPPLY FROM SOILS

THE SOIL MODEL AND ITS APPLICATION TO PLANT NUTRITION

P. H. NYE

Soil Science Laboratory, University of Oxford

The soil scientist is almost inevitably forced, in seeking to understand the mineral nutrition of vegetation growing in soil, to refine the scale of his enquiry, and eventually to concentrate his attention on those local zones of the soil that are most closely associated with the roots—just as the plant physiologist studies single cells for the light they can throw on whole organisms.

Confronted with something so complex as a complete ecosystem containing many species, like a forest, the best we have done so far is to measure its store of nutrients and follow their movements by measuring the nutrients returned to the soil in the litter and the rain. This tells us nothing about the competition for nutrients between the root systems, and little about the way in which the vegetation, by absorbing nutrients from the subsoil, impoverishes it, and at the same time improves the nutrient status of the topsoil. At the least we have to find out what proportion of the nutrients in the cycle are drawn from the subsoil.

With a simpler ecosystem it is sometimes possible to differentiate between nutrients absorbed from the subsoil and topsoil by using radio tracers. Dr Newbould will be describing this technique later in the symposium. The results of such an experiment are shown in Table 1, the object being to find the depth from which a grassland savanna was deriving its phosphorus (Nye and Foster 1961).

One might anticipate that the amounts taken from each horizon would be related to the rooting density and the phosphate status (to use a vague term) of the horizon, but in this experiment it would seem that the subsoil is providing more than might be expected. Possible explanations may lie in the moisture regime in the different horizons, the efficiency of the absorbing surface areas of the roots or possible interaction with other ions in the same horizon. This experiment is inadequate to tell us. In short we are led to enquire in detail about the processes that control the uptake of nutrients in each horizon of the soil. And the purpose of this session is to give some account of the limited way we soil scientists have come along the road towards true explanation of nutrient absorption.

Clearly, for detailed explanation of the processes involved in nutrient absorption, one must start with as simple a system as possible: which is a portion of an intact absorbing root in soil. If this can be thoroughly understood there is some hope eventually of integrating over the whole root system.

TABLE I

Relative uptake of soil phosphorus by savanna from different depths of the root zone

	Phosphorus from each depth per cent		Roots at each depth per cent by weight	
Depth (cm.)	*Grasses*	*Other spp.*	*Grasses*	*Other spp.*
0–25	70	64	81	66
25–50	19	23	14	27
50–75	11	13	5	7

	Phosphorus status of soil	
	Conc. P in 0·01 M $CaCl_2$ *(p.p.m.)*	*Exchangeable P (p.p.m.)*
0–25	0·018	18
25–50	0·003	7
50–75	too low to measure	

The overall rate of uptake over a length of root will depend upon a combination of soil factors and plant factors. So far as the soil is concerned, the root, at its simplest, may be conceived as a nutrient sink. As a nutrient ion is removed at the boundary and the concentration there is lowered, nutrients in the bulk of the soil will tend to diffuse towards the boundary. For a simple ion like NO_3^- which is not adsorbed by the soil, but exists only in the soil solution, the way the resulting diffusion zone builds up around the root is illustrated in Fig. 1. The radial distance of the zone is roughly equal to $\sqrt{Dt}$, where D is the diffusion coefficient of the ion in the soil. A typical value for D_{NO_3} in a moist soil is 10^{-6} cm^2 sec^{-1}.

Most nutrient ions—all the cations and phosphorus for instance—are adsorbed by the soil colloids, where there is usually anything from 10–1000 times as much of them as there is in the soil solution. These adsorbed ions are in rapid dynamic equilibrium with the soil solution. This pool of nutrient on the solid tends to buffer the soil solution against depletion. The relation between the total amount of diffusible nutrient ion in a soil (C)

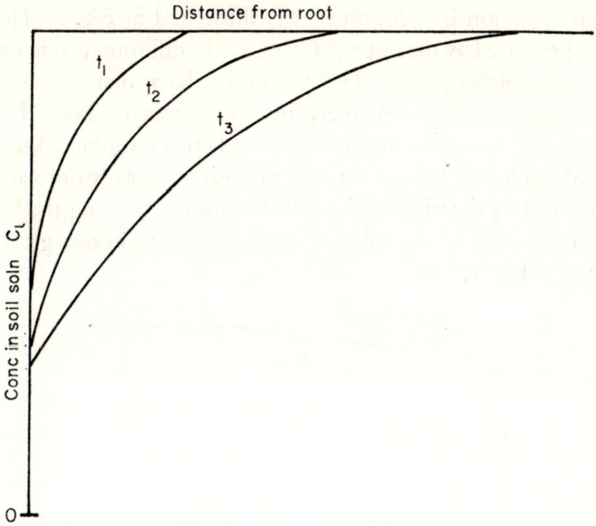

FIG. 1. Development of the zone of depletion of NO_3^- at a root surface at successive times.

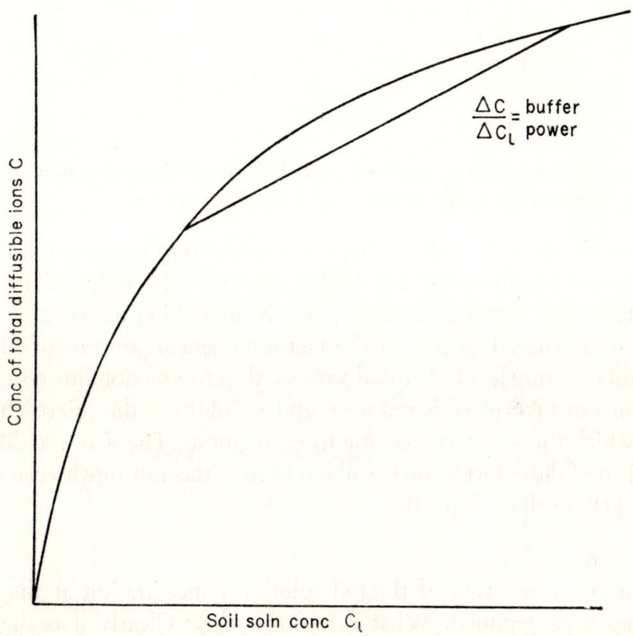

FIG. 2. The relation between the total amount of diffusible ion in the soil (C), and its concentration in the soil solution (C_l).

and its concentration in solution (C_l) is illustrated in Fig. 2. The buffer power is represented by the slope $\triangle C/\triangle C_l$. The diffusion coefficient varies inversely as the buffer power. The reason for this is that an ion is free to diffuse in solution but is relatively immobile when adsorbed. Thus it spends only a fraction of its time in a phase where it is mobile. Accordingly the zone of depletion for an ion like potassium spreads outwards more slowly than it does for nitrate; though the amount crossing the boundary for a given lowering of the solution concentration is in fact greater. This is illustrated in Fig. 3.

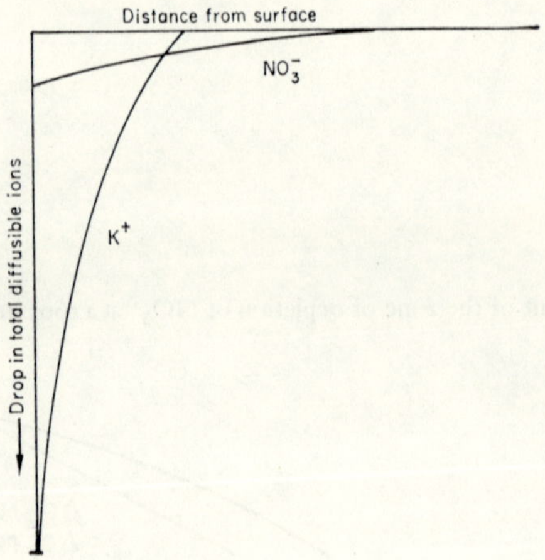

FIG. 3. Depletion zones for K^+ and NO_3^-. The lowering of the soil solution concentration is the same, but the buffer power of K^+ is ten times greater.

I have said enough to show that we must be very interested in the soil solution concentration and the buffer power of the soil, and Professor Arnold and Drs Sutton and Gunary will be describing this relationship in greater detail later. The other major factor influencing the rate of diffusion is the soil moisture level. As this decreases, the cross section through which diffusion can take place is reduced, and in addition the micro-channels along which the ions pass become more tortuous. The way the diffusion coefficient of the chloride ion in soil is related to the moisture level is shown in Fig. 4 (Rowell *et al.* 1967).

Plant factors

So far a given lowering of the soil solution concentration at the boundary has been assumed. What determines this? Clearly it is the result of a balance between the demand by the plant and the ability of the soil to maintain the supply. If demand is heavy and supply is slow the concentration at the surface will drop to a very low value. On the other hand if demand is light and diffusion is easy only a slight lowering is needed

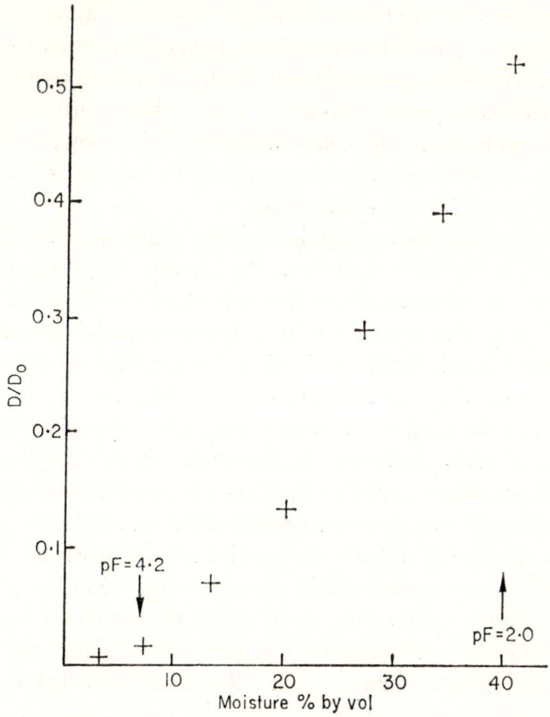

FIG. 4. Effect of soil moisture on the diffusion coefficient of Cl^-. D_0 is the diffusion coefficient in free solution.

to keep up with the demand. We must therefore have some measure of plant demand at the root surface. The simplest assumption we can make is that

$$F = \alpha C_l$$

where

F = flux across the surface of the root (g cm^{-2} sec^{-1})

α = a proportionality coefficient (not a constant) called the root absorbing power, and is a measure of the plant demand.

Considered purely as a problem in diffusion theory, if α, and r, the root radius, are known, and the diffusion characteristics of the soil are known, the course of uptake and the variation of C_l with time and radial distance can in principle be calculated (e.g. Nye 1966). If α is low the concentration at the root surface is maintained, and uptake is proportional to α. If α is high the concentration at the surface is reduced to a very low level and further increase in α will not result in an increased diffusion rate. Unfortunately we have very little knowledge as yet about the root adsorbing power of whole plants grown in soil. From short term work in nutrient culture on excised roots or portions of root we know that very often flux tends to be proportional to concentration if this is low, i.e. when α is independent of concentration, but approaches a maximum as the concentration is increased (Briggs *et al.* 1961).

Few comparable experiments have been made with whole plants in nutrient culture where the concentrations are anything like as low as those found in the soil solution; where they are (Russell *et al.* 1954; Loneragan and Asher 1967) the same sort of relation between uptake and concentration has been found. In soil, the determination of α is particularly difficult because it involves a knowledge of the concentration at the root boundary and the surface area of the root. Drew *et al.* (1969) have determined its value for single roots of onion plants growing for 10 days in soils ranging from very low to high potassium status (7·8 to 64·0 μg K per ml. in the soil solution). Under their conditions the concentration at the root surface was calculated to be about two-thirds of that in the solution at a distance from the root, and the value of α was constant within the rather large limits of error.

So far, only the uptake by a small length of intact root has been considered. Uptake by a whole plant or community of plants will be the sum of the uptakes from all these elements. We must know therefore how the absorbing surfaces increase in time and for how long each element of surface continues to function. For a rapidly growing plant exact knowledge of the absorption period of an individual root may not be so vital as might as first appear. If the relative growth rate of the total root surface is rapid it does not greatly matter if an element of root functions for only 10 days or for an infinity of days—its later contribution is swamped by the new roots being produced. What is essential, and what is almost totally lacking in the literature, is quantitative knowledge of the rate of increase of surface area of root system in soil. This involves simultaneous measurement of root length as well as root weight or volume. Dr Tinker will be describing a recent attempt to find the value of the root absorbing power for a number of nutrients in a soil-grown plant at different stages of its growth. This work illustrates these points. For the present we have to determine a value of the root absorbing power averaged over the whole plant: such questions as the effect of a high nutrient level in the top soil on the absorbing power of the roots in the subsoil have yet to be considered in detail.

The other plant factor that has to be taken into account in constructing a model of nutrient uptake is competition between roots, or—as it is

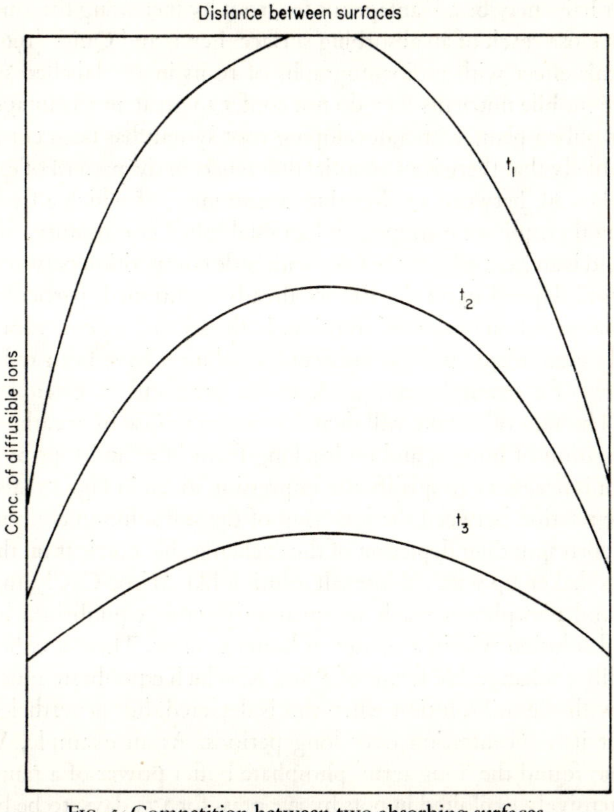

FIG. 5. Competition between two absorbing surfaces.

expressed in the soil—the overlap of their depletion zones. The successive stages in the depletion of nutrients between two absorbing surfaces is shown in Fig. 5. In principle, all stages of the pattern of depletion of a given volume of soil can be calculated if we know its diffusive characteristics and the absorbing power and distribution in space and time of the roots exploiting it. In practice, no such detailed study has been completed yet. However, Wiersum (1962) but not Cornforth (1968) has shown that a mobile nutrient like nitrate is taken up as rapidly by a plant with a coarse root system as it is by one with a fine root system, because competition sets in early. On the other hand a rather immobile nutrient such as phosphate is taken up more rapidly by the plant with the dense root system because each root exploits only a narrow zone around it.

Because of the very low mobility of *P* it has been calculated (Nye 1966) that root hairs may be advantageous to a root by increasing the amount of *P* within close reach of an absorbing surface. Lewis and Quirk (1967) have shown this effect with radioautographs of roots in P^{32} labelled soil. For the more mobile nutrients they do not confer so great an advantage.

So far only a plant with a developing root system has been considered. It seems likely that there is an essential difference, in the pattern of exploitation of the soil, between a colonizing community, of which a fresh sown agricultural crop is an example, and an established community, of which a grassland is an example. In the first, with little competition between roots, uptake will depend upon the factors already mentioned, particularly the initial concentration of the soil solution. In the second, a dense root system is already established, and the nutrients in solution have been reduced to low levels—for example, nitrate *N* under grassland in often less than 1 ppm. The rate of uptake will then depend on release by weathering, on mineralization of humus, and on leaching from litter and tops by rain.

Here it is necessary to qualify the impression given in Fig. 2 that there is a unique relation between the lowering of the soil solution concentration and the corresponding depletion of the exchangeable nutrient on the solid. If a soil is shaken up with a dilute salt solution like $M/100$ $CaCl_2$, the major cations and phosphorus reach an apparently stable equilibrium between solid and solution within a matter of hours at most. There are, however, less readily exchangeable forms of *P* and *K*, which equilibrate much more slowly with the soil solution when this is depleted, but nevertheless serve to buffer its concentration over long periods. As an example, Webber (1966) has found the 'long term' phosphate buffer power of a range of 10 soils, intensively exploited in pots by rye grass for 175 days, to be between 3 and 5 times greater than the short term buffer power. Professor Arnold will be discussing the slow release of *K*, and Drs Sutton and Gunary the slow release of *P*. The point to be made here is that when competition between roots is slight, uptake depends mainly on the soil solution concentration, and is rather insensitive to the buffer power—to a power less than $\frac{1}{2}$; but when competition is severe the buffer power and the total quantity of nutrient that can ultimately be extracted by the roots become all important. When more is known about the kinetics of this process of slow release, it can be incorporated in the simplified dynamic model presented here without great difficulty. But unless other speakers have something very new to tell us it is not yet possible to do this realistically.

So far only the diffusion of nutrients to root surfaces has been mentioned, and the mass flow of soil solution to the roots caused by transpiration has been neglected. Mass flow, or convection, modifies the diffusion process and will alter the concentration at the root surface and the consequent flux. It may be shown theoretically (Marriott and Nye 1968) that for $v/\alpha < 1$

$$\text{Flux} \simeq \alpha\, C_{l_1} \left(\frac{\beta + v}{\beta + \alpha}\right)$$

where v is the velocity of flow of solution at the root surface (ml. cm^{-2} sec^{-1}) and β is a term depending on the diffusion coefficient in the soil, the time and the root radius. The important point to note in this formula is that for any nutrient the flux should increase approximately linearly with the mass flow rate. This conclusion must be modified to the extent that α may depend on the transpiration rate—a matter to be covered by Professor Weatherley. In moist soils β for all ions is relatively large, and even at high transpiration rates v is small in comparison. But in drier soils mass flow may significantly increase flux, provided a high transpiration rate can still be achieved. For example, when $v = 10^{-6}$ ml cm^{-2} sec^{-1}, a reasonable average rate we have found in summer months, the following is the sort of increase of flux that may be expected because of the convection of nutrients to the roots.

Volume moisture content	0·40	0·16
p*F*	2·0	3·0
Increase in flux %	12	120

In the drier soil, diffusion acting alone would be much reduced, so that transpiration has a greater relative influence. Dr Tinker will be describing experiments to assess the significance of mass flow at different stages of a plant's development.

To sum up, it seems to me that we now understand the essential process by which nutrients are moved to the absorbing surfaces of roots, and the pattern of depletion in the soil. Many complications could be mentioned and no doubt others remain to be discovered. We are beginning to learn how to integrate these processes for the life span of a whole root system developing into a uniform soil. Eventually, even with the interwoven root system of a mixed community exploiting a succession of soil horizons, the course of removal of nutrients from the soil could be thoroughly described—provided we have the right plant parameters. It has been necessary here to stress the importance of the absorbing power of the roots because it provides us with an essential boundary condition without which the movement of nutrients through the soil to the root cannot be understood. It is however merely a parameter which could be further explained and analysed in terms of the growth of the plant and the demand for nutrients that this creates, but this trail leads away from the soil entirely and towards the leaves and the energy they receive and cannot be pursued here.

REFERENCES

BRIGGS G.E., HOPE A.B. and ROBERTSON R.N. (1961). *Electrolytes and Plant Cells.* Blackwell, Oxford.

CORNFORTH I.S. (1968). The effect of the size of soil aggregates on nutrient supply. *J. Agric. Sci.* **70**, 83–85.

DREW M.C., NYE P.H. and VAIDYANATHAN L.V. (1969). The supply of nutrient ions by diffusion to plant roots in soil. I. Absorption of potassium by cylindrical roots of onion (*Allium cepa*) and leek (*Allium porum*). *Plant and Soil* (in press).

LEWIS D.G. and QUIRK J.P. (1967). Phosphate diffusion in soil and uptake by plants. III. P^{31} movement and uptake by plants as indicated by P^{32} autoradiography. *Plant and Soil*, **26**, 445–453.

LONERAGAN J.F. and ASHER C.J. (1967). Response of plants to phosphate concentration in solution culture. II. Rate of phosphate absorption and its relation to growth. *Soil Sci.* **103**, 311–318.

MARRIOTT F.H.C. and NYE P.H. (1968). The importance of mass flow in the uptake of ions by roots from soil. *Trans. 9th Int. Cong. Soil Sci.* (*Adelaide*) **1**, 127–134.

NYE P.H. (1966). The effect of the nutrient intensity and buffering power of a soil, and the absorbing power, size and root-hairs of a root, on the nutrient absorption by diffusion. *Plant and Soil*, **25**, 81–105.

NYE P.H. and FOSTER W.N.M. (1961). The relative uptake of phosphate by crops and natural fallow from different parts of their root zone. *J. agric. Sci.*, **56**, 299–306.

ROWELL D.L., MARTIN M.W. and NYE P.H. (1967). The measurement and mechanism of ion diffusion in soils. III.The effect of moisture content and soil solution concentration on the self-diffusion of ions in soils. *J. Soil Sci.* **18**, 204–222.

RUSSELL R.S., MARTIN R.P. and BISHOP O.N. (1954). A study of the absorption and utilization of phosphate by young barley plants. III. The relation between the external concentration and the absorption of phosphate. *J. Expt. Bot.* **5**, 327–342.

WEBBER M.D. (1966). Studies on inorganic soil phosphate. Ph.D. Thesis, London.

WIERSUM L.K. (1962). Uptake of nitrogen and phosphorus in relation to soil structure and nutrient mobility. *Plant and Soil*, **16**, 62–70.

CATION EQUILIBRIA AND COMPETITION BETWEEN IONS

P.W. Arnold

School of Agriculture, The University,
Newcastle upon Tyne

The study of cation exchange equilibria between soils and soil solutions can be regarded as one small, though perhaps important, part of the complex problem of understanding soil–plant relationships. The intention in this paper is briefly to discuss certain aspects of cation exchange equilibria in so far as appropriate measurements can begin to characterize the ionic environment of roots and contribute towards the solution of ecological problems.

It has long been recognized that ion uptake by roots is concentration, or more strictly, activity dependent. In addition to external concentration, the influence of other ions and the rate at which metabolic processes utilize ions, are important. Viewed over the years there is no doubt that the study of plant nutrition via the soil solution has had a chequered history. The main difficulties are that, as well as being dilute, soil solution tends to vary in composition, even from day to day, and it is most sensitive to changes in the ratio of soil to liquid. Dilution of a soil–soil solution system containing mono- and divalent cations commonly displaces the equilibrium such that the adsorption of divalent ions increases and the adsorption of monovalent ions decreases. Changes in the nitrate content, for example, alter the amounts and proportions of other constituents in the soil solution and, for many years, no satisfactory means of interpreting analytical data could be found. Even as early as 1858 Liebig encountered such difficulties, which led him to conclude that soil solution studies should be abandoned. Despite further work over many years there was, especially after the middle 1920's, a notable absence of soil solution studies until quite recent times. From about 1920 for more than 30 years the main effort in soil cation–plant nutrition studies went into estimating the total amounts of labile cations in soils usually by displacement with concentrated salt solutions, or, less frequently, by isotopic tracer methods. These approaches to assessing the cation status of soils are too well-known for comment, but it should be emphasized that estimates of quantity alone need not correlate with uptake by the plant. Apart from the varying efficiency of the displacement or extraction procedures, estimates of quantity cannot be relied upon to provide a measure of the activity or intensity of a cation in the soil–soil solution system, which is one of the main factors governing absorption by roots. Over periods which are sufficiently long for the total reservoir of labile cations to be greatly depleted, it would seem reasonable to expect that uptake will depend on the total quantity which is, or becomes, labile in different soils. Again, for cations which are so weakly adsorbed by soil that much of the labile pool is in the soil solution (Tinker 1967), it might be expected that quantity would be a reliable index of availability. In soil–plant studies we are usually concerned with short periods of plant growth such as one growing season, or less, and it was Schofield (1955) who first seriously suggested that both intensity and capacity measurements should be considered together; the former largely governing immediate uptake, the latter endeavouring to maintain intensities.

For obvious practical reasons we must be satisfied with less than repeated checks on soil solution concentrations. Basically, the quest is for data which reliably reflect, in so far as this is possible, the effective cation status of soils. In this connection it must be emphasized that we cannot expect limited data, however precise and well-founded, to explain or account for the effects of variables which require additional consideration in the dynamic processes of soil–plant interaction. Indeed, there is a marked tendency among those dealing with soil–plant relationships to expect too much from a few simple measurements. On the other hand we must guard against discarding useful concepts simply because they do not come up to expectation when tested against a background of inadequately defined or uncontrolled variables.

The following discussion applies to soils which contain predominantly negatively charged colloids—which includes most, if not all, soils in this country—because special problems arise when more nearly equal amounts of negative and positive charge occur. The origin of the negative charge carried by soil colloids is now reasonably well understood; this includes the so-called permanent charge and the variable, or pH dependent charge (see Jackson 1964).

In considering how exchangeable, or labile, cations associate themselves with negatively charged soil colloids, it is to be expected that Coulombic forces play an all-important role in counteracting the kinetic energy of the cations. In addition there is invariably at least some specific interaction between colloid surface and adsorbed cations, which can be held very close to, or on, the surface with special affinity. The result is that adsorbed cations, most of which are in rapid exchange equilibrium with the soil solution, distribute themselves so that, although most are probably held close to the surface, their numbers decline with distance from the colloid surface, giving an enveloping atmosphere of cations, often referred to as a diffuse double layer. At points sufficiently remote from the particle surface the ionic concentration becomes uniform with equivalent amounts of cation and anion in the 'outer' soil solution. Labile anions, in contrast, will tend to be repelled by negatively charged surfaces so their concentration should rise to a maximum to the 'outer' solution. The net result of this is a difference in electrical potential between the solid surface and the 'outer' solution in which the electrolyte distribution is uniform. This difference in electrical potential will vary with the electrolyte concentration in the 'outer' solution.

When equilibrium is established between a soil, with its adsorbed cations, and an 'outer' solution, the chemical potential of each molecular species must be the same at all accessible parts of the system. Alternatively, it can be taken that the electrochemical potential of all ions of a particular species will have the same value regardless of their position, whether it be in the diffuse double layer—where electric potentials change with distance from the solid surface, or in the 'outer' solution. On either basis it emerges that a measure of the relative chemical potentials (μ) of cations, for example of M_1 and M_2 (where subscripts indicate valencies), in all parts of the system is given by

$$\mu_{M_1} - \tfrac{1}{2}\mu_{M_2} = RT \ln a_{M_1} - \tfrac{1}{2}RT \ln a_{M_2} + \text{constant terms} \qquad (1)$$

or

$$\mu_{M_1} - \tfrac{1}{2}\mu_{M_2} - \text{constant terms} = RT \ln \frac{a_{M_1}}{\sqrt{a_{M_2}}} \qquad (2)$$

where a_{M_1} and a_{M_2} are the activities of the ions in the soil solution, which can be calculated from measured concentrations. As stated by Beckett (1964a), when considering labile potassium and calcium in soil, the expression (2) 'depends on no assumptions about the soil'. Thus $RT \ln a_{M_1}/\sqrt{a_{M_2}}$ for an equilibrium system is uniquely proportional to the difference between the chemical potentials of labile cations M_1 and M_2 adsorbed on the soil. A formal treatment of the subject in terms of partial molar free energies was given by Woodruff (1955), and recently Barrow *et al.* (1965) clarified several of the principles concerned. However, in the present context there is no obvious advantage in expressing equilibrium data as energies of exchange; the basic data are the same in whatever terms they are expressed, and indeed parameters which could obscure some problems which arise, especially in attempts to use more than one type of equilibrium ion activity ratio, should be avoided.

Were it not for the validity of Schofield's (1947) Ratio Law it would be difficult, or impossible, to proceed further. It was shown by Schofield that for predominantly and highly negatively charged soils over a range of dilute electrolyte concentrations the values obtained for activity ratios of the type

$$\frac{a_{M_1}}{\sqrt{a_{M_2}}}, \frac{a_{M_1}}{\sqrt[3]{a_{M_3}}}$$

are independent of the total electrolyte (anion) concentration in the equilibrium solution. This means that the relative chemical potentials of adsorbed ions can be evaluated without taking arbitrary decisions about the method of measurement. Unique values are obtained at a particular temperature which depend only on the amounts of cations adsorbed; for this to be so, anions must effectively be excluded from at least the inner parts of the double layer so the environment of the vast majority of the adsorbed cations is not materially affected by changes in the solution concentration. However, in using equilibrium activity ratios we must accept the fact that we cannot go beyond measuring the potential of one ion relative to another.

Perhaps the best known application of ion activity ratios is in the calculation of the 'Lime Potential' of a soil. Whereas the pH of a soil, i.e. soil solution, varies with the electrolyte content, the activity ratio

$$\frac{\sqrt{a_{\mathrm{Ca}}}}{a_{\mathrm{H}}},$$

or expressed as negative logarithms, $\mathrm{pH}-\frac{1}{2}\,\mathrm{pCa}$, is a constant and can be equated to the calcium hydroxide potential of the soil (Schofield and Taylor 1955a, b).

For soil in which calcium is the dominant exchangeable cation, it is reasonable to use it as the reference ion, and thus a measure of the potential of another ion, M_1, relative to calcium in the particular soil can be obtained. In practice it is often convenient to use the sum of the calcium and magnesium as the reference 'ion' because Ca and Mg often behave in similar ways when adsorbed on soil, although this is not always the case. When the calcium (or calcium plus magnesium) potential in different soils is assumed to be fairly constant we can obtain a measure of the relative M_1 potentials among soils. Some of the problems which arise in extremely acid soils have been discussed by Tinker (1964a) who showed that certain activity ratios are critically dependent on the exact (small) amount of labile calcium present; he suggested aluminium as being a more appropriate reference cation in such soils.

In many soils the natural content of soluble anions in an equilibrium 'water extract' is enough to give analytically determinable amounts of the cations needed to calculate activity ratios such as $a_{M_1}/\sqrt{a_{M_2}}$, but the disadvantage is that water extracts are not always obtained clear enough for analysis and the act of obtaining a water extract, must, at least to some extent, deplete the soil of labile cations. Usually it is convenient to use Taylor's (1958) interpolation method for determining equilibrium cation activity ratios. When calcium is the dominant soil cation it is convenient to equilibrate soil with dilute calcium chloride solutions (usually within the range 0·02–0·005 M) containing both slightly more and slightly less M_1Cl than the roughly estimated equilibrium concentration; some M_1 will be either adsorbed or desorbed by the soil, but knowing the weight of soil and volume of solution, the composition of the solution which would not change on equilibration can be determined by interpolation. With appropriate activity coefficients (see, for example, Beckett 1965) the true equilibrium activity ratio $a_{M_1}/\sqrt{a_{\mathrm{Ca}}}$ is obtained. In practice it is desirable to use sufficient total electrolyte to keep soil colloids flocculated, and hence facilitate analysis.

When it comes to using ion activity ratios in availability studies, it is possible that uptake of some cations by plants is so small compared with the size of the labile pool that the total quantity in the pool is not measurably altered. Such could be the case for labile magnesium in clayey soils of moderate to high labile magnesium content. Thus, all other factors being equal, we would only need a suitable measure of the magnesium potential to characterize the prevailing magnesium status of the soil. It is, however, much more likely that as plant uptake ensues, there will be a fall in cation potential because the labile pool is depleted. The extent to which a cation potential falls will depend on the quantity removed from the soil–soil solution system and on the ability of the soil to resist or buffer the fall in potential. This concept, which is essentially that of Schofield's (1955) Intensity–Capacity idea, has been developed particularly by Beckett (1964b) to characterize the behaviour of labile potassium in soil. Using so-called 'immediate' (i.e. rapidly determined) Quantity/Intensity (Q/I) relationships, the Potential Buffering Capacity (PBC) of soil can be assessed, although it must be remembered that slow releases of initially non-labile cations, which can counteract depletions and hence increase the real 'buffer' capacity, are not included. Figure 1 illustrates how potassium potentials change with changing quantity of potassium adsorbed on two different soils, the changes in quantity being referred to the amount initially present. The slopes of the lines, which tend to be near-linear over the ranges normally encountered in agricultural soils, characterize the 'immediate' buffering capacity. For very small contents of labile potassium the Q/I relationship invariably shows pronounced curvature, due to the presence of adsorption sites which have special affinity for potassium ions. However, when dealing with most practical situations the slope of the Q/I relationship

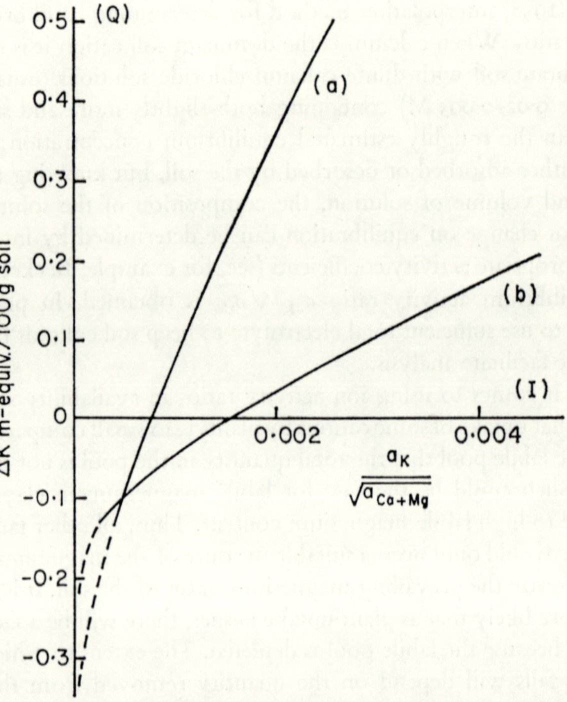

FIG. 1. Quantity-Intensity relationships of potassium in two contrasted soils.

can usually be represented by a single figure. Examples of how magnesium potentials are buffered in different soils have been reported by Salmon (1964a).

For soils which differ in their Q/I relationships it is predictable (Arnold 1962) and confirmed (Barrow *et al.* 1965) that cation potentials alone, or directly related measurements, are not satisfactory for forecasting even the relatively short-term cation supplying powers. Barrow (1966) studied potassium uptake under greenhouse conditions from soils which differed both in their initial potassium potentials (expressed as differences in free energy) and their 'immediate' PBCs, but which possessed negligible potassium releasing power from initially non-labile sources. In an analysis using both variables, he was able to account for most of the variation in the potassium contents of plants (*Trifolium subterraneum*).

In experiments on crop plants the usual practice is to keep constant as many factors as possible and to vary one, two or possibly three; under such circumstances good correlations between plant and soil measurements

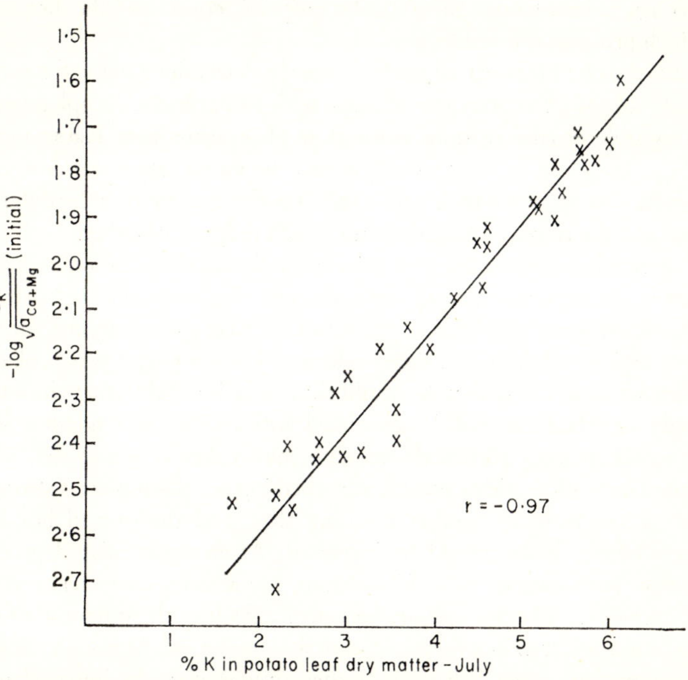

FIG. 2. Relationship between percentage potassium in potato leaf dry matter on 27 July and $-\log a_K/\sqrt{a_{Ca+Mg}}$ measured on field-condition soil immediately after planting.

often materialize. Using field plots in which only the potassium status of soil was intentionally varied, Arnold *et al.* (1968) found initial potassium potentials, expressed as negative logarithms, over a wide range of values were highly correlated with the potassium contents of potato leaf dry matter sampled in July. This is shown in Fig. 2 in which the potassium percentage in the leaf is plotted against $-\log a_K/\sqrt{a_{Ca+Mg}}$, $r = 0{\cdot}97^{***}$. However, over smaller ranges of potassium potential a near-linear relationship between $a_K/\sqrt{a_{Ca+Mg}}$ and the percentage potassium in leaf dry matter is obtained (Salmon 1964b). The data in Fig. 2 were obtained at two sites where the PBCs of the soils were almost identical. It remains to be seen how potassium potentials correlate with plant uptake under field conditions on soils which differ in their buffer capacity; the need, especially in the case of potassium, is to obtain improved estimates of buffering capacity because 'immediate' PBCs have limitations when initially non-labile sources of potassium are utilized by plants. In this connection Beckett (1967) has, at least from a theoretical standpoint, considered the directions in which progress is possible.

It can thus be summarized that a properly determined value for cation activity ratio is a characteristic of a soil with its particular complement of exchangeable or labile cations at the time of measurement. Ratios of the type $a_{M_1}/\sqrt{a_{M_2}}$ can be used to help define the initial cation status of soil–soil solution systems, for their values reflect both the saturations by different cations and the relative strengths with which they are adsorbed.

It is generally accepted that soils behave as polyfunctional cation exchange materials, meaning that not all the exchange 'sites' or exchange 'surfaces' are the same; some 'sites' have a preference, or have special affinity, for one type of cation, while others prefer another. Jarusov (1937) was probably the first to make clear that at equilibrium it is inevitable that the more strongly adsorbed ions will be associated with the stronger bonding sites and the less strongly adsorbed ions with the weaker bonding sites. Thus increased saturation with a weakly adsorbed ion would displace a strongly adsorbed ion from its weaker bonding sites, and the overall bonding strength of the latter would be increased. At any particular degree of saturation with a variety of labile cations, ion activity ratios serve as the most accessible, least ambiguous measures reflecting the outcome of the competition between ions for adsorption sites on the soil. Although calcium and magnesium often behave in rather similar ways in ion exchange equilibria between soil and soil solution—a fact which encourages the use of terms in which Ca and Mg appear as one species—these ions do not, in fact, always behave similarly. Differences in their behaviour seem to be greater when adsorbed on soil organic matter than on soil mineral matter. Thus, Salmon (1964a) found weaker adsorption of magnesium relative to calcium by peat with increasing magnesium saturation, so calcium becomes progressively more difficult to displace with increasing magnesium saturation.

In connection with the competition between cations during entry into plant roots growing in soil, there are two aspects to consider. If, in the first place, we are supplied only with information about the quantities of labile ions in the medium we should look both at the outcome of competition between ions in the soil–soil solution system, the results of which may differ greatly from soil to soil, and the final outcome of the competition for entry into the plant. In the latter the effects observed are the result of a number of processes taking place simultaneously. In the former, equilibrium cation activity ratios reflect the outcome of the competition. Clearly, it should not be forgotton that effects observed in the plant are, in part, the outcome of the interaction between cations in the soil–soil solution system.

In attempting to deal quantitatively with the problem of ion 'competition' for entry into plants, both Tinker (1964b) and Salmon (1964b) reached quite similar general conclusions. Salmon (1964b) investigated a method based on equilibrium ion activity ratios for characterizing the magnesium status of soils; it was especially interesting because it offered scope for studying the well-known antagonistic effect of potassium on magnesium uptake. At the outset it was assumed that for a given uptake of anions, half-moles of a divalent cation are as effective as moles of a monovalent cation, which for a given anion supply is probably true. Salmon's data showed that over a range of cation potentials the amount of a monovalent cation in plant dry matter depends directly on its molar activity, that is $(M_1)_{\text{plant}} \propto (a_{M_1})_{\text{solution}}$. However, the magnesium potential was calculated as $\sqrt{a_{\text{Mg}}/a_{\text{Ca+Mg}}}$ because from the limited data there was evidence that, other factors being equal, the amount of magnesium in the plant was proportional to $\sqrt{a_{\text{Mg}}/a_{\text{Ca+Mg}}}$, rather than $a_{\text{Mg}}/a_{\text{Ca+Mg}}$. Although this cannot easily be justified from a theoretical standpoint, it served as a basis from which to start developing 'unified' or combined activity ratios. An expression was derived involving both the magnesium potential, calculated as $\sqrt{a_{\text{Mg}}/a_{\text{Ca+Mg}}}$, and the antagonistic effect of competing ions, such as potassium, and the whole found to be well-correlated with magnesium concentration in the plant. Thus:

$$\text{Mg}_{(\text{plant})} \propto \sqrt{\frac{a_{\text{Mg}}}{a_{\text{Ca+Mg}}}} \cdot \frac{1}{1 + \sum \text{proportionality factor} \dfrac{\sqrt[z]{a_{\text{M}^{z+}}}}{\sqrt{a_{\text{Ca+Mg}}}}}$$

The significance of the proportionality factor was discussed by Salmon (1964b); its value, which must be determined by experiment, depends on a number of factors including the properties of the root, and the rate at which the potential of M^{z+} falls during uptake. It is difficult to visualize any substantially different approach for dealing with the complex problem of cation antagonism using simple experimental data. If we retain the use of the term 'available', which can mean no more than 'capable of being used', we may regard 'unified' activity ratios as beginning to provide a link between availability and utilization, in that some of the factors determining utilization are considered.

Ecologists are well aware that cation uptake depends on many factors which have not been mentioned, such as moisture supplies and how these fluctuate, anion supplies, and competition or synergism between plant roots and soil micro-organisms. In comparing the problems of ecologists with those who work on crop plants, it should be remembered that the latter usually work on cultivated soils in which only a few factors are intentionally varied, attempts being made to reduce the influence of other variables by draining, cultivating, irrigating and using blanket dressings of some plant nutrients; in this sense most ecological problems are likely to be more complex.

If equilibrium measurements continue to show promise in preparing the way for dealing with some aspects of the overall dynamic processes of nutrient uptake by plants, their worth must be partly due to other factors being either relatively constant (or unimportant), or quite well correlated with the measurements which are made. Although the laboratory procedures, as described in the literature, are usually straightforward, special consideration must always be given to the way in which soils are handled; for example, it is sometimes recommended that soils should not be air-dried before analysis, because not only are soils liable to undergo at least some irreversible changes on air-drying, but also appreciable leakage of cations from roots to soil may take place.

The hope is that the type of measurements described will prove useful beyond most soil type boundaries. However, for soils which are not predominantly negatively charged, the Ratio Law does not necessarily apply, mainly as a result of anions not being excluded from the inner part of the double layer, and the simple approach using ion activity ratios may be totally inadequate; for such soils there lies a whole field of enquiry which has barely started.

REFERENCES

ARNOLD P.W. (1962). The potassium status of some English soils considered as a problem of energy relationship. *Proc. Fertil. Soc.* **72**, 25–43.

ARNOLD P.W., TUNNEY H. and HUNTER F. (1968). Potassium status: soil measurements and crop performance. *9th Int. Congr. Soil Sci.* (*Adelaide*) **2**, 613–620.

BARROW N.J., OZANNE P.G. and SHAW T.C. (1965). Nutrient potential and capacity. I. The concepts of nutrient potential and capacity and their application to soil potassium and phosphorus. *Aust. J. Agric. Res.* **16**, 61–76.

BARROW N.J. (1966). Nutrient potential and capacity. II. Relationship between potassium potential and buffering capacity and the supply of potassium to plants. *Aust. J. Agric. Res.* **17**, 849–861.

★BECKETT P.H.T. (1964a). Studies on soil potassium. I. Confirmation of the ratio law: measurement of potassium potential. *J. Soil Sci.* **15**, 1–8.

★BECKETT P.H.T. (1964b). Studies on soil potassium. II. The 'immediate' Q/I relations of labile potassium in the soil. *J. Soil Sci.* **15**, 9–23.

BECKETT P.H.T. (1965). Activity coefficients for studies on soil potassium. *Agrochimica*, **9**, 150–153.

BECKETT P.H.T. (1967). Potassium potentials. in Tech. Bull. Min. Agric. Fish. Food, No. 14. *Soil potassium and magnesium*, pp. 32–38, H.M.S.O. London.

JACKSON M.L. (1964). Chemical composition of soils. *Chemistry of the soil* (Ed. F.E. Bear), pp. 71–141. New York.

JARUSOV S.S. (1937). Mobility of exchangeable cations in the soil. *Soil Sci.* **43**, 285–303.

★SALMON R.C. (1964a). Cation exchange reactions. *J. Soil Sci.* **15**, 273–283.

SALMON R.C. (1964b). Cation-activity ratios in equilibrium soil solutions and the availability of magnesium. *Soil Sci.* **98**, 213–221.

SCHOFIELD R.K. (1947). A ratio law governing the equilibrium of cations in soil solution. *Proc. 11th intern. Congr. Pure and Applied Chem. (London)*, **3**, 257–261.

SCHOFIELD R.K. (1955). Can a precise meaning be given to 'available' soil phosphorus? *Soils and Fert.* **18**, 373–375.

SCHOFIELD R.K. and TAYLOR A.W. (1955a). Measurement of the activities of bases in soils. *J. Soil Sci.* **6**, 137–146.

★SCHOFIELD R.K. and TAYLOR A.W. (1955b). The measurement of soil pH. *Soil Sci. Soc. Amer. Proc.* **19**, 164–167.

★Taylor A.W. (1958). Some equilibrium solution studies on Rothamsted soils. *Soil Sci. Soc. Amer. Proc.* **22**, 511–513.

TINKER P.B. (1964a). Studies on soil potassium. III. Cation activity ratios in acid Nigerian soils. *J. Soil Sci.* **15**, 24–34.

TINKER P.B. (1964b). Studies on soil potassium. IV. Equilibrium cation activity ratios and responses to potassium fertilizer of Nigerian oil palms. *J. Soil Sci.* **15**, 35–41.

★TINKER P.B. (1967). The relationship of sodium in the soil to uptake of sodium by sugar beet in the greenhouse and to yield responses in the field. *Trans. Meet. Commns. II and IV int. Soc. Soil Sci.* 1966, 223–231.

WOODRUFF C.M. (1955). Energies of replacement of calcium and potassium in soils. *Soil Sci. Soc. Amer. Proc.* **19**, 167–171.

Experimental Methods

Articles marked (★) contain useful information about experimental procedures. In addition, reference should be made to Beckett, P.H.T. and Craig, J.B. (1964). The determination of potassium potentials. *8th Int. Congr. Soil Sci. (Bucharest)*, **2**, 249–255.

PHOSPHATE EQUILIBRIA IN SOIL

C. D. SUTTON and D. GUNARY
Levington Research Station, Ipswich, Suffolk

The process of uptake of phosphate by a plant involves the transfer of a phosphate ion from the soil solution near to the root surface into the cytoplasm of the root cells. Whatever the mechanism of this transfer is, it occurs so rapidly that normally the limiting factor in uptake is the number of phosphate ions that are at the root surface.

When a length of new root starts to exploit fresh soil, the number of phosphate ions at the root is proportional to the concentration of phosphate in the soil solution. In most soils this concentration is extremely low so that the supply of ions near to the root will be quickly exhausted. Diffusion towards the root of ions from more distant solution will augment the supply but, as diffusion is so slow, this is still inadequate for normal growth.

However, as the solution phase near the root is depleted, phosphate ions from the solid phase will be drawn into solution. Thus, to assess the phosphate supplying power of a particular soil, the rate at which this desorption process occurs and the quantity of phosphate per unit volume of soil that can be released must also be known. The quantity of phosphate that can be released into solution is usually called the labile fraction. This quantity may be measured by isotopic dilution techniques and is usually called the *E*-value for laboratory measurements and the *L*-value where growing plants are involved.

These four soil properties, phosphate concentration in solution, diffusion rate, release rate and the total quantity of releasable or labile phosphate are interrelated. The way in which this occurs for labile phosphate and phosphate concentration in solution is described by phosphate adsorption isotherms, so that to visualize this relationship one must also take into account the ability of the soil to adsorb phosphate.

The adsorption of phosphate can be likened to the temporary trapping of ping-pong balls as they bounce across an egg tray. For a given total number of balls, the number trapped at any one time will be a function of the total number of pockets in the tray. Hence this latter number will determine the percentage of the balls likely to be free at any one time, and the rate of movement across the tray. That is to say, for a soil of given labile phosphate content, the phosphate adsorption capacity (the maximum which is approached by the adsorption isotherm) will determine the concentration in solution and the diffusion and release rates. The way in which phosphate supply depends on the inter-relationship of all four soil properties can be visualized by using the analogy of the supply of water from a well (Fig. 1).

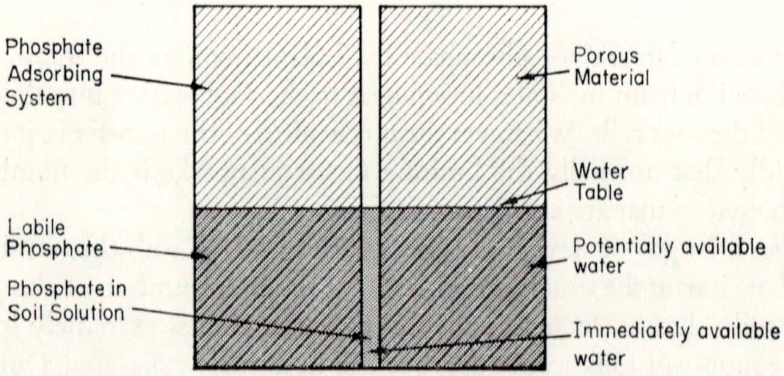

FIG. 1. Well analogy of phosphate behaviour.

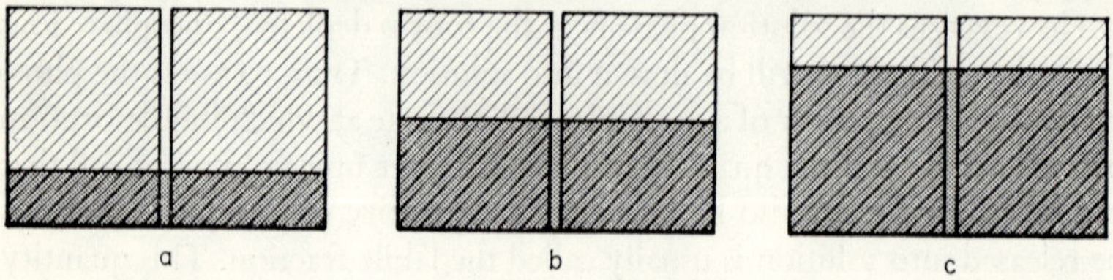

FIG. 2. When *P*-adsorption capacity is similar, only labile phosphate need be known to describe the system.

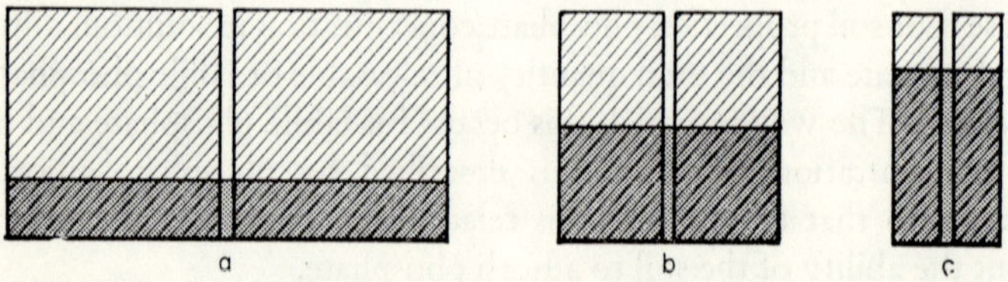

FIG. 3. With different adsorption capacities, two parameters are needed to describe the system.

In this analogy the water in the well represents phosphate in solution, water in the porous material surrounding the well represents labile phosphate, while the volume of porous material represents the size of the adsorption system. By comparing a, b and c in Fig. 2 it will be seen that for a given adsorption system, it is only necessary to specify the quantity of labile phosphate present. The concentration in solution, the rate of diffusion and the rate of replenishment then follow since the higher the fraction of the adsorption capacity that is filled the higher will be the water level in the analogy and obviously the greater is the 'head of pressure' for replacing what has been removed. With different adsorption systems (Fig. 3) a knowledge of two parameters is necessary for a complete description. Thus if the size of the adsorption system and the amount of labile phosphate are measured the concentration in solution and release rate need not then be determined. Alternatively, if labile phosphate and concentration in solution are measured, then the size of the adsorption system can be inferred.

The above relationships are, however, only half the story, dealing only

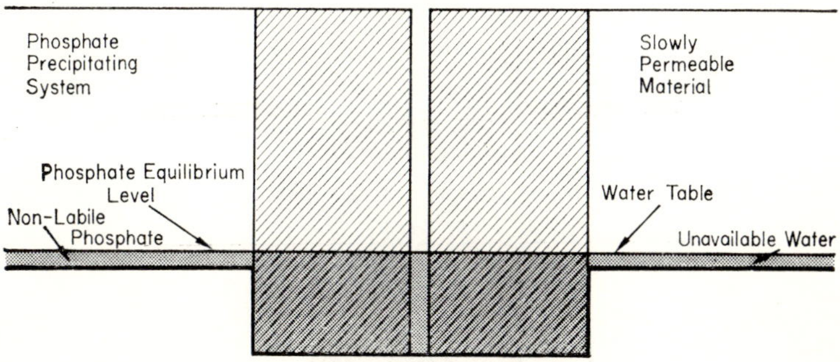

FIG. 4. Well analogy of phosphate behaviour, extended to include non-labile phosphate and the equilibrium phosphate level.

with the equilibrium between *P* in solution and labile *P*, whereas that between labile and non-labile *P*:

$$P_{\text{solution}} \rightleftharpoons P_{\text{labile}} \rightleftharpoons P_{\text{non-labile}}$$

must also be taken into account in the long term.

To visualize the complete system the well analogy can be extended (Fig. 4). As before, the centre portion represents the $P_{\text{solution}}/P_{\text{labile}}$ part of the equilibrium but now the porous material is surrounded by an infinite extent of slowly permeable material.

Starting with the whole system empty, the first water added will flow from the central well into the porous material, behaving as before. As more water is added the water table will rise until it reaches the lip of the trough where it will spill over and seep into the slowly permeable material. If further water is added, the level in the well may be temporarily raised, but eventually it will sink back to the lip of the trough. Conversely, water removed from the trough will be slowly replaced by seepage from the slowly permeable material, so that the lip of the trough represents an equilibrium level (Fig. 5).

This means that at very low levels of phosphate in the soil it can all be labile (Larsen, 1964). At higher levels, the solubility product of some phosphate compound will be exceeded—and a crystalline phase will build up in such a way that phosphate inside the crystal lattice is no longer able to enter solution freely. Only when phosphate depletion occurs is the phosphate likely to be released by dissolution.

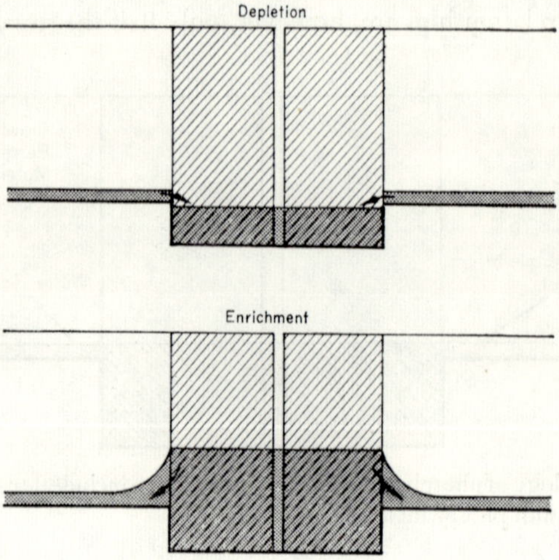

FIG. 5. An equilibrium phosphate level is re-established after depletion or enrichment

The picture of phosphate supply is, therefore, of a reservoir of crystalline phosphate that is in equilibrium with the remaining adsorbed phosphate. The quantity of this adsorbed labile phosphate in relation to the size of the adsorption system will determine the amount and mobility of phosphate in solution and the ability of the soil to maintain the concentration in solution as phosphate is removed by the plant. Soils with the same content of labile phosphate will have a buffering capacity directly dependent on the size of the adsorption system.

Effect of soil type

The relative importance and nature of the various components in the soil phosphate system will vary with soil type. Since there are a large number of variables depending on parent material and the past history of the soil it is only possible to generalize. However, a broad distinction can be made between acid and calcareous soils.

In calcareous soils and in soils in which calcium is the dominant exchangeable cation, say above pH 6, the non-labile phosphate will ultimately be a crystalline calcium phosphate. The most stable of these is hydroxyapatite, and it is the solubility of this compound which is likely to determine the quantity of labile phosphate which is present at equilibrium. In these soils, added or freshly released labile phosphate will be converted to non-labile forms at a rate which will depend on the reactivity of the calcium ions. Calcareous soils are not all alike in this respect and it might be expected that soft chalky soils will be more reactive than those developed from hard limestone. The labile phosphate on these soils, though its chemical nature is less clear, is probably adsorbed on the surface of calcium phosphate minerals or linked to clay surfaces through 'calcium bridges'. The phosphate adsorption capacity is, therefore, related to the clay content of these soils and to the content and surface area of calcium-rich minerals.

In acid soils the phosphate adsorption capacity is associated with the nature and quantity of the clay. This is because in these soils phosphate is adsorbed by hydrated iron and aluminium oxides which are particularly active on clay surfaces. The quantity of active iron and aluminium is not only a function of clay content but depends on the parent material from which the soil is developed. Thus basaltic soils which are rich in active iron have a phosphate adsorption capacity substantially higher than that of many other soils of similar clay content. Acid sands usually have a low phosphate adsorption capacity and in extreme cases phosphate has been shown to leach from these soils. The non-labile phosphates which occur in very acid soils are iron and aluminium phosphates, and it is the solubility of these which will determine the quantity of labile phosphate present at equilibrium.

Measurements necessary to predict phosphate supply

With such a large series of reactions involved in phosphate uptake it is not surprising that the unequivocal measurement of soil phosphate status is difficult. The classical methods based on acid extraction immediately upset the equilibria and are of little value, particularly where comparisons between soils of different type are involved.

Because of the interdependence of the various soil properties it is not necessary to measure all of them adequately to describe the soil's ability to supply phosphate. For example, phosphate uptake by reygrass was measured in a pot experiment and correlated with a number of soil parameters (Gunary and Sutton 1967). By taking combined account of phosphate concentration in solution and the amount of labile phosphate, it was possible to explain 70–80% of the total variation in phosphate uptake between soils (Table 1). Thus phosphate supply may be described by only two parameters, since for one growing season the mobilization of non-labile phosphate may be ignored.

TABLE 1

Percentage of variation in phosphate uptake accounted for by selected parameters

	Initial uptake	*Long-term uptake*
Phosphate concentration—log [*P*]	49·4	59·2
L-value	56·5	55·7
Multiple regression of log [*P*] and *L* value	73·5	79·6
Resin phosphate[1]	69·9	69·4

[1] Hislop and Cooke (1968).

With only small loss in accuracy, it is possible to predict phosphate supplying power by making only one measurement—the amount extractable by anion exchange resin. Since this resin removes phosphate ions from the solution phase, but relies on desorption of phosphate from the solid phase for the bulk of its supply—as does the plant—it is not surprising that it can account for about 70% of the variation in uptake.

Conclusions

To sum up, active plant roots can quickly deplete the phosphate in solution in their immediate vicinity. Their subsequent phosphate supply then depends on release from the solid phase and diffusion to the root surface. The rate of diffusion sets a limit to the zone of soil which can contribute phosphate to the root, but within that small zone extensive removal of the labile phosphate may well occur.

Finally it must be emphasized that with such an immobile nutrient as phosphate, the supplying power of the soil means the supplying power to a unit length of active root. The total length of this active root system must, therefore, be taken into account in determining the total phosphate uptake by a particular plant. Hence differences in uptake between plant species in the same soil are likely if their root systems have different lengths of active tissue. It also follows, and this is perhaps of more significance for ecologists, that while a low level of soil supply may be quite adequate for say a perennial grass with a well established root system, it will be totally inadequate for the tap root system of a tiny seedling.

Recommended methods

1. *Phosphate concentration in solution.*
By shaking soil for 18 hr with 0·01 M $CaCl_2$ at a soil : solution ratio of 1 : 20 and determining the phosphate concentration in the extract using the method of Fogg and Wilkinson (1958).

2. *Labile phosphate*
(a) *L* value
By growing ryegrass in soil labelled with 10 μC carrier free $H_3{}^{32}PO_4$ per kg soil and determining the specific activity of the phosphate taken up at each cut up to a total of six cuts. Gunary and Sutton (1967).

(b) *E* value
By determining the quantity of soil phosphate exchangeable with carrier free $H_3{}^{32}PO_4$ after end over end shaking for an appropriate period. For short term comparisons between soils 18 hr shaking and a soil: solution ratio of 1 : 20 are used.

For 'total exchangeable' phosphate a 14-day period and a soil : solution ratio of 1 : 50 are used. Gunary and Sutton (1967).

3. *Phosphate adsorption capacity*
The method of Olsen and Watanabe (1957) is modified by having a tangent to the curved Langmuir plot at equilibrium phosphate concentration of 10 p.p.m. P.

Considerable simplification of this technique could give a rough approximation which may be adequate for comparative purposes.

4. *Resin extractable phosphate*
By shaking 2 ml $<$ 0·4 mm soil with 5 ml $>$ 0·5 mm De Acidite FF resin in 100 ml water for 16 hr at 25° C and determining the quantity of phosphate adsorbed by the resin after separation from the soil. Cooke and Hislop (1963), Hislop and Cooke (1968).

REFERENCES

COOKE I.J. and HISLOP J. (1963). Use of anion exchange resin for the assessment of available soil phosphate. *Soil Sci.*, **96**, 308–312.

FOGG D.N. and WILKINSON N.T. (1958). The colorimetric determination of phosphorus. *Analyst, Lond.* **83**, 406–414.

GUNARY D. and SUTTON C.D. (1967). Soil factors affecting plant uptake of phosphate. *J. Soil Sci.*, **18**, 167–173.

HISLOP J. and COOKE I.J. (1968). Anion exchange resin as a means of assessing soil phosphate status: a laboratory technique. *Soil Sci.* **105**, 8–11.

LARSEN S. (1964). On the relationship between labile and non-labile phosphate. *Acta. Agr. Scand.* **14**, 249–253.

OLSEN S.R. and WATANABE F.S. (1957). A method to determine a phosphorus adsorption maximum of soil as measured by the Langmuir Isotherm. *Proc. Soil Sci. Soc. Am.* **21**, 144–149.

THE TRANSPORT OF IONS IN THE SOIL AROUND PLANT ROOTS

P. B. TINKER
Department of Agriculture, Oxford University

INTRODUCTION

The basic processes operating in the soil around plant roots have been dealt with in the first paper in this section. I propose to amplify some of these ideas, as they apply to field crops, and then, as a practical example, to describe work in which Mr Brewster in our department investigated nutrient transport to isolated whole plants in field conditions. We had two main aims in this work: firstly to estimate the relative importance of mass flow and diffusion at different stages of growth, and secondly to determine whether any significant changes in concentration occurred at the plant root surfaces. We believe that these methods may be applied in principle to ecological problems, though the practical difficulties may well be greater. The treatment is simple, but it is difficult to see how more refined methods could be applied in the field at present.

In the general field situation, both mass flow and diffusion operate. Mr Nye has dealt in detail with diffusion, so I will concentrate on mass flow. The term could cover two processes: the movement of nutrients into an existing depletion zone around a root by general movement of soil water up or down a profile, or the movement of nutrients up to the root surface in the transpiration-induced flow of water to the root. The former has not yet been fully treated, though it may be of considerable interest for ions such as nitrate, and 'mass flow' normally refers to the second process exclusively. I emphasize that this is movement *up to* the root—we do not discuss the effect of transpiration on movement *into* the root.

Early attempts at analysing soil transport processes, mainly by Barber *et al.* (1962), used approximate calculations based on transpiration ratios, average crop compositions and average soil solution compositions. They indicated that mass flow was important for NO_3, SO_4, Ca and Mg, and we have no reason at present to alter these broad conclusions. However, uncertainty arises from the attempt to average all processes over a whole growing season. Transpiration, and hence mass flow, will be greater after mid-summer, but much of the final nutrient content may have been absorbed by then (e.g. Knowles and Watkins 1931). It seems quite possible that some nutrients in the 'mass flow' group mentioned above reach the root by a predominantly diffusive mechanism early in the season, and that much of the mass flow later is wasted as a nutrient-supplying mechanism for some crops. It is also assumed that water and nutrients enter, in constant ratio, through the same root portions, but mass flow must become less effective if water uptake and nutrient uptake zones on the root are separated. Finally, there are considerable difficulties in defining a 'mass flow contribution' even to a single uniform piece of root over a short time.

The obvious course is to define an 'apparent mass flow contribution' VC_{li}, where V is the flux intensity of water into the root (equals flow velocity at root surface) and C_{li} is the initial soil solution concentration. If uptake flux intensity F is exactly equal to this, mass flow equals uptake, and no problem arises. If VC_{li} is greater than F, the concentration increases next to the root, and diffusion carries ions away from the root. If VC_{li} is smaller than F, there is a concentration decrease at the root surface, and diffusion carries ions towards the root. It is reasonable to say that mass flow supplies the root entirely in the first case, and mass flow and diffusion in the second case, but in neither can we satisfactorily distinguish the mass flow and diffusive fluxes from each other. When there is a diffusive flux there is a concentration gradient, and the mass flow operates on different concentrations at different points. The mechanisms can in theory be separated at any one point, but their relative contributions vary with position.

The correct theoretical approach has been given by Nye and Spiers (1964), but their differential equation is not generally soluble. It can be solved for the special case of linear, as opposed to radial, flow (Nye 1966), and this type of system is being investigated by us, but it does not give any direct help in dealing with natural systems, where flow is almost always radially towards the root.

For our attempt to investigate the average situation around roots in the field, we have had to accept the simple assumption of the 'apparent mass flow contribution', which is implicit in the equation originally presented by Passioura (1963). This is

$$F = VC_{li} + \frac{(C_i - C_r)Dg}{a} = \alpha C_{lr} \qquad (1)$$

when

F = nutrient ion flux into root, in m/cm²/sec.

α = root uptake coefficient; we assume that uptake is proportional to external concentration. This is discussed later.

V = water flux into root, ml/cm²/sec, or cm/sec.

C_i = initial soil concentration of ion, in m/cm³.

C_r = concentration at root surface; subscript l indicates concentrations in liquid phase.
D = diffusion coefficient in soil for that ion.
g = a function of Dt/a^2, which diminishes with t, and is usually not very far from 1.
a = root radius.

(Note root uptake per unit length $= 2\pi aF$).

The original equation was restricted to non-adsorbed ions such as nitrate, but the form given above should be correct for all ions for which equilibrium between solid and liquid phases in the soil is rapid.

The equation implicitly separates the diffusive and mass flow fluxes, and states the mass flow contribution as VC_{li}, when it might just as logically be VC_{lr}, or any concentration between these two. If VC_{lr} and VC_{li} are widely different, an averaged value may give better results.

Some doubts attach to the diffusion term in the equation also. This is a standard solution for a 'constant boundary' condition, i.e. C_r is a constant. This is unlikely, and also contradicts the assumption $F = \alpha C_{lr}$, as Passioura (1965) has pointed out. Again, the error is probably acceptable. C_{lr} varies rather slowly after short times for most ions and Drew (1966) found both boundary conditions gave very similar results for uptake of ions.

Nye and Marriott (1968) have compared the results obtained from this equation (1) with numerical computer solutions of the correct differential equation, and show that the equation is acceptable in most normal situations. We can therefore use it with some confidence, despite its known imperfections, as an approximation to the true conditions.

To clarify the effects of mass flow on uptake we may rewrite the equation as (Nye and Marriott, 1968);

$$F = VC_{li} + \frac{D'(C_{li} - C_{lr})g}{a} = \alpha C_{lr} \tag{2}$$

where

$$D = D_l v_l f_l \frac{dC_l}{dC} \sim D_l v_l f_l \frac{C_l}{C} = D' \frac{C_l}{C}$$

The properties of the solid phase then only appear in g.

Then

$$F = C_{li}\left\{\frac{V}{1 + (D'g/a\alpha)} + \frac{D'g/a}{1 + (D'g/a\alpha)}\right\}, \tag{3}$$

which is essentially the same formula in Mr Nye's paper. The effects of V on F are in Fig. 1. The apparent mass flow contribution therefore increases much more rapidly than F, and overtakes the total flux at $V = \alpha$. Distances

X and *Y* represent respectively positive and negative apparent diffusive contributions. The effect on flux of increasing *V* thus depends upon the term

$$\frac{VC_{li}}{1+(D'g/a\alpha)},$$

which may be quite small, and much smaller than the apparent mass flow contribution VC_{li}. It is possible that effects of water flux on root uptake properties (Brouwer 1954) will be more important than the simple mass flow effect.

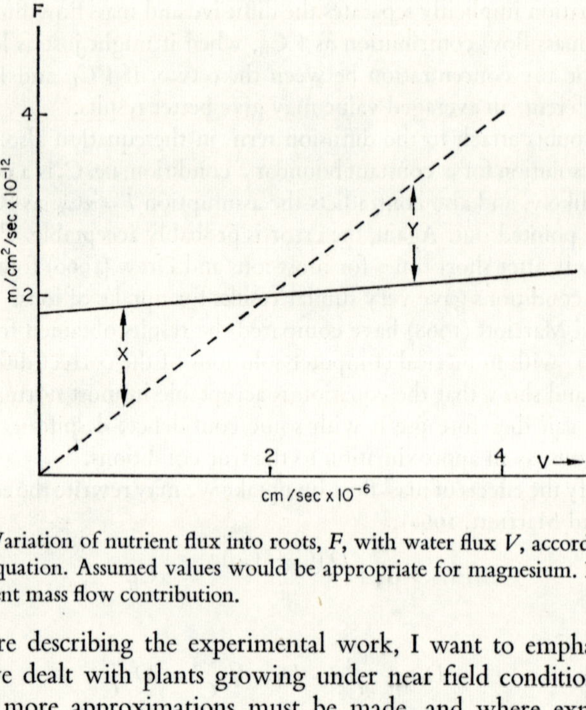

FIG. 1. Variation of nutrient flux into roots, *F*, with water flux *V*, according to Passioura equation. Assumed values would be appropriate for magnesium. Dotted line is apparent mass flow contribution.

Before describing the experimental work, I want to emphasize why we have dealt with plants growing under near field conditions, where clearly more approximations must be made, and where experimental control is less exact, than in greenhouse or growth chamber. There is no question that transport processes must occur in soil—the critical problem is whether they are important in the nutrition of plants. If the transport processes are so efficient, relative to the value of *F* required by the plant, that C_{lr} differs little from C_{li}, then the transport stage has no important influence in nutrient uptake, and may well be neglected. This is not a problem which can be answered finally for any one soil, nutrient or climate, as it depends upon plant growth rates and root uptake efficiencies. Very little information on fluxes into plant roots is available now, and no generalizations can safely be made until many such measurements have been made.

Experimental

The aim was to approximate to field conditions, while ensuring a fairly simple system. Leek seedlings (chosen because of their simple root system) were planted out into pots filled with a clay loam soil which had previously been dried, sieved and uniformly fertilized with *N*, *P* and *K* as powdered pure chemicals. The amounts of fertilizer were selected to correspond roughly with the rates which might be applied to a field crop. The largest pots contained about 30 kg of soil, to give a large reservoir of water. The soil was wetted with sufficient water to give a tension of 100 cm, and the pots were allowed to equilibrate for 2 weeks. No more water was added. After the leeks were planted, the soil was covered with polystyrene granules and a sheet of polythene, with a small hole to accept the base of the leek. This reduced loss of water from the soil surface to trivial amounts. The pots were sunk into the soil out in the field, and transpiration was measured by periodic weighings. They were protected from direct rain by a light plastic-covered frame. Soil temperatures in pots and field were always closely similar.

On 4 occasions a set of 10 pots was harvested; wet and dry weights and nutrient contents of roots and tops, and length and radius of roots (and hence surface area) were determined. Soil samples were taken at intervals to allow displacement and analysis of soil solution (Moss 1963). Other measurements made were of exchange isotherms of ions, and quantities of exchangeable ions. No root hairs were seen at any harvest.

Results

Only results for potassium will be discussed here, and any complications which can arise from curved isotherms (which cause concentration-dependent diffusion coefficients) are ignored in these calculations, which are mainly intended to illustrate the problems.

Figures 2 and 3 show the changes with time in total potassium content of plant, total root length, total weight of fresh matter and cumulative transpiration per plant. As expected, all these curves approximate to exponential form. We require the rate of uptake of potassium and the

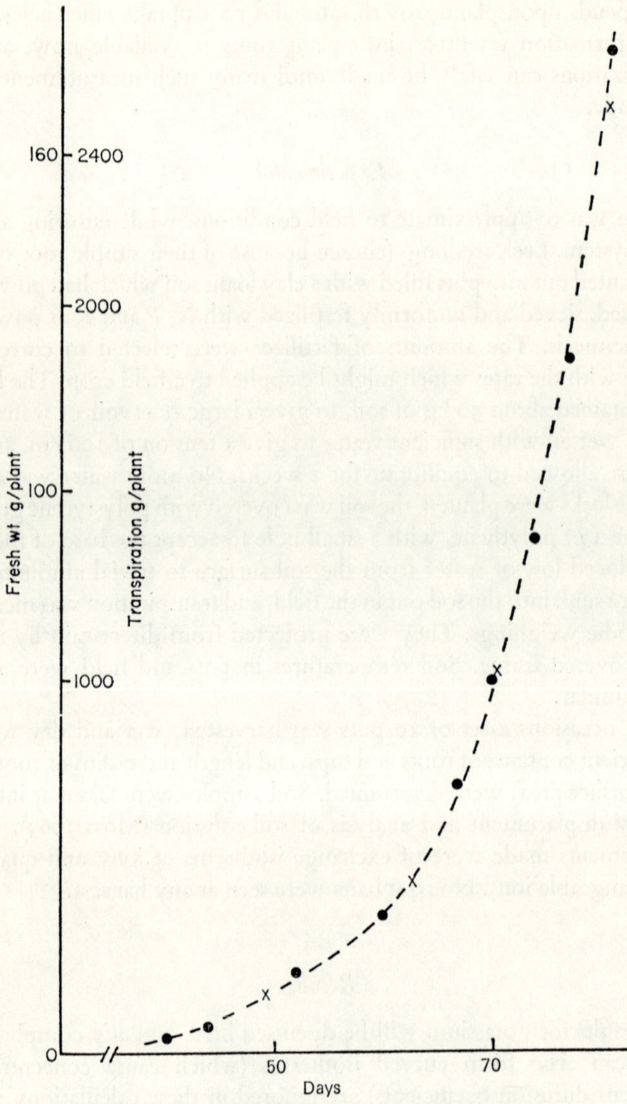

FIG. 2. Transpiration and plant fresh weight, changes with time.
Transpiration ● Fresh weight = X

mean rate of transpiration, at different times, from these curves. Radford (1967), in a discussion of the treatment of growth analysis results, suggested fitting equations, which are easily differentiated when the constants have

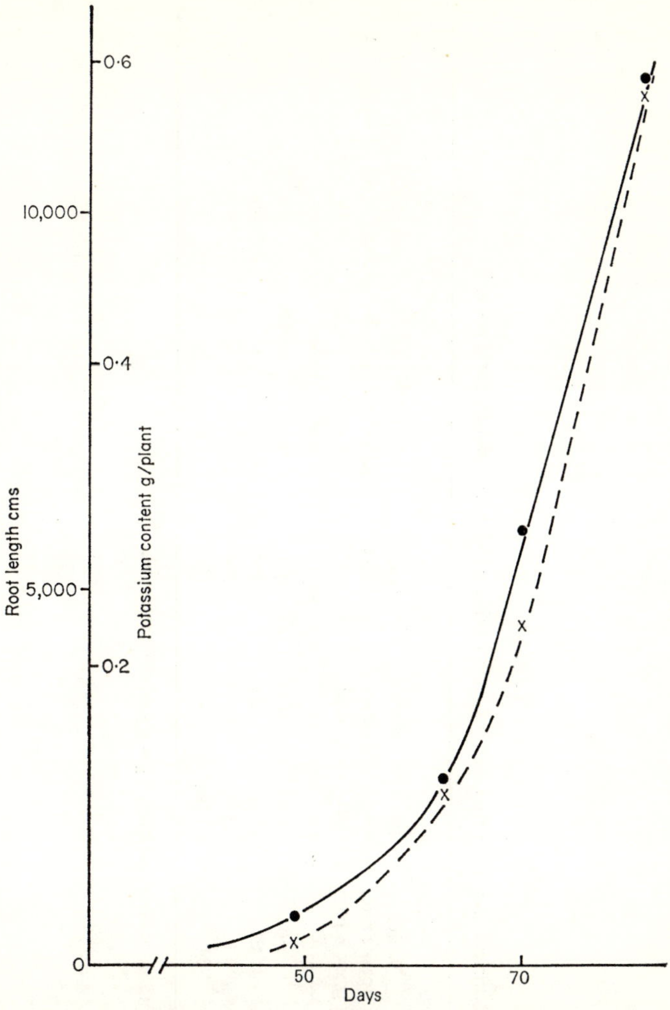

FIG. 3. Root length per plant and potassium content, changes with time.
Root length ● Potassium content = X

been found. With only 4 harvests curve fitting cannot be very accurate, and in this case the slopes have been estimated by drawing tangents to the smooth curve through all points. Mean values of F and V are then found, knowing total root length and mean radius at the harvest times (Table 1).

TABLE 1

Nutrient and water flux data for roots of single leek plants during early growth

Age of plant, days	Potassium uptake rate, moles/sec $\times 10^{-9}$	Water uptake rate, g/sec $\times 10^{-4}$	Root length cm.	Surface area cm^2	Mean F moles/cm^2/sec $\times 10^{-12}$	Mean V g/cm^2/sec $\times 10^{-7}$	$\frac{VC_{li}}{F}$ %	Mean C_{lr} moles/ml $\times 10^{-7}$	$\bar{\alpha}$ cm/sec $\times 10^{-5}$
49	0·8	1·9	630	150	5·6	13	10	2·4	2·3
62	3·1	3·1	2370	580	5·4	5·3	4	2·2	2·4
70	6·7	10	5340	1260	5·4	7·9	7	2·4	2·2
83	9·8	31	11750	1850	5·3	16	13	3·3	1·6

Mean bulk soil solution potassium concentration $C_{li} = 4{\cdot}5 \times 10^{-7}$ moles/ml.

Transpiration was closely dependent upon the amount of leaf material, probably because the mean monthly potential evapo-transpiration does not alter greatly over the experimental period of May to July (Monteith 1966). The weather caused short term fluctuations, and the maximum value of V measured was 2×10^{-6} cm/sec for a 3-day period. Transpiration only occurs during the day, and the calculated value of V may therefore be about a half of the daily maximum values. Possibly V may be even greater than this for short bursts of transpiration. Passioura and Frere (1967) consider 5×10^{-6} cm/sec to be the largest value normally attained, which agrees fairly well with these results.

The measured soil solution concentration, C_{li}, of K, remained fairly constant throughout (Table 1), but this is not a necessary condition for the calculations. In the calculation of the mean concentration at the root surface, C_{lr}, by equation (2), D_l for potassium was taken as $1{\cdot}98 \times 10^{-5}$ cm²/sec, v_l, the volume fraction of liquid, was close to 0·3 during the whole experiment and f_l was 0·3 at a bulk density of 1·0. The root radius a varied from 0·039 to 0·025 cm, and g was estimated at about 0·5 mean for an uptake period of over 8 days (Passioura 1963).

Discussion

The results show a remarkably uniform situation at the 4 harvests. The root length kept pace with the development and nutrient requirements of the rest of the plant so closely that the *mean* flux of potassium over the root surface varied within less than 10%. Even if the whole root system was not absorbing, the active part presumably formed a reasonably constant fraction of the whole during the period, and the comparison between the F values still applies. Mean V, the water flux, varied rather more; it tended to increase with time, as one would expect from the initially increasing day length and rising temperatures, except for harvest 1. Uptake rates for this first harvest could not be estimated accurately. The apparent mass flow contribution VC_{li} therefore increased with time after harvest 2, but never exceeded 13% of the whole flux on average. It may possibly have attained up to 50% over short periods, but in practice it seems reasonable to ignore it, as we have seen that the real effect of mass flow on uptake is probably smaller than the 'apparent mass flow contribution'. The uniformity of the results at the four harvests is of course a result of the steady, uninterrupted vegetative growth of the plants in this period.

In calculating mean F values, we are assuming that the uptake is uniformly distributed over the entire root surface. On this assumption, equation (2) gives the mean concentration at the root surface C_{lr} (Table 1); this was always around half of the initial soil solution concentration. Any other distribution of the total flux over the root surface would give a still greater effective depression of concentration at the root surface, and diffusion is clearly of considerable importance for these plants. Any studies on their potassium nutrition which assumed the roots to be in contact with the original soil would be seriously incorrect. The initial aim of the work was thus achieved.

The above calculation does not depend upon any assumption about plant properties except that all root surfaces were continuously and equally active in uptake. This is very unlikely (though all roots appeared fresh at all harvests) and we really require information on the change of uptake with age of each piece of root, and experiments on this are planned. The simplest assumption would be that uptake ceased fairly sharply after a certain time; knowing this 'root duration' one could then easily estimate the length of active root from Fig. 3. We have no definite information on this, but Drew (1966) found onion roots to absorb potassium ions for at least 16 days, and Brouwer (1954) reported uptake by bean roots up to 15 cm from the tip, which indicates periods of the same order. With root durations of around 16 days, and exponential growth of the type found here, most roots must be active at any time, the lowest fraction being 0·75 at the fourth harvest. The problem would be more serious for non-exponential growth, and more information on the duration and variation of active uptake by a root, under field conditions, is urgently needed.

Any proof that only part of the root system was active would reinforce the conclusions on the importance of diffusion, as C_{lr} at the active roots would have to be still smaller to give the greater flux over the smaller area. In view of these uncertainties, we cannot give the exact value of 'mean C_{lr}' any weight; its importance lies in the inference that there must be a considerable concentration drop at the absorbing root surface. This concentration decrease, which is greatest at the most active surfaces, will tend to give a more uniform flux to roots in soil than in water culture. Many refinements may be suggested to this very simple model, but nearly all of these seem to require a still greater concentration decrease. The sole exception appears to be mechanisms which would liberate potassium immediately around the root from non-exchangeable sources, but there is no evidence for this at present.

A value for the 'mean root uptake coefficient', $\bar{\alpha}$, can be calculated from the 'mean C_{lr}' value and the mean flux. It is important to note that though mean F is an algebraic mean of the real F values over all parts of the system, this is not so for mean C_{lr} and $\bar{\alpha}$, as α and F are not proportional. If large parts of the total root surface do not absorb ions, it is not possible to give any exact significance to $\bar{\alpha}$, though it may still be of value as a rough measure of the efficiency of a root system. If we could say that a certain fraction of the total root was active in ion uptake, we could calculate $\bar{\alpha}$ over the active part only ($\bar{\alpha}_a$). As the active fraction decreased, so $\bar{\alpha}_a$ would increase, reaching infinity in the limit where C_{lr} is nil—all the active part of the root is then a zero sink.

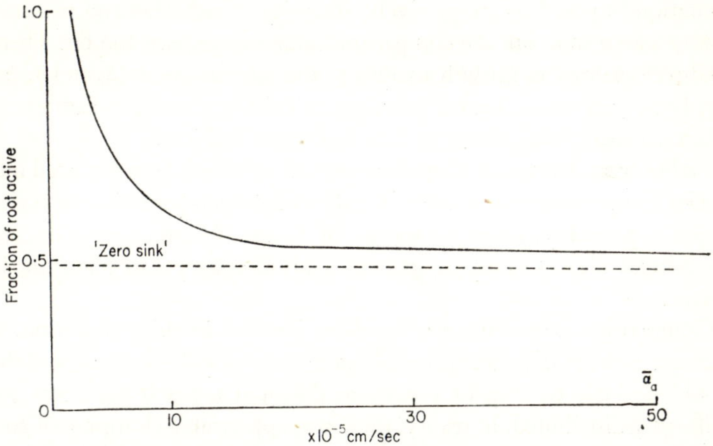

FIG. 4. Change in the calculated root uptake coefficient $\bar{\alpha}_a$ with changes in the fraction of the total root length which absorbs potassium ions. The 'Zero sink' line indicates the condition that $\bar{\alpha}_a$ is infinity, and the concentration at the surface of the absorbing roots is zero.

The situation for this experiment is expressed in Fig. 4. It is therefore unlikely that much less than half of the root surface was active in uptake of potassium in these plants, as the soil simply could not supply the measured total flux to a smaller surface.

The mean root uptake coefficient $\bar{\alpha}_a$ is thus dependent upon the amount of root which absorbs potassium. It is nevertheless interesting that the $\bar{\alpha}$ values found, of around 2×10^{-5} cm/sec, are comparable to those found by Drew (1966) for segments of single roots of seedling onions (3×10^{-5}) and leeks (6×10^{-5}) in growth chamber experiments. For the present, we suggest $\bar{\alpha}$ may be considered as a parameter to indicate the ease with which a plant is obtaining its requirements of a nutrient, and comparisons of its value for different species in the same soil, or the same species in different soils, or in the same soils with different growth conditions, may be interesting.

The procedure described has obvious parallels with growth analysis. In both cases, we are concerned with the function of an external absorbing surface of the plant, and are attempting to analyse this function in terms of absorbing area and effectiveness. Thus uptake rate of nutrients corresponds to growth rate, root surface to leaf surface, and F to net assimilation rate. Both procedures calculate mean fluxes of material, even though it is likely that this varies widely from point to point. In both cases, there is an outside limitation imposed by, respectively, intensity of radiation and maximum flux to a zero sink, but the comparison cannot be pressed too far. There is no direct analogy in growth analysis to the calculation of C_{lr} and $\bar{\alpha}$, or to supply by mass flow. Leaves intercept an existing flux of radiant energy, whereas roots create their own fluxes of water and ions.

Earlier work has touched on this point, as Welbank (1962) related inter-species competition to the rate of uptake of nitrogen per unit root weight, which is probably a rough measure of F, and Watson (1965) discussed the general idea of applying growth analysis concepts to plant nutrient uptake.

Competition between roots has been ignored in this calculation. On average, each cm of root enjoyed about 2·5 cm^3 of soil at harvest, and there would be little overlap of potassium diffusion zones if the roots were uniformly distributed. In reality, there was appreciable clumping of roots, and there was certainly some competition between them, though it was unlikely to be important. No correction is at present possible for this, but the effect would be to still further lower the true C_{lr}. The possibility of calculating uptake with competition between roots is now being investigated; this could hardly be neglected in any situation where inter-plant or inter-species competition was important. Competition will of course vary for different nutrients, depending upon their mobility, and the possible importance of this factor must be assessed in each particular situation.

This approach, which we might call nutrient flux analysis, may be of value in ecological problems. There are differences between species (e.g. Langer 1966) in nutrient and soil requirements, but little apppears to be known about the mechanism of such effects. Analysis of uptake patterns in terms of root surface areas, mean fluxes across root surfaces, and possibly mean root uptake coefficients, might explain some such differences. Information on the life of each root and the extent of the absorbing region of the total root, would make this approach very much more exact, and the uptake of nutrients in mixed stands might then be dealt with.

REFERENCES

BARBER S.A., WALKER J.M. and VASEY E.H. (1962). Principles of ion movement through the soil to the plant root. *Trans. Int. Soil Sci. Comm. II & IV, N. Zealand*, 121–124.

BROUWER R. (1954). The regulating influence of transpiration and suction tension on the water and salt uptake by the roots of intact Vicia Faba plants. *Acta Botan. Neerl.* **3**, 264–312.

DREW M.C. (1966). Uptake of nutrients by plant roots growing in the soil. Thesis, Oxford.

KNOWLES F. and WATKINS J.E. (1931). The assimilation and translocation of plant nutrients in wheat during growth. *J. Agric. Sci. Camb.* **21**, 612–637.

LANGER R.H.M. (1966) Nutrition of grasses and cereals, in The Growth of Cereals and Grasses (Ed. F.L. Milthorpe and J.D. Ivins), pp. 213–226. London.

MONTEITH J.L. (1966). Photosynthesis and transpiration of crops. *Exp. Agric.* **2**, 1–14.

MOSS C. (1963). Some aspects of the cation status of soil moisture. Part I. *Pl. Soil*, **18**, 99–113.

NYE P.H. (1966). Changes in the concentration of nutrients in the soil near planar absorbing surfaces when simultaneous mass flow and diffusion occur. *Trans. Int. Soil Sci. Comm. II & IV, Aberdeen*, 317–328.

NYE P. H. and MARRIOTT F.H.C. (1968). The importance of mass flow in the uptake of ions by roots from soil. *Trans. 9th Congr. Int. Soil Sci. Soc.*, **1**, 127–134.

NYE P.H. and SPIERS J.A. (1964). Simultaneous diffusion and mass flow to plant roots. *Trans. 8th Congr. Int. Soil Sci. Soc.* **3**, 535–542.

PASSIOURA J.B. (1963). A mathematical model for the uptake of ions from the soil solution. *Pl. Soil*, **18**, 225–238.

PASSIOURA J.B. (1965). Letter to editor. *Pl. Soil*, **22**, 317.

PASSIOURA J.B. and FRERE M.H. (1967). Numerical analysis of the convection and diffusion of solutes to roots. *J. Aust. Soil Res.* **5**, 149–160.

RADFORD P.J. (1967). Growth analysis formulae—their use and abuse. *Crop Sci.* **7**, 171–175.

WATSON D.J. (1965). Some features of crop nutrition, in The growth of the potato. (Ed. D. Ivins and F.L. Milthorpe), pp. 223–247. London.

WELBANK P.J. (1962). The effects of competition with *Agropyron repens* and of nitrogen and water supply on the nitrogen content of *Impatiens parviflora*. *Ann. Bot.* **26**, 370–373.

DISCUSSION ON MINERAL NUTRIENT SUPPLY FROM SOILS

Recorded by Dr D. GUNARY

The discussion was concerned largely with the nature and possible limitations of the model introduced by Mr Nye and the related experiments described by Dr Tinker.

Commenting on a suggestion that the rate of elongation of the roots into fresh soil should be taken into account, Mr Nye considered that if this referred to the so-called 'root interception', it was unimportant. This is because firstly, the root cap is not an absorbing system and, secondly, as it moves into new soil the cap pushes material away, so that nutrients must then pass back to the root by diffusion and mass flow.

In the model the absorptive system of the root was represented as a uniformly active cylinder. This was questioned by several speakers, for a number of reasons. Mr Nye stated that root hairs were deliberately avoided in the model represented and the experimental plants were selected because they have no root hairs. He thought that in practice, the presence of root hairs will change the picture appreciably only for slow moving ions such as phosphate, where uptake should be increased since a large part of the exploited volume of soil is within the root hair zone. Thus autoradiographs using ^{32}P-labelled soil have demonstrated zones of depletion in the soil which correspond to the root hair zone.

That the root is not, in fact, a uniformly absorbing cylinder was illustrated in examples given by Dr Rovira (Fig. 1). These showed that with phosphorus, uptake and retention in the apical 6–7 cm of a root was two to three times greater than the uptake further back and with sulphur this effect was even more striking, uptake being confined to the apical 7 cm. Dr Scott Russell supported this view and stressed that the ratio in which different ions are taken up varied with different members of the root system. Professor Harley considered that in some cases the root was so unimportant as to be an artefact. This is because practically all plants in the natural condition are mycorrhizic and it is through mycorrhiza that nutrient uptake from soil occurs.

The simple model presented by Mr Nye and Dr Tinker was thought to make no provision for certain important changes which may occur in the soil near to the root system. In spite of the fact that Gardner's theoretical calculations predict that the soil will not dry out appreciably during experiments of the type conducted by Dr Tinker, Professor Weatherly had experimental evidence to suggest that appreciable drying out does occur. In fine sand, water tension at the root had been increased from less

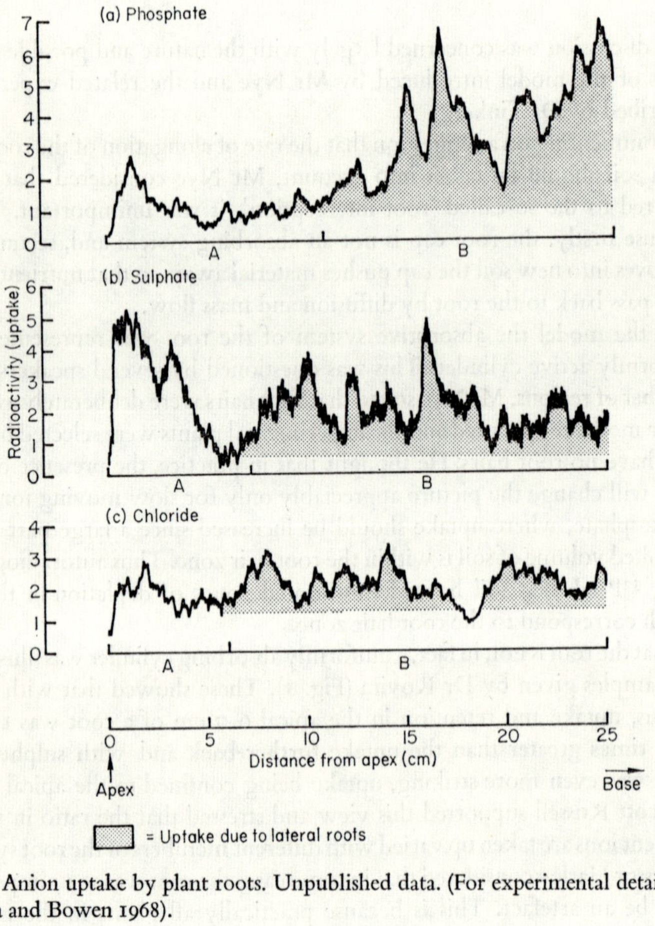

FIG. 1. Anion uptake by plant roots. Unpublished data. (For experimental details see Rovira and Bowen 1968).

than one bar to about ten bars. Although this effect had been observed in a fine sand at moderate transpiration rates, similar effects had been observed in other soils, including a clay soil and with not very high transpiration rates. Dr Tinker thought that this was not unexpected in fine sand where the hydraulic conductivity decreases sharply as drying out begins. However, he was unable to suggest any reasons for such an observation in a clay soil. If drying out does occur Dr Tinker agreed that the diffusion coefficient would decrease sharply so it would be necessary to consider not only a concentration dependent but a position dependent diffusion coefficient.

It was suggested by Dr Scott Russell that the mass of root exudates present in the soil might modify the system. Defining their nature Dr Rovira explained that round the root tip, the most likely exudates are polysaccharides, whereas further back in the zone of elongation organic acids, amino acids and sugars have been identified. It is probable, however, that other, as yet unidentified, exudates may be present in significant quantities. Quantitative determinations of these materials in the soil system is badly needed. Commenting on their significance, Mr Nye drew a distinction between those exudates which will form chelates, and polysaccharides, amino acids and organic acids which have no chelating properties. So far as the diffusion zone is concerned, the latter group are likely to be associated with high bacterial numbers, but under aerobic conditions the large amount of CO_2 produced by these bacteria will diffuse away so rapidly that it will hardly affect diffusion of nutrients. In contrast, chelates could be influential; for example, those which complex iron might bring about an increase in the concentration of phosphate in solution. Again, quantitative evidence is needed.

Mr Barley suggested that another important change which may occur in the soil near to the root is the deformation of the soil by the root so that the porosity and diffusivity of the shell of soil round the root may be different from the bulk.

The value of whole plant work was appreciated by Professor Ohlrogge, but he pointed out that the use of models may ignore incontrovertible facts such as the micro-heterogeneity of the soil. Mr Nye said that, save in exceptional circumstances, such as when fertilizer is used, roots ramify randomly and are thus good 'averagers' over the whole soil system. In criticizing the model approach, Mr Barley went so far as to suggest that so many qualifying assumptions have to be made in order to couch the model in mathematical terms, that too much realism is being sacrificed for internal precision. Ecologists must therefore beware not to confuse conclusions based on models with reality. Dr Tinker replied that at least they were trying to apply the model to real plants!

Amplifying an aspect of his paper Professor Arnold quoted the work on ion antagonism by Tinker and by Salmon, in which a compound or 'unified' activity ratio is used to indicate the 'intensity' or potential of a nutrient in the soil. This 'unified' activity ratio then depends upon the activity ratios of other cations in a way which can express ion antagonism, but a proportionality factor is needed which may vary between plant species.

It was suggested by Professor Mengel that it might be simpler to measure the intensity and buffer capacity of the soil for potash alone without using ion activity ratios. Although this is experimentally simpler, Professor Arnold pointed out that it was subject to a great deal of variation depending on experimental conditions. In contrast, the ratio of K activity/Ca activity is a unique figure, almost independent of technique, which may be related to the average potassium concentration over a range of conditions.

REFERENCE

Rovira A.D. and Bowen G.D. (1968). Anion uptake by plant roots: Distribution of anions and effects of micro-organisms. *Trans 9th Congr. Int. Soil Sci. Soc.*, Adelaide, **2**, 209–217.

SECTION THREE

MINERAL NUTRITION OF THE WHOLE PLANT SYSTEM

ECOLOGICAL INFERENCES FROM LABORATORY EXPERIMENTS ON MINERAL NUTRITION

I. H. RORISON
Department of Botany, University of Sheffield

INTRODUCTION

This paper first considers the problems involved in measuring the response of plants to mineral nutrients when they are grown as isolated individuals in the greenhouse or growth-room. There follows a consideration of the extent to which such measurements can be used to interpret the behaviour of the same plant species when they are growing under natural or semi-natural field conditions.

This consideration is thought timely because many of us who are interested in nutritional aspects of plant ecology find ourselves spending more time in the laboratory than in the field and we tend to use physiological methods in our work. Our attention has been drawn to problems arising from field studies of species distribution in relation to edaphic factors. We have brought the problems into the laboratory and have tested, by intensive experimentation in controlled environments, what we have inferred from the results of previous extensive experiments. For the most part we are studying aspects of the physiology of the whole plant, using dry weight yield as an index of the plant's response. On the other hand we work with relatively few species and we largely ignore the consequences of seasonal changes in the environment and of the presence of other individuals of the same or other species. Two major questions therefore arise:

1. Are we advancing knowledge of ecological situations by our present methods and, equally important,

2. are we forging a mutually advantageous link between the field ecologist and the cellular physiologist who is interested in mechanisms?

PHYSIOLOGICAL EXPERIMENTS

Methods employed by ecologists are often derived from crop physiology and it is a measure of our dependence upon them that all the remaining speakers in Section III are crop physiologists. Their predecessors have had the tremendous advantages of resources of men and laboratories, and of sources of selected and uniform seed. They have discovered excellent indicator species (e.g. Wallace 1951) for a range of nutritional situations and the physiological response of these species may be used as a guide when considering naturally occurring species, material of which is either difficult to collect or to handle experimentally.

Despite this, there has been a tendency among ecologists to be either unaware of, or to ignore, the results of crop experiments. While it is valuable to confirm for semi-natural species what is already known for good crop indicator species it would seem desirable to advance our knowledge by a distillation of results from both fields.

The crop-specialist may place different emphasis on his experimentation because he is more likely to be interested in yield and quality of certain plant parts, than in species survival within a community. Relative growth rates of crop plants are invariably much higher than of those found in more natural conditions, and their responses to nutrient additions continue to concentrations well above the optimum for wild species, many of which are adapted to edaphically extreme habitats. The crop physiologist is usually concerned with short generation-times while field ecologists may pursue studies of longevity running into tens of years at one end of the time-scale, and seedling establishment taking as many weeks at the other. Cell physiologists studying primary processes within the plant may be down to time-scales of as many minutes (Jennings, this volume).

It is important to stress that all experiments take time and that this enables us to study the *rates* of processes rather than only the final results of their operation. Nor should we forget the time-relations of relevant processes outside the plant, such as the seasonal variations in climate which affect in turn the activity of soil microorganisms and the availability of soil nutrients to higher plants (Barber, Harley, both in this volume).

The significance of space as a factor is often only considered in the laboratory in terms of container size and its possible limitation of growth within an experimental period. In the field there may or may not be competition between individuals for air volume and soil volume. The former is a relatively uniform medium but the latter is (Nye, this volume) more heterogeneous both physically and chemically. Different life-forms and the plasticity of individuals both in root and shoot growth affect their various successes in an environment. According to De Witt (1960), Harper (1961) and others, an individual which is subject to the pressures of a plant community may respond very differently from an otherwise identical individual growing alone. Taken literally this statement could be a major deterrent to making an autecological approach to population studies. But study of the nutrient response of individuals of each species in a community can provide results of ecological significance and is certainly the most obvious way of keeping experiments simple, and manageable enough to provide reproducible results. Before such a statement can be elaborated four matters of methodology will be considered which may affect both the results of physiological experiments and also their ecological application. They are: choice of species; the phase of life cycle to be studied; selection of factors to be measured; and, methods of experimentation.

1. *Choice of species*

A study of plant distribution in the field shows wide variation in edaphic tolerance from species to species. In grasslands in the Sheffield region *Deschampsia flexuosa* and *Scabiosa columbaria* show narrow and contrasting pH ranges while *Festuca ovina* and *Rumex acetosa* are widely tolerant (Fig. 1). (Reference to *The Atlas of the British Flora* (1962) confirms this distribution at a national level.) All these species grow best in fertile soils of intermediate pH but their ecological amplitude is constant in the field, *D. flexuosa* being a strict and often dominant calcifuge, *S. columbaria* a strict calcicole but never dominant, while *F. ovina* may dominate grassland on soils ranging from strongly calcareous to markedly acid. The differences between ecological and physiological response are shown diagrammatically in the manner of Ellenberg (1958) in Fig. 2. Ellenberg (1958) reports that there may be regional differences in the behaviour of species—that plants which in Northern Europe are calcicole may be more or less indifferent to pH in the climate of Southern Europe. One general explanation given is that a species at the limit of its area of distribution can survive competition only in edaphic extremes of which it is tolerant but most of its potential competitors are not. Certainly strict calcicoles and calcifuges in the Sheffield area are slow-growing species which tolerate extreme soil conditions but fail to survive competition with species of higher growth rate when introduced into more fertile soils (Rorison 1967). Conflicting reports of species distribution may be attributed to two factors: that many early statements were based on observation without strict examination of soil pH in the immediate vicinity of the plant roots; and the occurrence of physiological races which, though morphologically identical, may be both geographically and physiologically distinct. It is therefore important to use care in choosing species for experiments and to be guided by the precise questions to be answered. In order to investigate edaphic factors characterizing acid and calcareous soils we began by using strict calcicoles and calcifuges belonging to different genera (Rorison 1960; Grime 1963), but there are advantages in using related species (Clymo 1962; Clarkson, this volume) or, better still, physiological races of the same species. These latter would not only keep morphological and anatomical differences to a minimum but also make it possible to study the evolutionary mechanisms controlling adaptations of tolerance, as illustrated by the results of Bradshaw; Goodman; and Turner (in this volume). Unfortunately, few species have enough

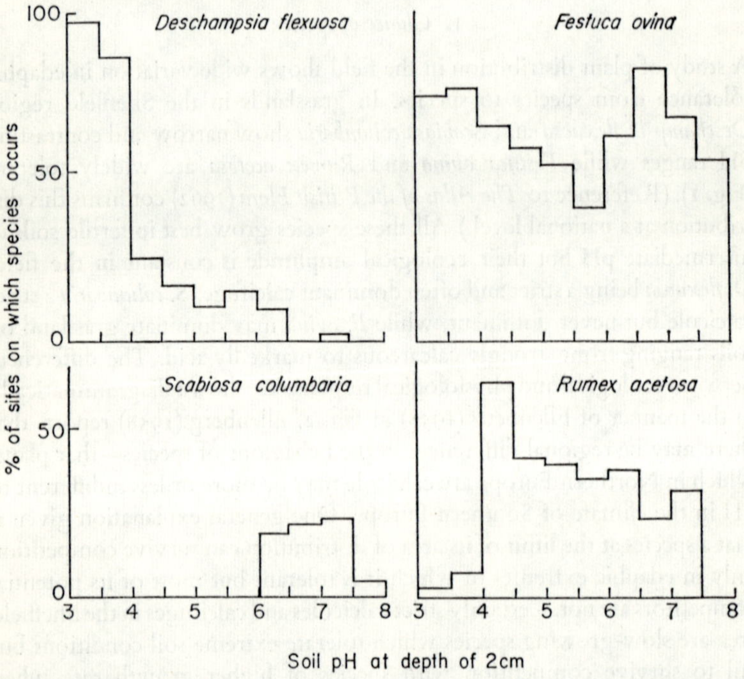

FIG. 1. Histograms showing the frequency of occurrence of 4 species over a range of soil surface pHs. Data provided from 340 random m² quadrats in established grassland from 41 sites in the Sheffield area.

races to cover the whole range from strongly acidic to strongly calcareous soils and thus provide ideal material for studying the calcicole–calcifuge problem. *F. ovina* (Fig. 1), whose distribution is bimodal in the Sheffield survey data, has possibilities which we are investigating (see also Snaydon and Bradshaw 1961). To assay soils for nutrient availability, a widely tolerant species has many advantages over a group of narrowly tolerant species or races which would be needed to span the same edaphic range. *R. acetosa* is a possible candidate here (Fig. 1).

2. *Phase of life cycle studied*

The most sensitive phase in a plant's life is usually considered to be during germination and seedling establishment (Salisbury 1920; Rorison 1960a, and Harper 1965, *inter alia*). If the parent plant can survive the rigours of the habitat and produce viable seed, now is the most crucial time in the

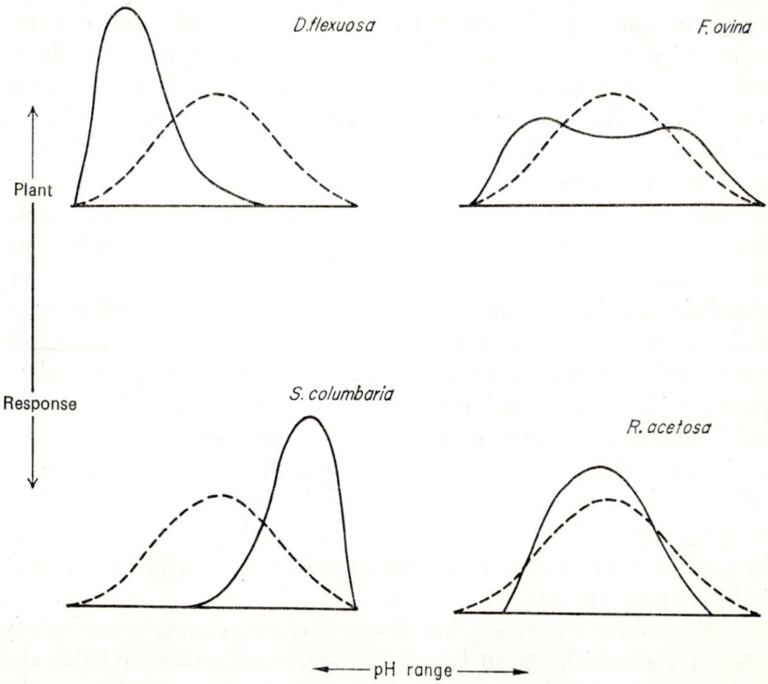

FIG. 2. A diagrammatic representation (see Ellenberg 1958) of the physiological optimum curve (- - - -) and ecological optimum curve (——) of the 4 species whose frequency in relation to soil pH is shown in Fig. 1.

chain of events, when response is most sensitive and external influences may be most accurately measured.

With any species, two factors must be considered when experimenting with nutritional factors at the seedling stage: seed size and seed reserves. If differences in seed size involve differences only in weight and volume but not in stage of development then relative growth rates are often equal but absolute rates vary with differences in initial capital (Black 1957; Rorison 1961). In competitive situations individuals from large seeds have therefore a better chance of survival than those from small seeds. If difference in seed size involves varying degrees of differentiation and maturity this may lead not only to different initial growth rates but also to differences in speed and time of germination, and these may blur experimental responses to prescribed ranges of nutrient conditions unless taken into account. In the field situation it may be that, because of discontinuous germination, the few stragglers which survive unusual climatic or other disturbance in the community are in some seasons the only effective propagules. It can therefore be important to determine the responses of seedlings derived from the whole spectrum of seed produced by a species and not only from the largest and most rapidly developed. Seed reserves may play a significant role in nutrient studies, although in known cases involving mineral nutrients their influence is detectable for no more than three to four weeks. For example, in *Onobrychis sativa* the onset of aluminium toxicity is measurable in terms of dry weight only after three to four weeks although visual symptoms and differences in root: shoot ratios are apparent earlier. Distribution of embryo reserve phosphorus occurs largely within the first three to four weeks. With plants surviving in distilled water, petiole collapse and phosphorus withdrawal from leaves to cotyledons all occur during the 4th week. (Rorison unpublished.) Larsen and Sutton (1963) have shown that seed reserve phosphorus of ryegrass plants used to determine 'L' values effects results in the first 4 to 6 weeks and that it is advisable to continue until a steady state is reached around 10 weeks. Seed reserves must therefore be accounted for, or experiments must run beyond the phase where their influence is significant.

The use of tillers or cuttings instead of seeds is very common, particularly with species which spread largely by vegetative means, and this also creates problems if their nutritional history is not known. Their supply, particularly of major nutrients which are mobile within the plant, such as nitrogen, potassium and phosphorus, may be adequate for prolonged growth whatever the subsequent treatment of the material. Accounts of some work using transplants reports raising cuttings or seedlings in full nutrient conditions (sometimes unidentified potting compost) for several months. This may enable visibly uniform individuals to be selected but must necessarily bias the results of any nutritional investigation. One is effectively building up the plants' resistance to toxicity and deficiency effects, and even responses to varying additions of nutrients, in terms of both uptake and growth, may be strongly affected by pre-treatments (Nassery 1968). Just as pre-treatment at high levels builds up a plant's resistance, so pre-treatment at low levels may lower resistance to deficiencies and toxicities and may also affect the extent of absorption of ions supplied subsequently, as Loughman and Russell (1957) found for phosphorus.

3. *Selection of factors to measure*

Most of our ecological problems involve a study of adaptation to extreme edaphic conditions. We are looking therefore at responses to nutrient stress rather than at levels for optimum growth. The selection of factors to measure is based on a distillation of widespread field experimentation (Clapham, this volume) and the aim is to carry out single, reproducible experiments to test the response of plants to selected nutrients and toxins. When studying seedling establishment we measure response in terms of growth and nutrient uptake. In doing so we should remember that far more information can be gained from measuring rates of reaction than merely recording yield data.

A simple diagram (Fig. 3) illustrates three of the possible ways in which yield X may be achieved in time Y. Speed of emergence influences the early part of the curve, and later the waning influence of seed reserves may become apparent under conditions of stress (upper curve). By the time Y, growth may be proceeding at one of several very different rates. For the benefit of further experiments therefore yield X alone does not tell us whether a vital stage has been reached nor does it allow an estimate of further response. A series of dry weights transformed to natural logarithms allow us to plot relative growth rates directly and to compare the effects of treatments in an easy and straightforward manner. A high overall relative growth rate is of great importance for a seedling to become established in a space. Its survival and replacement in the habitat will depend *inter alia* on its subsequent division of labour which Harper (1967) has termed reproductive strategy.

Nutrient uptake as measured by chemical analyses of root and shoot fractions provides two valuable pieces of information: (a) what the plant has managed to extract from its growth medium in a given time, and (b) how much net translocation there has been from root to shoot. Many workers present only results of foliar analyses, particularly when they are using soils from which it is difficult to extract entire and uncontaminated root systems. While foliar analysis is a useful supplementary, and may be a useful guide for husbandry purposes, it can be misleading for dealing with mineral toxicities and deficiencies in infertile soils. This is because the drastic effects of such elements as aluminium are primarily in the root

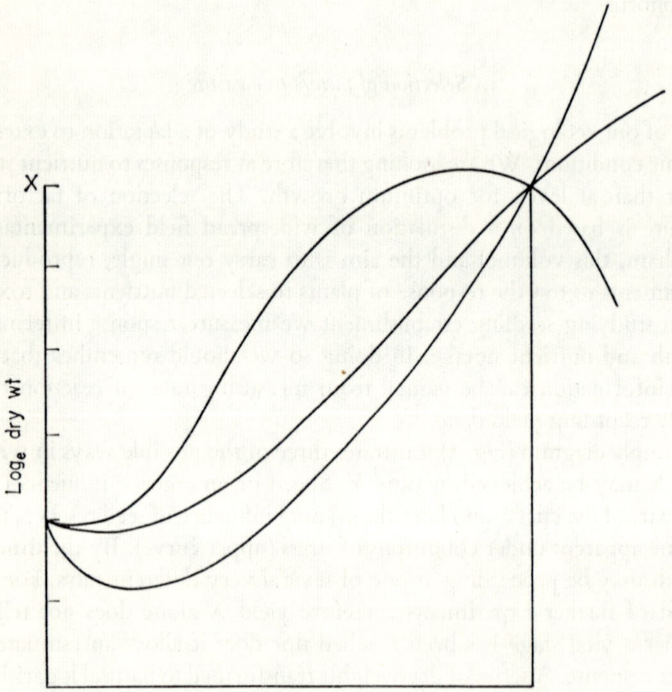

FIG. 3. A diagrammatic representation of a range of relative growth rates which achieve the same yield *X* at time *Y*.

system, where the initial and crucial physiological upsets occur. At the same time, the foliar content of the toxic elements may not differ appreciably between tolerant and susceptible species, because of limited translocation.

It is vital therefore to study root systems with care. It is as important to measure root geometry as it is to determine patterns of absorption and translocation. The problem is further extended because nutrient availability varies with soil depth in response to both profile development and differences in soil moisture content (Newbould, this volume).

4. *Methods of experimentation*

(a) The use of controlled environment facilities. One of the most significant technical advances for this subject has been the development of growth rooms and cabinets (Evans 1963). It is now possible, in most laboratories, to carry out a series of experiments under controlled and identical environmental conditions. Results are then comparable and cumulative. The system has two major advantages; the effects of any required number of combinations of nutrient factors may be tested, and interactions between nutritional and certain climatic factors may be measured.

(b) Selection of growth media—soil, other solids, or solutions? The use of soil in experiments involves many difficulties. In the first place the levels and relative proportions of the various nutrient ions can never be known precisely because of the complexity of the system and the well-known shortcomings of the chemical analysis of soils. And if soil is moved from the field to the garden, greenhouse or growth-room its structure and the availability of mineral nutrients in it may be significantly altered. Alternatives to soil are solid inert media such as sand and vermiculite, or solution cultures. Both offer a means of presenting the plant with a known concentration of nutrients, and in addition the solid media offer a form of mechanical support which approximates to soil conditions. However, it is difficult to attain uniform drainage and nutrient percolation through most solid media or to maintain a uniform pH. If very low levels of nutrients are added it is likely that a high proportion of the ions will be adsorbed on sand and pot surfaces.

Solution culture has certain advantages: it is possible to maintain pH and nutrient levels near the root surface, and to sample and adjust both accurately; sterile conditions may be maintained; whole plants may be removed, measured and returned without injury. With some species, root systems, including root hairs, develop in approximately similar patterns in both solid and liquid media.

(c) The choice of nutrient concentrations provides further problems. The levels of nutrients used in the majority of experiments (Hewitt 1966) are based on the responses of arable crops and are extremely high compared with conditions in more natural soils. Wallace *et al.* (1967) have shown that maize and tobacco grow and develop satisfactorily when calcium is supplied at as low a level as 2 ppm, but only if other nutrient ions are also at low levels: balanced growth appears to depend on an appropriate balance of mineral nutrients in the rooting medium.

It is also important to realize that the true pattern of response can only be understood if a sufficiently wide range of nutrient levels is used (Antonovics *et al.* 1967). This is made clear in a recent experiment (Rorison 1968) on the response to phosphate of four species from different habitats (Fig. 4).

The range of nutrient levels used, of from 10^{-7} to 10^{-3} M, is very wide in terms of biological response. Above 10^{-3} M phosphorus the growth of even some crop plants is depressed and yet most nutrient solutions listed by Hewitt (1966) have phosphorus levels of more than 10^{-3} M. This is particularly important to remember when considering species adapted to soils with low available phosphorus levels.

The logarithmic yield data in Fig. 4 shows that *R. acetosa* and *S. columbaria* reach an optimum response at or below 10^{-3} M. *Urtica dioica* is the only species likely to respond to concentrations greater than 10^{-3} M and *D. flexuosa* is the only species likely to survive at concentrations lower than 10^{-7} M. The problem is how to keep concentrations of 10^{-7} M in solution and not adsorbed on the walls of containers. Constant flow would seem to be the only way to maintain a supply at very low concentrations. Asher *et al.* (1965, 1966) describe a system which gave nutrient solutions with potassium concentrations as low as 10^{-8} M and they obtained an optimum growth response with a range of plants at *c.* 10^{-4} M potassium. To achieve this optimum response, 275 L were circulated through each 2 L container (of 8 seedlings) per day, a total of 1072·5 mg K per day. This is equivalent to presenting a 0·03 M concentration of K in a 1-litre pot each day. Constant flow has the advantage of bathing the roots of plants with a steadily maintained level of nutrients which under idealized conditions in the soil would be maintained by either nutrient movement from soil to root surface and/or of root movement through the soil. Gradients however, are inevitable in any natural system and periodic renewal with its simplicity and ease of maintenance is adequate for most ecological experiments particularly at the seedling stage of growth. Aeration ensures both the provision of oxygen and a minimum development of concentration gradients between root and solution. The most important factor to record is the total amount of each nutrient provided per plant per unit of time. If this is known then comparison between the results of different workers is possible.

(d) Variation in results obtained by chemical analysis can be caused not only by small differences in experimental conditions experienced by the same species but also by the use, by different operators, of different analytical methods and measuring instruments. Several surveys have been carried out (e.g. Bowen 1967) in which ground samples of the same material were

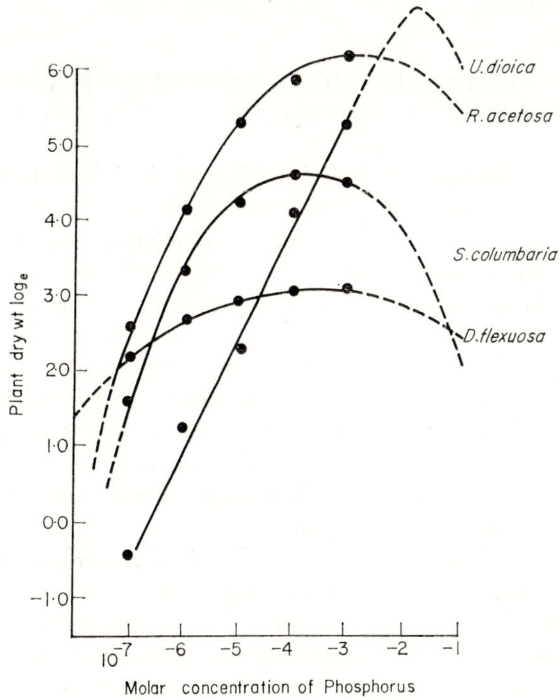

FIG. 4. Dry weights achieved in 6 weeks. Data transformed to natural logarithms and plotted against external phosphorus concentration. Dotted extension lines suggest trends beyond the range of external phosphorus concentrations used in the present experiment.

distributed to several analytical laboratories and analysed by several techniques for a wide range of elements. The results ranged from strictly comparable to the grossly variable. Fortunately, consistent results were obtained from different laboratories for Ca, Fe, Mn, Mo, N, P (e.g. Ca 41,400 ± 2230 ppm) but small and significantly different results for Cu, K, Mg, P and Zn when different techniques were used. Gross differences, attributed to contamination, are quoted for aluminium from 6·4 ± 0·7 ppm by colorimetric methods to 80·1 ± 11·9 by spectrometric methods.

We have therefore variation in results stemming from variable experimental conditions, variable plant material, including differing species response, and variability in analytical technique. It is important to read the mass of papers on mineral nutrition with these variations in mind. We may reconcile conflicting theories in some cases by allowing for differing techniques and the use of species of differing tolerances.

PHYSIOLOGICAL EXPERIMENTS RELATING TO ECOLOGICAL SITUATIONS

Having noted some of the technical problems of physiological experiments and having suggested some aspects requiring particular care and attention, can we justify their use to interpret plant distribution in terms of adaptation to nutrient factors? A general trend follows the addition of nutrients to semi-natural communities (Milton 1940; Willis 1963; Thurston, this volume); there is a decrease in the number of species, an increase in vegetation cover, and a replacement of herbs by grasses. Is it reasonable to assume that those species with the greatest potential response to nutrient additions will, given a suitable life-form, smother the less responsive species? May we therefore look to growth rate, growth form, and response to nutrients in physiological studies for clues to the response of individuals in groups as exemplified by three field situations:

(1) where seedlings establish themselves in favourable conditions of mineral nutrition and in the absence of competitors;

(2) where mineral factors are decisive in the establishment or otherwise of certain species, whether with or without competition; and

(3) where individuals are competing with other individuals of the same or different species.

1. *Seedling establishment in the absence of competitors*

A major factor in the recolonization of grassland by perennial species in the Sheffield area is the availability of gaps. Most sites are sloping and soil movement and the actions of animals and human beings ensure that patches of bare earth are exposed throughout the year. Initially there will be no competition above ground and virtually none below ground. Seeds falling into the gaps under favourable climatic conditions will behave like individuals sown into pots of the same soil albeit with the absence of animal predation (see Darwin, in Harper 1967). Their survival will depend on the efficiency with which they hold the open territory against surrounding invaders. They may, for example, produce close-packed rosettes of leaves or grow tall, but equally important, they must produce a root system adequate to exploit the available soil volume. To maintain their numbers in the population, they must produce one successor.

There have been few studies of longevity of perennials. Rough estimates for *S. columbaria* and *Poterium sanguisorba* vary from 10–30 years. In the field *S. columbaria* probably has a total production of around 5000 seeds in 10 years. Under favourable conditions in the experimental gardens at Sheffield plants from seed collected in Derbyshire have produced *c.* 50,000 seeds in 4 years. So the chances of *S. columbaria* reproducing itself in a gap are reasonably good. A knowledge of the nutritional responses and growth rate of this species measured in the laboratory should therefore be of value in predicting its ability to colonize a gap in its natural habitat or indeed any open space.

2. *Mineral factors which are decisive*

There are soils which are unfavourable to some species whether competition from other individuals occurs or not. Competition may accelerate the disappearance of the species but its demise is sure and takes place under laboratory conditions. Such a case is the failure of *S. columbaria* to become established on certain mineral acid soils owing to aluminium toxicity. It may not be the ultimate cause of death but its effect of inhibiting root growth and rendering seedlings susceptable to desiccation (by drought or frost) is the initial and decisive factor in their elimination from acidic sites.

Some ruderal species may be excluded from soils by the low level of availability of certain essential nutrients. It has been shown in laboratory experiments that *U. dioica* fails to survive in media with low ($< 10^{-5}$ M) phosphorus concentrations. The increase in growth rate obtained with phosphorus addition is great and is maintained beyond 10^{-3} M (Rorison 1968). So, although like several ruderal species it might thrive in a gap whose soil had a high nutritional status, its relative growth rate would be so slow under conditions of nutrient stress that it would be a poor competitor for other invading species and would probably fail to survive even in the absence of competition.

3. *Competitive situations*

We next come to situations where different species are competing for mineral nutrients. It has been shown that the response of individuals in mixed populations may be different from their response when grown singly or in pure stands (De Witt 1960; Harper 1961), and we have heard from Mr Van den Bergh (this volume) that under sub-optimal nutrient conditions, a high-yielding species may be less competitive than a low-

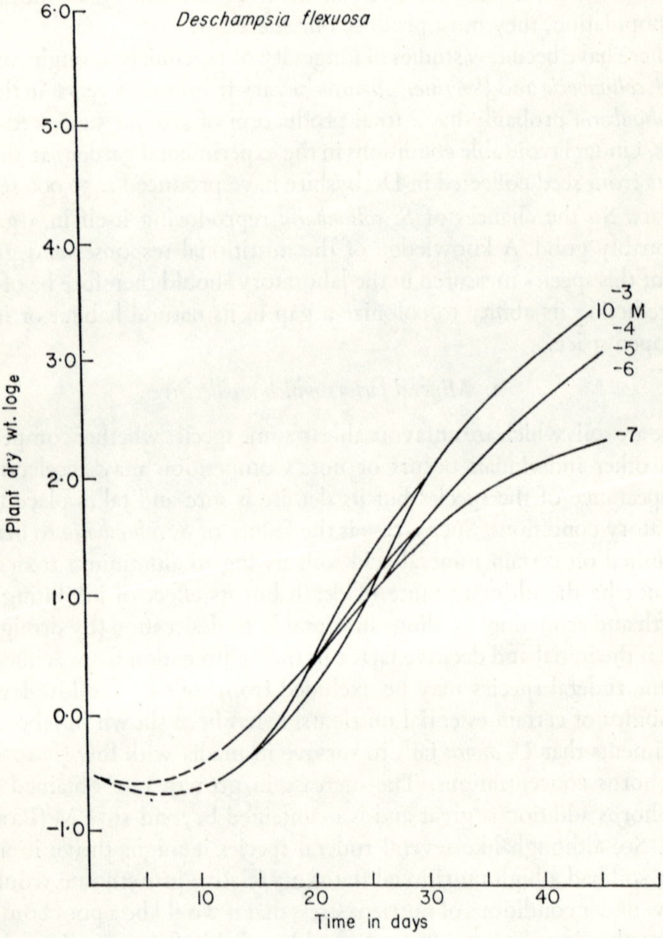

FIG. 5a.

FIG. 5. Dry weight response (transformed to natural logarithms) of (a) *Deschampsia flexuosa* and (b) *Rumex acetosa* when grown in nutrient solution with initial phosphorus concentrations of 10^{-7}, 10^{-6}, 10^{-5}, 10^{-4} and 10^{-3} M.

yielding species. This is a result which is unexpected if predictions have been made only from monoculture experiments designed to determine the optimum response of each species to a particular nutrient factor. If however the monoculture experiments are designed to determine the *range of*

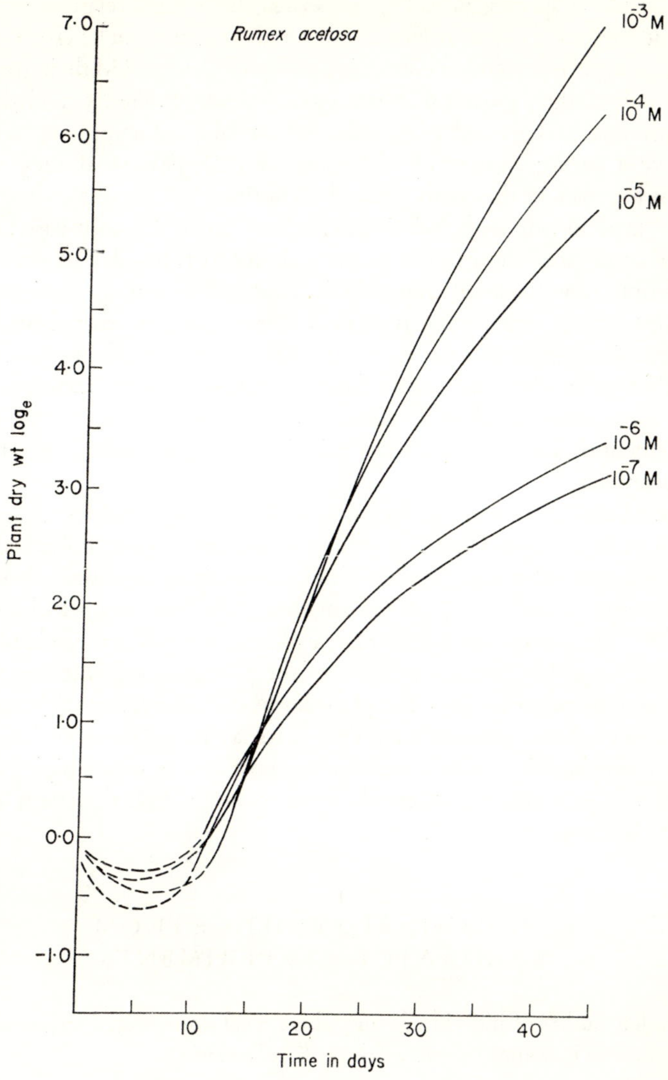

FIG. 5b.

tolerance of each species to a particular nutrient, it would be less unexpected. For example, (Rorison 1968), *D. flexuosa* has a low relative growth rate, tolerates very low nutrient levels and responds only slightly to increase (in nutrient level), e.g. response to phosphorus levels (Fig. 5a). By comparison several ruderal species, including *R. acetosa*, have high relative growth rates but they are also very responsive to variation in nutrient levels (Fig. 5b) so that in a nutrient solution containing phosphorus at 10^{-7} M their growth rate is only slightly greater than that of *D. flexuosa* (cf. Figs. 5a and 5b).

With this background knowledge we would not expect *R. acetosa* to have a strong competive advantage over *D. flexuosa* if they were grown together in these sub-optimal conditions. It is suggested therefore that if experiments with individuals are sufficiently broadly based they permit some prediction of competitive ability. It is possible to test these predictions experimentally using a limited number of even-aged individuals and this type of experiment is of undoubted value for crop plants. It is however inadequate in semi-natural grassland, for example, because seed of different species, each with different germination characteristics, is introduced among established plants of varying degrees of maturity. Even if competition for nutrients, water and light is not immediate, it is likely to occur later.

Model experiments to simulate such field conditions would be altogether too complex to handle and experiments with even-aged individuals are an inadequate substitute. Studies of non-competitive conditions therefore remain our main source of information. We have well-documented evidence to say, for example, that *D. flexuosa* is confined in the field to very acidic soils (a) because it is tolerant of the prevailing edaphic conditions such as aluminium toxicity and phosphorus deficiency, and (b) because it has a low relative growth rate. That is to say its tolerance enables it to survive in conditions which are toxic to those species whose high growth rates at higher pHs would enable them to exclude *D. flexuosa* (Hackett 1967; Rorison 1967, 1968).

ADDITIONAL EVIDENCE FROM LABORATORY EXPERIMENTS

Detailed studies of individuals may also reveal unsuspected responses such as occurred in some current work on *S. columbaria*.

Scabiosa columbaria has a low relative growth rate but it also shows very poor growth response, at low phosphorus levels. How does it survive in the field where analysis of the calcareous soils on which it occurs often show very low levels of available phosphorus? A clue came in the analysis of plant material for phosphorus (Fig. 6). It was found that after six weeks the phosphorus contents of both roots and shoots of *S. columbaria* seedlings varied with levels of phosphorus in the external solution. The most marked increases occurred in the roots and were greater than the increases in concentration recorded for the other species grown, e.g. *R. acetosa* (Fig. 6). The experiment was repeated by Dr Nassery (1968) of this department. He confirmed that in high external concentrations more phosphorus was accumulated in the roots of *S. columbaria* than in the roots of *R. acetosa*. In

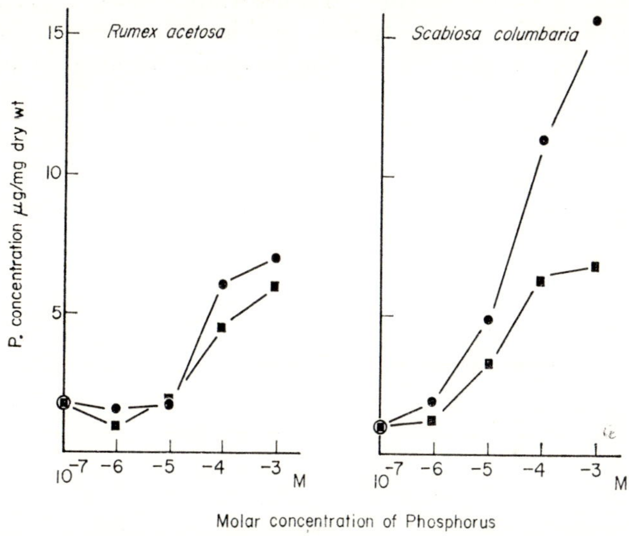

FIG. 6. Phosphorus concentrations in roots ● and shoots ■ after 6 weeks growth.

addition, he determined the amount of inorganic phosphate, acid-soluble phosphate, and acid-insoluble phosphate in the plants. He showed that in the roots of *S. columbaria* the amount of the inorganic fraction (P_i) is greater than the acid-insoluble fraction at all concentrations of phosphorus. In *R. acetosa* the quantity of P_i is greater than the acid insoluble fraction only at an external concentration of 10^{-3} M phosphorus, and then only slightly greater.

The percentages of P_i in *S. columbaria* and *R. acetosa* compare as follows:

	P molarities in nutrient solutions		
	10^{-6}	10^{-3}	
S. columbaria	63	80	Percentage
R. acetosa	18	42	P_i

We then considered whether this ability of *S. columbaria* to accumulate phosphorus in excess of immediate requirements could be of ecological significance. Although the soils of its natural habitat are usually low in available phosphorus, seasonal flushes are reported to occur (Davison 1964). Increased accumulation by *S. columbaria* at flush times might be adequate to maintain steady growth throughout the season—if the accumulation of P_i could be utilized.

It was an interesting contrast to *R. acetosa* which although responding to increases in phosphorus supply with rapid increases in growth, did not show the same tendency to accumulate inorganic phosphorus in the root.

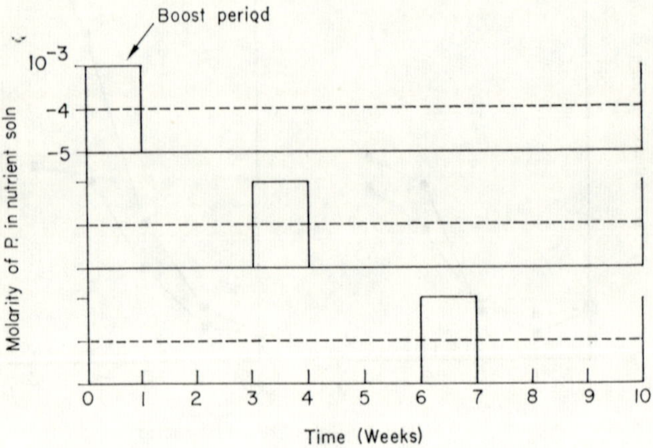

FIG. 7. An outline scheme for supplying short-term increases (□) in excess of the basic phosphorus supply (——) so that in 10 weeks plants are presented with the same total amount of phosphorus in solution as the control system (- - - -).

We have continued this investigation in two ways. We have returned to the field to measure seasonal variation in phosphorus availability in a range of soils. We are doing this by means of chemical analyses of soil extracts and of seedlings grown to assay the soils. Fractionation of the phosphorus contents of the seedlings should indicate whether inorganic phosphorus is accumulated in the roots of *S. columbaria* during and immediately after the occurrence of flushes.

Meanwhile in the laboratory we have been measuring the effect of short-term increases in the supply of phosphorus to *S. columbaria* and *R. acetosa* seedlings. The range of phosphorus levels was the same as in the previous experiment (10^{-3}–10^{-7} M) but the method of presentation was different. Figure 7 illustrates the procedure. The seedlings were grown for

10 weeks in a controlled environment. One set of plants was grown in nutrient solution containing an initial concentration of 10^{-4} M phosphorus. With 10-weekly changes a plant in a ½-litre pot was thus presented with a total of 15,500 μg P. This was the control. Another set of plants was also given a total of 15,500 μg P, but in two parts. For nine weeks they were grown in 10^{-5} M concentrations of phosphorus and for one week in 10^{-3} M (in fact $0{\cdot}9 \times 10^{-3}$ M.P). The one week's increase was given either during the 1st or the 4th week or the 7th. These treatments were also given with controls of 10^{-5} M and 10^{-6} M phosphorus. We were thus able to measure the response of both species to an increased phosphorus supply which has been given at different stages in their seedling development, and after growth at different initial phosphorus levels, and we hope to be able to say how much and for how long each species responds to an increase in phosphorus supply (measured as dry weight increases), and what effect the stage of development and of pretreatment has on the response of each species.

This incomplete experiment has been outlined because it illustrates the type of questions that we are asking and the type of experiments we design to answer them. We are well aware that the reasoning has its limitations and in particular that we are following up single factors which alone may or may not be of great importance in determining the distribution of the species we examine and to which the plant may respond differently when another factor (or factors) is varied. Perhaps we shall find that the ability of *S. columbaria* to accumulate phosphorus is an important adaptation for survival in its natural habitat. In any event we are collecting information about its response to phosphorus levels *per se*. And we consider this is particularly valuable for us because we can use the data as a background to studies of aluminium toxicity which, in part, is concerned with impaired phosphorus absorption.

CONCLUSION

It is suggested that laboratory experiments of the kind described in this paper and by subsequent contributors to Section III, are valuable not only as contributions to our knowledge of the physiology of the plant species involved but also because they provide indications of ways in which mineral nutrient conditions might influence species distribution in the field. They will have been undertaken in the light of the evidence from field studies and from extensive comparative experiments of the kind described by

Grime and Hodgson (this volume) and will complement that evidence. In favourable circumstances results from the two approaches will combine to support an explanation of the field situation in terms of the growth-response of the species in question to particular levels, steady or changing, of one or more mineral nutrients. This may well be the point at which the cell physiologist or biochemist can be interested in an investigation of the primary processes involved, and so an important bridge may be built between the territories of the field ecologist and the cell biologist.

It is also the point at which the ecologist should consider a return to the field with experiments designed to test the validity of explanations formulated in the laboratory.

REFERENCES

ANTONOVICS J., LOVETT J. and BRADSHAW A.D. (1967). The evolution of adaptation to nutritional factors in populations of herbage plants. *Isotopes in plant nutrition and physiology*. International Atomic Agency, Vienna.

ASHER C.J., OZANNE P.G. and LONERAGAN J.F. (1965). A method for controlling the ionic environment of plant roots. *Soil Sci.* **100**, 149–156.

ASHER C.J. and OZANNE P.G. (1967). Growth and potassium content of plants in solution cultures maintained at constant potassium concentrations. *Soil Sci.* **103**, No. 3, 155–161.

BLACK, J.N. (1957). Seed size as a factor in the growth of Subterranean Clover (*Trifolium subterraneum* L.) under spaced and sward conditions. *Aust. J. agric. Res.* **8**, 335–351.

BOWEN H.J.M. (1967). Comparative elemental analyses of a standard plant material. *Analyst*, **92**, 124–131.

CLYMO R.S. (1962). An experimental approach to part of the calcicole problem. *J. Ecol.* **50**, 707–731.

DAVISON A.W. (1964). Some factors affecting seedling establishment in calcareous soils. *Ph.D. thesis, University of Sheffield.*

DE WIT C.T. (1960). On competition. *Versl. Landb. Onderz.*, Wageningen, No. 66.8.

ELLENBERG H. (1958). Mineralstoffe für die pflanzliche Besiedlung des Bodens. A. Bodenreaktion (einschliesslich Kalkfrage). *Handb. PflPhysiol* **IV**, 638–709.

EVANS L.T. (1963). Extrapolation from controlled environments to the field. *Environmental control of plant growth.* (Ed. L.T. Evans.) Canberra, pp 421–435.

GRIME J.P. (1963). An ecological investigation at a junction between two plant communities in Coombsdale on the Derbyshire Limestone. *J. Ecol.* **51**, 391–402.

HACKETT C. (1967). Ecological aspects of the nutrition of *Deschampsia flexuosa* (L). Trin. III. Investigation of phosphorus requirement and response to aluminium in water culture and a study of growth in soil. *J. Ecol.* **55**, 831–840.

HARPER J.L. (1961). Approaches to the study of plant competition. *Symp. Soc. Exptl. Biol.* **15**, 1–39.

HARPER J.L. (1965). Establishment, aggression and cohabitation in weedy species. *The genetics of colonizing species.* (Ed. H.G.Baker and G.L.Stebbins), pp 243–268. New York.

HARPER J.L. (1967). A Darwinian approach to plant ecology. *J. Ecol.* **55**, 247–271.

HEWITT E.J. (1966). *Sand and water culture methods used in the study of plant nutrition.* 2nd Ed. Commonwealth Agric. Bureaux. Tech. Comm. No. 22.

LARSEN S. and SUTTON C.D. (1963). The influence of soil volume on the absorption of soil phosphorus by plants and on the determination of labile phosphorus. *Pl. Soil* **18**, 77–84.

LOUGHMAN B.C. and RUSSELL R.S. (1957). The absorption and utilization of phosphate by young barley plants. IV. The initial stages of phosphate metabolism in roots. *J. exp. Bot.* **8**, 280–293.

MILTON W.E.J. (1940). The effect of manuring, grazing and cutting on the yield, botanical and chemical composition of natural hill pastures. *I. Yield and botanical section. J. Ecol.* **28**, 326–356.

NASSERY H. (1968). Phosphorus absorption by plants from habitats of different phosphorus status. Ph.D. Thesis, University of Sheffield.

RORISON I.H. (1960a). Some experimental aspects of the calcicole-calcifuge problem. I. The effects of competition and mineral nutrition upon seedling growth in the field. *J. Ecol.* **48**, 585–599.

RORISON I.H. (1960b). The calcicole-calcifuge problem. II. The effects of mineral nutrition on seedling growth in solution culture. *J. Ecol.* **48**, 679–688.

RORISON I.H. (1961). The growth of Sainfoin seedlings. University of Nottingham School of Agric. Rept. 1960, 44–46.

RORISON I.H. (1967). A seedling bioassay on some soils in the Sheffield area. *J. Ecol.* **55**, 725–741.

RORISON I.H. (1968). The response to phosphorus of some ecologically distinct plant species. I. Growth rates and phosphorus absorption. *New Phytol.* **67**, 913–923.

SALISBURY E.J. (1920). The significance of the calcicolous habit. *J. Ecol.* **8**, 202–215.

SNAYDON R.W. and BRADSHAW A.D. (1961). Differential response to calcium within the species *Festuca ovina* L. *New Phytol.* **60**, 219–234.

WALLACE A., FROLICH, E. and LUNT, O.R. (1967). Calcium requirements of higher plants. *Nature* **209**, 634.

WALLACE T. (1951). The diagnosis of mineral deficiencies in plants by visual symptoms. London. H.M.S.O.

WILLIS A.J. (1963). Braunton Burrows: the effects on the vegetation of the addition of mineral nutrients to the dune soils. *J. Ecol.* **51**, 353–374.

THE ABSORPTION OF NUTRIENTS BY PLANTS FROM DIFFERENT ZONES IN THE SOIL

P. NEWBOULD

Agricultural Research Council Radiobiological Laboratory,
Letcombe Regis, Wantage, Berkshire

INTRODUCTION

Quantitative information on the relative extent to which nutrients are absorbed from different zones in the soil, and on factors which affect it is meagre. This lack of data reflects the limitations of procedures hitherto available for studying this problem, namely, the measurement of nutrient uptake when fertilizers are placed at different depths in soil or observations of the distribution of roots which have been separated from soil. The former method is subject to considerable limitations because high local concentrations of nutrient can change root form and function. Conclusions based on root distribution are uncertain both because of the difficulty of making accurate observations and lack of knowledge on the capacity of different types of roots to absorb nutrients.

The injection of radioactive tracers into defined zones of soil can provide a more direct method for measuring uptake from different parts of root systems under natural conditions (Hall *et al.* 1953, Nye and Foster 1960, 1961) provided that the procedure used is such that the uptake of tracer from different depths varies with the quantity of nutrient absorbed from them. It has been established that this requirement can be satisfied (Newbould and Taylor 1968; Newbould *et al.* 1968a). In this paper, reference is made to work with this technique which will illustrate marked gradients in the absorption of phosphate and calcium by plants from different depths in soil and the effect on them of water supply, soil type and the position in which fertilizers are placed. The experiments were carried out with perennial ryegrass and barley grown in monoculture. These crops were chosen because of their agricultural importance but in addition they provide suitable models from which to start to unravel the more complex situations of mixed communities either of artificial or natural origin.

The results discussed here are from experiments carried out in collaboration with the Grassland Research Institute, Hurley and the Chemistry Department, Rothamsted. Some preliminary details of the experiments have been included in the Annual Reports of this Laboratory (Newbould *et al.* 1967) and fuller descriptions are in preparation. In the present discussion an attempt has been made to select aspects of the results which appear likely to be representative of widespread situations.

METHODS

The requirements which must be satisfied to obtain reliable information with the tracer procedure have been described in detail elsewhere (Newbould and Taylor 1968; Newbould *et al.* 1968a). They can be briefly summarized as:

1. The injection of the tracer must not alter the concentration of labile ions; the use of 'carrier-free' isotope ensures this.

2. The tracer must equilibrate in a constant manner with the labile nutrient in the soil at all sites of injection; the use of a carefully standardized injection procedure and the choice of soils with an homogeneous exchange complex at all depths should ensure this. Experiments to verify this may be necessary at sites with non-uniform soils.

3. Allowance must be made for variation between different depths in the quantities of labile nutrients with which the tracer can equilibrate.

4. The sites of injection must be random relative to the distribution of roots; the placing of injections at each depth in a constant geometric pattern ensures this.

5. The injection procedure should not alter the pattern of uptake by mechanical injury to the roots or in other ways.

To inject the tracer, a polythene tube (external diameter 6·8 mm) through which a tightly fitting steel rod has been inserted is pushed into the soil to the required depth. The steel rod is then withdrawn and the polythene tube raised 2 cm to leave a small cavity in the soil. The solution containing the carrier-free tracer is injected into this cavity through a needle made from flexible polythene capillary tubing (external diameter 2·5 mm) using an automatic syringe (Cornwall Luer-Lok).

The quantity of tracer subsequently detected in the above-ground parts of the plants can be used to assess the relative contribution of roots at different depths in the soil to the nutrition of the plants according to the equation:

$$N_p(1-n) = T_p(1-n) \times K \frac{N_s(1-n)}{T_s}$$

Where $N_p(1-n)$ and $T_p(1-n)$ are respectively the quantities of nutrient and tracer absorbed by the plant from several depths in the soil $(1-n)$, T_s is the quantity of tracer injected at each depth, $N_s(1-n)$ is the quantity of readily available or labile nutrient in the soil and K is a constant representing the extent to which the tracer is diluted at each depth. When comparing absorption from two or more zones it is convenient to express uptake from each zone in relative units. The convention was adopted of expressing these units as the quantity of tracer present in the shoots per m² relative to the quantity of tracer injected at each site of injection multiplied by the quantity of labile nutrient in the soil (g labile nutrient per 10 g soil). It is much more difficult to estimate the actual quantities of nutrient absorbed from different depths and this has not been considered here.

In the soils used in these experiments variations with depth in the quantity of labile nutrients ($N_s(1-n)$ in above equation) were found to be small. This simplifies interpretation, as except when fertilizers were added the observed pattern of uptake cannot be attributed to nutrient gradients in the soil.

For experiments with ryegrass, 25 injections are made in square array to any one depth; while for barley, zones of soil are labelled by making 12 injections at each of two depths in the same plot (e.g. 12 at 5 cm and 12 at 10 cm) in rectangular array spanning three rows of crop. These procedures caused the above ground parts of the crops from 1 m² of ground to be labelled with adequate uniformity; the coefficient of variation of uptake of tracer from different depths has been found to be similar to, or less than, the coefficient of variation of yield.

The uptake of phosphate and calcium has been examined by these methods. From the practical viewpoint knowledge on uptake from different depths is most relevant for nutrients which diffuse slowly in soil and for this reason the absorption of phosphate is of particular interest. The inclusion of calcium provides information on whether the pattern of uptake throughout root systems is similar for different nutrients and moreover is a big contrast to phosphate since it is absorbed in much larger quantities. Phosphorus-32 is a convenient tracer for phosphate since its energetic β-radiation makes it simple to assay samples of dried and ground plant tissues with a large plastic β-scintillation counter. Because of the low energy of the β-radiation of calcium-45 there is no equally convenient radioactive isotope of calcium; however strontium-89 can be used because the ratio of strontium to calcium in the soil solution is closely related to that in which the ions are absorbed by plants (Russell and Newbould 1966).

EFFECT OF SOIL WATER AND NITROGEN FERTILIZER ON UPTAKE BY PERENNIAL RYEGRASS

The effect of water and nitrogen on the nutrition of ryegrass can be illustrated by some results of an experiment conducted on a 2-year-old sward of perennial ryegrass. The form of the experiment was influenced by an earlier observation that in periods of limited water supply nitrogen may have a greater effect on the growth of the crop when placed at 45 or more cm in the soil as opposed to application in the traditional manner on the soil surface (Garwood 1965).

Early in the spring after the area was top dressed with nitrochalk (7·8 g N/m^2), two watering regimes were imposed on a sward of perennial ryegrass which was protected from rain with plastic covers, namely:

A. No water was added during the early part of the experiment.

B. Water was added whenever the calculated water deficit reached 25·4 mm.

The water content of the soil in 6 cm layers down to about 70 cm was determined by a method using β-radiation described previously (Newbould *et al.* 1968b). The method depends on the attenuation of β-radiation by water held in absorbent nylon pads; it allows the water content of narrow zones of soil to be measured and unlike the nylon/stainless steel electrical resistance method is not affected by the presence of soluble salts.

The experiment was started 50 days after the sward had been covered. By that time the soil to a depth of about 30 cm in plots of regime A had a moisture content of 10·7 g H_2O/100 g soil as compared with a value of about 19 (i.e. approximately field capacity) in plots of regime B. The grass was cut to 2·5 cm and ammonium nitrate equivalent to 11·2 g N/m^2 was added to replicate plots by one of the following methods—placed on the soil surface, injected into soil at 7·5 or into soil at 45 cm below the surface. At the same time radioactive tracers were injected at 7·5, 45 or 60 cm below the soil surface to enable the uptake of phosphate and calcium to be measured.

Ryegrass was sampled to 2·5 cm above the ground surface at approximately monthly intervals for four months. The water content of soil for the period preceding each sampling occasion (Fig. 1) has been averaged from weekly measurements (Newbould *et al.* 1968b). In regime A, no water was added until after the second sampling occasion and during this period the water content decreased progressively throughout the soil profile. In regime B, 102 mm of water was applied in the same period and the soil was on average 40–50% wetter than in regime A. Shortly after the second sampling 152 mm of water was applied to regime A and 51 mm to regime B and the plastic covers were removed from the plots of both regimes leaving them exposed to natural rainfall; subsequently the water status of soil in regime A rose to that in regime B.

The drying out of the soil brought about by water regime A resulted in a lower yield of dry matter for all nitrogen treatments (Table 1) though

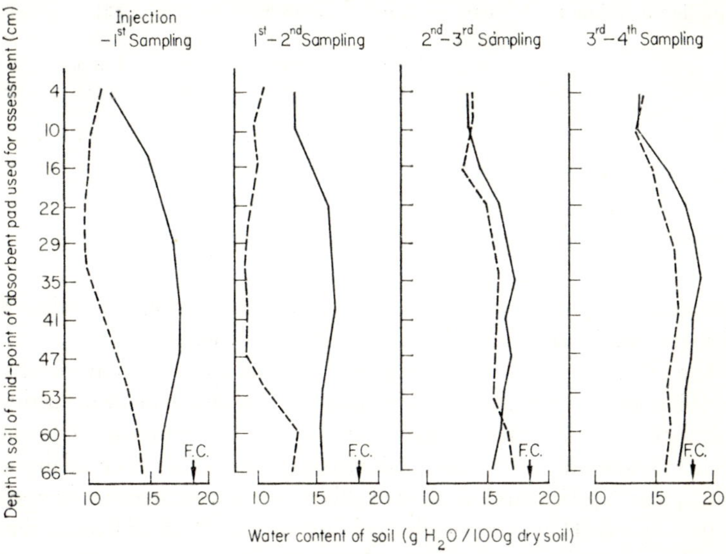

FIG. 1. The mean water content of soil (g H_2O/100 g soil) for the period between sampling occasions at eleven depths below perennial ryegrass swards with contrasting water regimes (see text for their description). ---- Regime A, —— Regime B. F. C. is water content at Field Capacity.

this was only statistically significant for the surface and 45 cm placements. Broadly similar effects were observed in the total content of nitrogen, phosphate and calcium in the herbage. The placement of nitrogen at 45 cm into soil which had about 20% more water than at depths nearer the surface in regime A gave twice as much dry weight at the first sampling occasion than if the nitrogen were placed on the soil surface; the total content of nitrogen, phosphate and calcium was enhanced in the same fashion.

The heavy irrigation of the soil in water regime A after the second sampling was soon reflected in increased dry matter yield and uptake of

TABLE I

The effect of water regime and depth of nitrogen placement in soil on the yield (g dry weight/m^2) of perennial ryegrass

Depth of nitrogen placement	Surface		7·5 cm		45 cm	
Water regime*	A	B	A	B	A	B
Sampling occasion 1	63·0	214·9	80·0	226·2	121·6	234·2
2	40·6	96·7	40·1	105·2	38·7	112·1
3	159·5	60·0	186·6	62·5	144·6	68·0
4	68·8	34·9	78·0	43·8	68·5	40·8
Total	331·9	406·5	384·7	437·7	373·4	455·1

L.S.D. (P = 0·05) for totals = 56·6

* See text and Fig. 1 for a description of the water regimes.

nitrogen, phosphate and calcium; the plants were now able to utilize the nitrogen previously unused because of the lack of water.

The relative uptake of phosphate and calcium from different depths in the soil is shown in Fig. 2; the wide variations with depth made logarithmic transformation necessary for statistical analysis, thus significant differences cannot be shown in the figure. Discussion of the results both of this and subsequent experiments is limited to effects which were statistically significant (P=0·05). The extent to which phosphate was absorbed from different depths was markedly affected by the water regime. Averaged over the season about three times as much phosphate was absorbed from 7·5 cm in regime B as in regime A; uptake from 45 cm was also greater. Uptake from near the soil surface in regime A was particularly restricted by low water supply before the first two harvests; the subsequent addition of water led to much increased uptake. The relationship between effects on nutrient uptake and on yield of dry matter are obvious. During the period of water shortage the absorption of phosphate was considerably greater when nitrogen was placed at 45 cm as opposed to nearer the soil surface; this effect ceased when the water supply later became ample.

The absorption of calcium from near the soil surface was reduced by a smaller fraction than was observed with phosphate when the water supply was limited. Moreover, in contrast to the situation with phosphate the absorption of calcium from 45 cm was generally greater with water regime A.

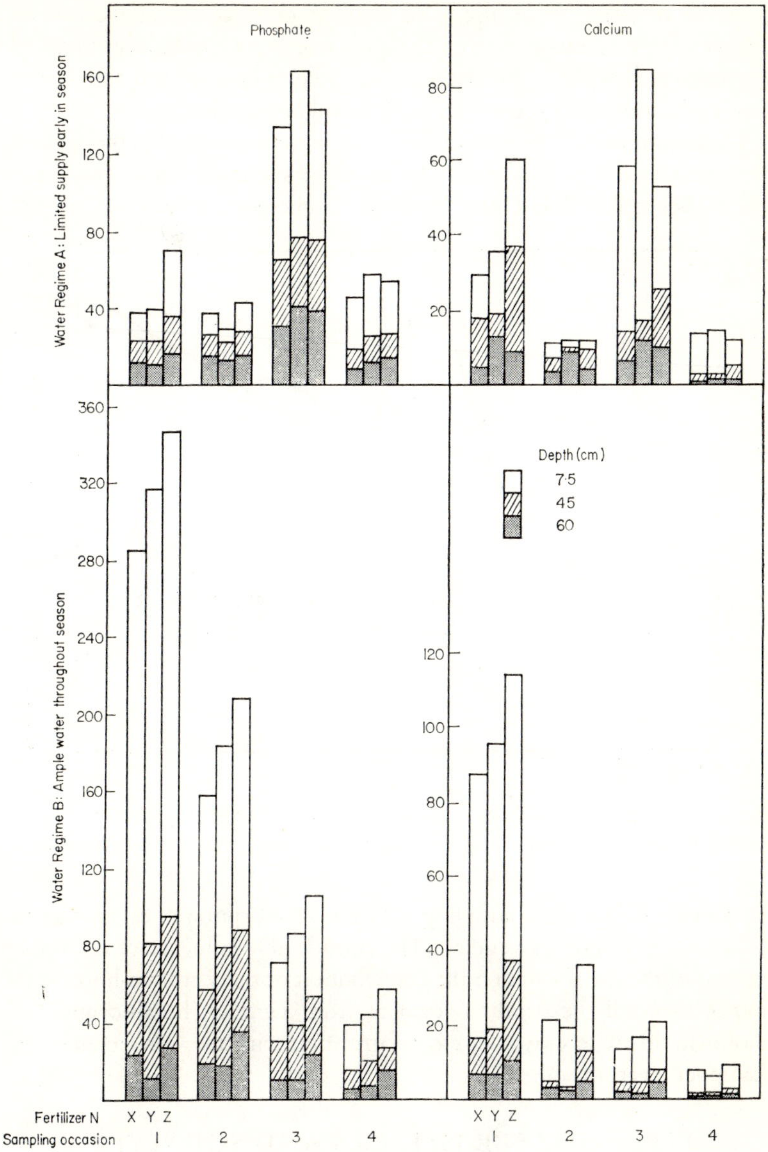

FIG. 2. Seasonal trends in the uptake of phosphate and calcium from three depths in soil by perennial ryegrass under two water regimes with fertilizer nitrogen applied to the surface (X), at 7·5 cm (Y) or 45 cm (Z) below the surface. Results are given in arbitrary units for each element.

In soil with an ample supply of water (regime B) the application of nitrogen at 45 cm rather than on the surface or at 7·5 cm slightly enhanced the relative absorption of phosphate and calcium from 45 and 60 cm; this was significant at all except the fourth sampling occasion.

The importance of the quantity of water in the surface soil to the growth of ryegrass is also well illustrated by another experiment on a sandy loam soil at Begbroke Hill, Oxon in which water content was determined

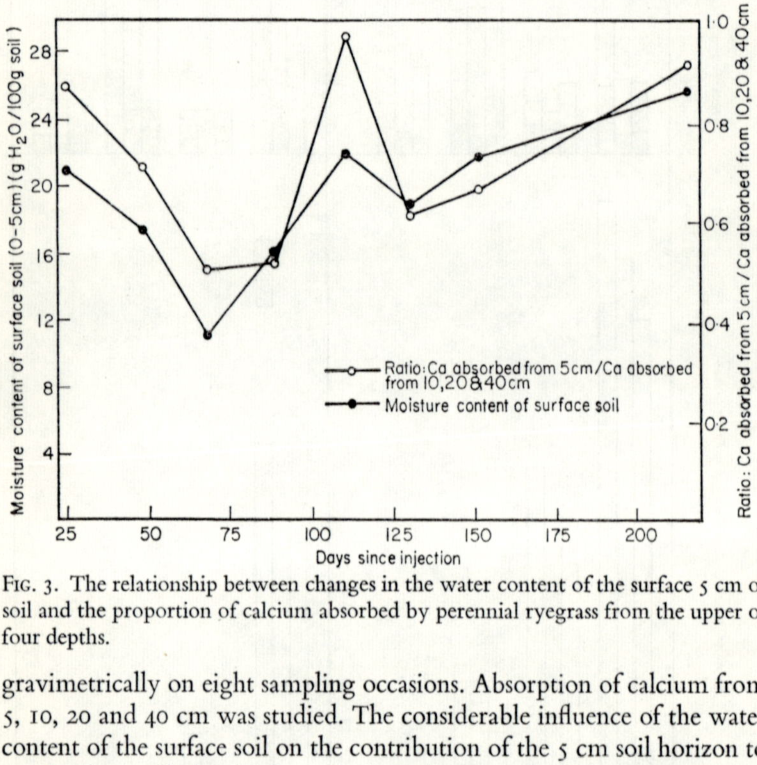

FIG. 3. The relationship between changes in the water content of the surface 5 cm of soil and the proportion of calcium absorbed by perennial ryegrass from the upper of four depths.

gravimetrically on eight sampling occasions. Absorption of calcium from 5, 10, 20 and 40 cm was studied. The considerable influence of the water content of the surface soil on the contribution of the 5 cm soil horizon to uptake from all the depths is evident from Fig. 3. The absorption of phosphate likewise showed a clear correlation with the water content of the upper layers of soil.

EFFECT OF FERTILIZERS AND SOIL TYPE ON UPTAKE BY BARLEY

The effect of nutrient gradients in soils of different type on the distribution of roots and their absorption of nutrients is illustrated by experiments on barley conducted in collaboration with the Chemistry Department, Rothamsted at Rothamsted and Woburn. At the former site, the soil was a clay with flints while at the latter it was a sand; brief descriptions of the properties of the soils are given in Table 2.

No fertilizer was added to some plots, others received an NPK dressing (12·5, 10·4 and 10·9 g/m² of N, P and K respectively) in the spring. After the fertilizer had been thoroughly mixed into the top 13 cm of soil by rotovation, spring barley (Maris Badger) was sown. The unfertilized plots were cultivated in a similar manner. Tracers were injected at 5–10, 15–20 and

TABLE 2

The effect of fertilizer applied to the top 13 cm of two soil types on the relative transfer of phosphate and calcium to the shoots of barley from different depths in soil

Site	Highfield IV Rothamsted			Stackyard field Woburn		
Soil type	Clay with flints			Sand		
Range of soil properties over the depths investigated;						
Isotopically exchangeable phosphate (mg P/100 g soil)	2·9–5·9			4·4–8·5		
Calcium extracted by N NH_4Ac (mg Ca/100 g soil)	308–414			89–117		
pH (in N/10 KCl)	5·3–5·5			5·8–6·1		
Rainfall during experiment (mm) (April–August)	352			276		
Fertilizer treatment*	None	NPK	NPK/None	None	NPK	NPK/None
Yield (g/m²) Shoots	614	895	1·5	175	718	4·1
Roots (0–45 cm)	39	51	1·3	18	46	2·6
Phosphate						
Total content (g/m²)	1·3	2·0	1·5	0·5	1·3	2·6
Absorption from different depths relative to that from 5–10 cm. without fertilizer						
5–10 cm	100	158		100	141	
15–20 cm	84	63		68	101	
25–30 cm	21	16		22	58	
35–40 cm	Not measured			15	32	

* NPK represents the application of a complete fertilizer equivalent to 12·5, 10·4 and 10·9 g/m² of N, P and K, respectively.

TABLE 2—*continued*

Fertilizer treatment	None	NPK	NPK/None	None	NPK	NPK/None
Calcium						
Total content (g/m^2)	1·3	2·4	1·8	0·4	3·0	7·5
Absorption from different depths relative to that from 5–10 cm without fertilizer						
5–10 cm	100	287		100	526	
15–20 cm	99	121		160	567	
25–30 cm	68	53		85	437	
35–40 cm	Not measured			75	395	
Roots						
Distribution relative to that at 5–10 cm without fertilizer						
5–10 cm	100	123		100	202	
15–20 cm	17	32		59	157	
25–30 cm	8	9		8	77	
35–40 cm	9	10		10	52	

25–30 cm at both sites and also at 35–40 cm below the soil surface at Woburn, shortly after the crop had emerged. The above ground parts of the crop were sampled at maturity about 5 months later.

The approximate distribution of roots down the soil profile was also examined. Soil cores were divided into zones corresponding to the depths of injection; roots were separated by the method of Schuurman and Goedewagen (1965), and their dry weight determined.

At both sites, fertilizer increased the dry weight of the shoots and the total quantity of phosphate and calcium transferred to them (Table 2); this effect was greater at Woburn, the yield there being extremely low in the absence of fertilizer. The relative extent to which phosphate was absorbed from different depths in the soil differed little between the unfertilized plots in the two experiments, but relatively more calcium was absorbed from 15–20 and 25–30 cm below the soil surface at Woburn. The total quantity of roots was much less in the unfertilized plots at Woburn but relatively more of them were present at 15–20 cm indicating a deeper rooting pattern in the sandy soil.

The effect of fertilizer on the relative uptake of the two nutrients differed markedly between the two sites. On the clay with flints soil, at Rothamsted, absorption of phosphate and calcium was enhanced only from the zone to which fertilizers were applied, absorption from below this region being either unaltered or significantly depressed. In contrast, on the sandy soil at Woburn, absorption of both elements was enhanced from all depths; the effect being greater for calcium than phosphate. In general the application of fertilizer caused an increased fraction of the root system to be in those zones from which the relative uptake of nutrients was enhanced but no close correlation was evident.

The factors responsible for the greater depth of rooting of barley at Woburn than Rothamsted cannot be assessed in experiments of this type. This difference is not eliminated by the addition of fertilizers which caused the total yield and nutrient content of both roots and shoots at Woburn to become comparable with that at Rothamsted. Thus, the contrast in root distribution did not appear due to differences either in the total supply of major nutrients or the availability of carbohydrates for root growth. Nor does it seem that seasonal differences in climate can provide an explanation since similar results were obtained at Rothamsted in two seasons between which the rainfall differed by 70 mm. It is concluded, therefore, that the difference in the patterns of rooting between the two sites was caused by soil factors other than the total supply of major nutrients. The contrasting texture of the soil suggests that mechanical impedance in the clay at Rothamsted may have been significantly greater than on the Woburn sand. Evidence that the length of root axes can vary inversely with the bulk density in soil (Schuurman 1965) suggests that this question deserves consideration. In addition, the possible effect of contrasting gradients in moisture, oxygen, minor nutrients or the leaching of nitrogen between the two sites cannot be excluded.

Experiments of a similar type were carried out with kale which not only absorbs considerably greater quantities of both nutrients but draws a larger proportion of them from greater depths in the soil. As with barley, comparable differences between the effects of fertilizer at the two sites were apparent.

RELATIONSHIP BETWEEN PHOSPHATE AND CALCIUM

It has already been noted that there were only small gradients in labile phosphate and calcium between the depths of soil studied. Accordingly, in the absence of a physiological difference in the performance of roots at different depths in soil the ions should be absorbed in the same ratio from all depths. This did not occur in any experiment; for example, it is apparent from Table 2 that at Woburn especially, phosphate and calcium were absorbed in very contrasting ratios from 5–10 and 15–20 cm. Fig. 4 shows equally big variations in an experiment with ryegrass which was sampled on six successive occasions when the absorption of phosphate and calcium from four depths was compared. Early in the season the ratio in which phosphate and calcium was absorbed from near the soil surface was significantly higher than from greater depths; late in the season this effect disappeared. Since soil analysis showed only small changes in the quantity

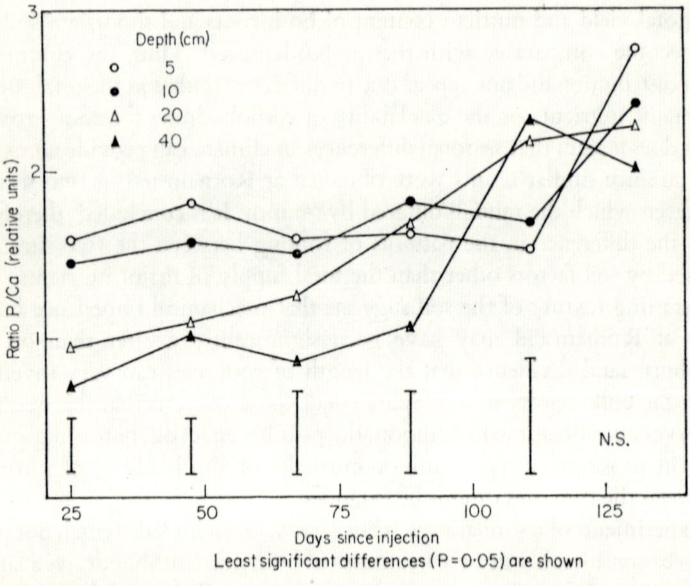

FIG. 4. Seasonal trends in the ratio of phosphate to calcium absorbed from four depths in soil by perennial ryegrass.

of labile nutrients between the beginning and end of the season the results in Fig. 4 provide further evidence that the difference in ratio is not due to soil factors. As these ions contrast markedly both in their metabolic function and the manner in which they are redistributed in plants these variations in the relative extent to which they are absorbed cannot be regarded as unexpected.

The occurrence of considerable variations in the ratio in which ions are absorbed at any one time by different parts of intact root systems has previously been demonstrated in laboratory studies in which the performance of different parts of a root system are studied separately (Russell and Sanderson 1967). One important implication of this situation is that measurements of uptake of any one nutrient from different zones cannot be regarded as a guide to root distribution or 'activity' in these zones. The interest and difficulties in making such assessments are obvious. The technique described in the present paper gives little information on this aspect.

DISCUSSION

The experiments described here show that the relative extent to which plants grown in monoculture draw nutrients from different depths in the soil can vary in a complex manner. Perennial ryegrass and barley both absorb the major fraction of their nutrient requirements from close to the soil surface but the supply of moisture, and local enhancement of nutrient supply cause considerable quantitative variations. Other soil factors also exert an influence causing significant differences between soils of contrasting type. The investigation of these questions is made more difficult by the fact that at any one time the ratio in which different nutrients are absorbed may vary significantly from one part of the root system to another due to plant physiological factors. Thus, the uptake of any one nutrient cannot be regarded as a precise guide to the uptake of others, even when gradients in the external supply do not complicate relationships. For the same reason the distribution of living roots could be only approximately inferred from the results of uptake studies. The comparative study of nutrient uptake and root growth in different zones is thus of particular interest. The inadequacies of mechanical procedures such as those used in the present work for identifying living roots in the soil, and especially their fine actively absorbing members, is a major limitation to the study of this question. The extent to which this difficulty can be overcome by 'labelling' living roots with gamma ray emitting isotopes which are freely translocated from shoots is under study (Russell and Ellis 1968).

Complex though they are, the patterns of root activity demonstrated in these experiments are simple in comparison with those which are to be expected in mixed communities. The effect of competition must then be taken into account equally with respect to root and shoot development; effects on shoots may be particularly important because of the changes which competition can cause in the supply of carbohydrates available for root development of different species. None the less it seems possible that the experimental methods described here can contribute to the analysis of these situations.

ACKNOWLEDGEMENTS

It is a pleasure to acknowledge collaboration from many sources in the conduct of these experiments:

Mr T.E. Williams and Mr E.A. Garwood of the Grassland Research Institute, Hurley; Mr F.V. Widdowson of the Chemistry Department, Rothamsted; Dr F.B. Ellis, Mr B.T. Barnes and Mr K.R. Howse together with members of the Chemistry, Electronics and Statistics sections of the Radiobiological Laboratory. The author is greatly indebted to Dr R. Scott Russell for much useful discussion and for his comments on the manuscript.

REFERENCES

GARWOOD E.A. (1965). Growth, water use and nutrient uptake from the sub-soil. The Grassland Research Institute, *Experiments in Progress* **17**, 51.

HALL N.S., CHANDLER W.F., VAN BAVEL C.H.M., REID P.M. and ANDERSON J.H. (1953). A tracer technique to measure growth and activity of plant root systems. *Nth Carolina Agr. Exp. Sta. Tech. Bull.* 101.

NEWBOULD P. and TAYLOR R. (1968). Uptake of nutrients from different depths in soil by plants. *Proc. 8th Int. Congr. Soil Sci.* **IV**, 731–742.

NEWBOULD P., HOWSE K.R. and ELLIS F.B. (1967). Effect of fertilizers on the relative absorption of phosphate and calcium by barley and kale from different depths in soil. Agricultural Research Council Radiobiological Laboratory Annual Report 1966 *ARCRL* **17**, 25–28.

NEWBOULD P., TAYLOR R. and HOWSE K.R. (1968a). *The absorption of nutrients from different depths in soil by ryegrass swards.* (In press.)

NEWBOULD P., MERCER E.R. and LAY P.M. (1968b). Estimation of changes in the water status of soils under field conditions from the attenuation of β-radiation by water held in absorbent nylon pads. *Expl. Agric.* **4**, 167–177.

NYE P.H. and FOSTER W.N.M. (1960). The use of radioisotopes to study plant feeding zones in natural soil. *Proc. 7th Int. Congr. Soil Sci.* **II**, 215–222.

NYE P.H. and FOSTER W.N.M. (1961). The relative uptake of phosphorus by crops and natural fallow from different parts of their root zone. *J. Agric. Sci.* **56**, 299–306.

RUSSELL R.S. and NEWBOULD P. (1966). Entry of strontium-90 into plants from the soil. Chapter 11 In: *Radioactivity and Human Diet.* Ed. by R.S. Russell, 213–245 Pergamon Press, Oxford.

RUSSELL R.S. and SANDERSON J. (1967). Nutrient uptake by different parts of the intact roots of plants. *J. exp. Bot.* **18**, 491–508.

RUSSELL R.S. and ELLIS F.B. (1968). Estimation of the distribution of plant roots in soil. *Nature (Lond.)*, **217**, 582–583.

SCHUURMAN J.J. (1965). Influence of soil density on root development and growth of oats. *Plant and Soil*, **22**, 352–374.

SCHUURMAN J.J. and GOEDEWAAGEN M.A.J. (1965). *Methods for the examination of root systems and roots.* Pudoc. Wageningen.

THE INFLUENCE OF THE MICROFLORA ON THE ACCUMULATION OF IONS BY PLANTS

D. A. Barber

Agricultural Research Council Radiobiological Laboratory,
Letcombe Regis, Wantage, Berkshire

INTRODUCTION

Plants growing in soil are surrounded by very large numbers of micro-organisms. Bacterial numbers of up to 10^{10}/g of soil have been recorded, although on average, in most soils there are likely to be about 2×10^9 bacteria/g (Clark 1967). However, with the exception of the nitrogen fixing organisms and ectotrophic mycorrhiza and to a lesser extent endotrophic mycorrhiza the possible effects of this very extensive microflora have been very largely ignored by plant physiologists concerned with the accumulation of inorganic nutrients by plants. On the other hand, soil microbiologists interested in the micro-organisms rather than in the plant have amassed much data on the activities of the flora. They have shown that the numbers and types of bacteria and fungi in the soil fluctuate greatly depending upon the proximity to roots, depths, moisture content, nutrient status, plant species and many other factors. For example Berezova (1965) has shown that in the rhizosphere of potatoes there are over sixty million bacteria/g of soil capable of breaking down calcium triphosphate and twenty million able to break down organic phosphate, while these numbers fall to two and one million respectively in soil a centimetre or so away from the roots. Other types of organisms showed similar differences in distribution. These findings encourage speculation as to the significance of the microflora in the inorganic nutrition of plants. A comparative study of the nutritional relationships of plants grown in the presence and absence of micro-organisms at the ambient laboratory level has therefore been undertaken. The first phase of this work concerned with plants growing in water culture is here described.

The choice of water culture for the initial investigation was determined by two factors; not only is it a much simpler system but the majority of current concepts regarding the accumulation of inorganic nutrients by plants are derived from such studies rather than from plants growing in soil.

For a number of reasons the phosphate ion was chosen for investigation; firstly, it occurs in the soil solution in very low concentrations, secondly, micro-organisms have a great avidity for it and thirdly, results of experiments on the uptake and distribution of phosphate from dilute solutions (Russell and Martin 1953; Loughman and Russell 1957) under clean but not sterile conditions appeared compatible with a marked effect of micro-organisms.

METHODS

In all experiments the uptake and distribution of phosphate in barley plants (*Hordeum vulgare* var. Maris Badger) grown in the absence of micro-organisms has been compared with that of plants under non-sterile conditions. The experimental procedure is described in detail elsewhere (Barber 1967a) but the salient facts are as follows.

The seeds are sterilized and thoroughly washed in sterile water and after germination the seedlings are individually screened for the presence of micro-organisms by growth for 4 days on nutrient agar. Obviously clean seedlings are then transferred to totally enclosed culture vessels and are either grown in the complete absence of micro-organisms by the use of sterilized apparatus, air and culture solutions or allowed to become infected with indigenous laboratory organisms by omitting these precautions.

The uptake and distribution of ^{32}P-labelled phosphate was determined from solutions of potassium phosphate of varying concentrations and pH (Barber and Loughman 1967; Barber 1967b). The effect of the carbohydrate status of the plant on the extent of microbial activity was determined by comparing the absorption of phosphate by plants previously kept for twenty-four hours in light or in darkness. Before plants were harvested and assayed for ^{32}P, samples of solution were withdrawn from the sterile cultures and plated on malt and nutrient agar; contaminated replicates were discarded.

RESULTS

1. *Effects of external concentration*

Table 1 shows the effects of micro-organisms on the uptake of phosphate in 1 hour from solutions of increasing concentration. Except at the lowest concentration examined (0·001 ppm P) micro-organisms had little effect

TABLE 1

Effect of micro-organisms on the uptake and distribution of ^{32}P-labelled phosphate in young barley plants after absorption for one hour in solutions of potassium phosphate of different concentrations

		Solution concentration ppmP				
		0·001	0·01	0·1	1·0	10·0
Total uptake μg P	Sterile	0·302	1·39	8·00	42·9	78·7
	N. Sterile	0·091	2·24	8·22	48·0	84·9
% in shoot	Sterile	0·7	3·9	22·8	27·9	29·7
	N. Sterile	0·05	2·0	9·0	17·7	20·0
Distribution in roots (as % total root P):						
Soluble esters + inorganic P	Sterile	85·0	86·4	91·9	91·5	89·3
	N. Sterile	33·3	31·2	73·6	78·9	86·7
Nucleic acids	Sterile	7·8	6·5	3·2	5·8	6·0
	N. Sterile	49·3	41·3	16·2	16·9	9·5
Phospholipids + phosphoproteins	Sterile	7·1	7·0	4·7	2·7	4·6
	N. Sterile	17·3	27·4	10·2	3·9	3·4

on the total amount of phosphate taken up but the distribution of the absorbed ion was greatly affected. When micro-organisms were present, a considerably larger quantity of the absorbed phosphate was incorporated into nucleic acids, phospholipids and phosphoproteins in the roots and less was transferred to the shoots. These results would be expected if micro-organisms in and on the roots absorbed phosphate at the expense of the root cells thus reducing the amount available for transport to the shoot. Because of the very short generation time of the micro-organisms they would rapidly incorporate phosphate into complex organic compounds and since it is not possible to separate the roots from the attendant microflora (Jenny and Grossenbacher 1963; Parkinson 1965; Scott 1966) analyses of roots grown under non-sterile conditions will reflect the dual nature of the material. It would be expected that such effects should be reduced when the external concentration of phosphate is high as the requirements of both plants and micro-organisms would then be satisfied.

Some visual evidence of the effects of bacterial infection are provided by autoradiographs which show that under non-sterile conditions there are localized sites of accumulation of ^{32}P-labelled phosphate around the periphery of transverse sections of roots (Barber, Sanderson and Russell 1968).

Beyond this, evidence that micro-organisms in the culture solution may reduce the effective ionic concentrations is provided by the fact that at the end of the experiment more phosphate can be stripped off the walls of the vessels which contained the non-sterile plants. In the subsequent experiments, some factors other than concentration which influence the extent of the microbial effects are examined.

2. *Effects of pH*

Many bacteria are very sensitive to hydrion concentration, their metabolic rate being much reduced at low pH. Since in water culture studies bacteria tend to predominate over fungi it has been shown, not surprisingly, that the magnitude of the effects of micro-organisms demonstrated above is considerably influenced by pH (Barber 1967b). These effects are quite distinct from those caused by the change in the valency of the dominant phosphate ion in solution from the monovalent $H_2PO_4^-$ at low pH to the divalent HPO_4^{--} at high pH.

Table 2 shows that both in the presence and absence of micro-organisms, the total amount of phosphate taken up from all concentrations was greatest when the pH was low and that more was transferred to the shoots of plants supplied with the highest concentrations, irrespective of pH and microbial infection. However, although micro-organisms had a considerable influence on the transfer of phosphate to the shoots, except at the higher concentration this effect was much less pronounced at pH 4 than at pH 6 or 8. A similar effect of pH was apparent on the extent of microbial activity on the roots as indicated by the quantity of phosphate present as soluble esters and inorganic phosphate.

It is evident therefore that when micro-organisms are present, differences in the accumulation of phosphate at high and low pH do not necessarily reflect the functioning of separate carrier mechanisms for the two ion species (Hagen and Hopkins 1955).

3. *Effects of photosynthesis*

Since the microflora on and around the roots is dependent very largely for its supply of organic nutrients on exudation from the roots and breakdown of root tissues (Rovira 1965) factors affecting the synthetic processes of the plant are likely therefore to affect the metabolism of these organisms through their effect on exudation. For example, it has been shown by

TABLE 2

Effect of pH on the uptake and distribution of ³²P-labelled phosphate in young barley plants grown under sterile and non-sterile conditions. The plants were exposed to solutions of potassium phosphate of various concentrations at pH 4, 6 and 8 for 3 hours

		Concentration of phosphate ppm P			
pH	Conditions of plants	0·001	0·01	0·1	1·0
	Total uptake μgP/plant				
4	Sterile	0·40	3·12	24·68	95·7
	N. Sterile	0·156	2·81	28·06	100·6
6	Sterile	0·32	3·40	26·87	79·5
	N. Sterile	0·15	1·53	22·49	77·6
8	Sterile	0·23	0·94	11·09	53·8
	N. Sterile	0·13	1·34	14·61	64·2
	P transported to shoot as % total plant P				
4	Sterile	7·5	8·7	50·2	48·6
	N. Sterile	3·8	6·4	34·3	49·6
	Ratio Sterile/N. Sterile	2·0	1·3	1·5	1·0
6	Sterile	6·5	15·3	36·0	50·7
	N. Sterile	0·7	3·3	24·1	44·8
	Ratio Sterile/N. Sterile	9·3	4·6	1·5	1·1
8	Sterile	3·5	5·3	17·1	48·9
	N. Sterile	0·8	1·5	8·2	38·1
	Ratio Sterile/N. Sterile	4·4	3·5	2·1	1·3
	Soluble esters and inorganic P as % root P				
4	Sterile	81·1	82·5	86·3	87·7
	N. Sterile	43·5	65·3	62·8	75·2
	Ratio Sterile/N. Sterile	1·9	1·3	1·4	1·2
6	Sterile	80·0	78·9	81·2	88·0
	N. Sterile	37·2	46·6	58·3	71·3
	Ratio Sterile/N. Sterile	2·1	1·7	1·4	1·2
8	Sterile	76·5	82·6	82·3	85·2
	N. Sterile	29·3	22·3	52·3	70·6
	Ratio Sterile/N. Sterile	2·6	3·7	1·6	1·2

Rovira (1959) for tomato and clover plants in solution culture that a 60% decrease in light intensity caused considerable changes in the relative quantitites of different amino acids in the exudate and decreased the overall quantity exuded quite markedly. It is possible therefore that the accumulation of phosphate by micro-organisms on roots would be dependent to some extent on the carbohydrate status of the plant. Accordingly, the uptake and distribution of phosphate by 'starved' plants which had been grown for 24 hours in the dark was compared with that of 'normal' plants under sterile and non-sterile conditions. Uptake of phosphate was determined from a solution of potassium phosphate containing 0·01 ppm P at pH 6 for 3 hours in the dark so that no photosynthate was produced during this period.

TABLE 3

Effect of carbohydrate status on the uptake and distribution of ^{32}P-labelled phosphate in young barley plants grown under sterile and non-sterile conditions. The plants were exposed to a solution of potassium phosphate containing 0·01 ppm P at pH 6 for 3 hours in the dark after 24 hours either in the light ('normal' plants) or in darkness ('starved' plants)

Carbohydrate status	Normal		Starved	
Condition of plants	Sterile	N. Sterile	Sterile	N. Sterile
Phosphate in roots				
μg P/plant	1·96	1·43	0·62	1·33
% in soluble esters/inorganic P	71·5	25·7	71·5	42·4
Phosphate in shoots				
μg P/plant	0·12	0·001	0·18	0·17
% total plant content	5·6	0·1	22·4	12·7
Phosphate removed from culture vessel walls with acidified carrier: μg P	0·087	0·377	0·065	0·187

As in the previous experiments, micro-organisms caused a marked reduction in the fraction of the phosphate in roots which was present as soluble esters and inorganic P (Table 3). This effect was much greater in the 'normal' than in the 'starved' plants suggesting that there was less microbial activity in the latter. Some independent evidence of this was provided by the reduction in the quantity of phosphate which could be removed from the walls of the culture vessels with acidified carrier at the end of the experiment.

An interesting feature of the present results is that although under sterile conditions starvation reduced root content, it did not affect the fraction present in soluble esters; this suggests that the decrease in substrate limited uptake but had little effect on the partition of such phosphate as was absorbed between the soluble esters and nucleic acids. It is of interest also that in the absence of micro-organisms the actual quantity of phosphate transferred to the shoots was somewhat increased by starvation. In both these respects the uninfected plants contrasted markedly with those under non-sterile conditions thus providing further evidence that effects on the utilization of phosphate caused by metabolic factors in plants can be readily concealed by microbial activity.

DISCUSSION

As in any other aspect of plant nutrition it is difficult to extrapolate from the results of studies in solution culture to the possible situation of plants growing in soil. In the present work this is particularly so as apart from the different conditions for plant growth, micro-organisms in the soil are likely to be more numerous and diverse in their action. The very simple microflora in the present experiments associated with the plants grown under non-sterile conditions appeared merely to compete for nutrients. In contrast, the microflora of the rhizosphere may be expected to influence nutrition and the growth of plants in a variety of ways. The micro-organisms may act beneficially by transforming unavailable mineral and organic compounds in the soil into forms available for uptake by higher plants, by breaking down root secretions which have a toxic effect on plant growth, by producing substances which may accelerate plant growth and development and by producing antibiotics which may suppress the activities of parasitic organisms. Alternatively micro-organisms may act in a deleterious manner by competing with plants for available nutrients, by producing substances that may suppress growth and retard development and by parasitizing plants.

It is therefore of particular interest to compare results obtained in the present work with findings in studies in which soil micro-organisms have been used. In an investigation of the effects of a mixed suspension of soil micro-organisms on the absorption of phosphate by tomato and clover seedlings in solution culture Bowen and Rovira (1966) observed an increase in both uptake and translocation of phosphate, indicating possibly that when a wide spectrum of organisms is present the stimulatory effects of microbial secretions may predominate over any competitive effect. In sand culture on the other hand Akhromeiko and Shestakova (1958) showed that the addition of rhizosphere organisms reduced the uptake by oak and ash seedlings of ^{32}P-labelled phosphate supplied as sodium dihydrogen phosphate. When, however, phosphate is supplied as insoluble compounds the opposite effect can sometimes be demonstrated. Gerretsen (1948) found that the addition of soil micro-organisms to previously sterile cultures caused insoluble phosphate containing compounds to be broken down with a resultant increase in the uptake of phosphorus by the plant. None the less, despite the widespread occurrence of organisms in the soil which can release phosphate from minerals when grown in isolation on agar plates, no marked increase in the uptake of phosphate by plants growing in soil has been demonstrated which can be ascribed to such microbial activity (Cooper 1959; Myśków 1961). Endophytic mycorrhiza alone have been unequivocally shown to increase uptake of phosphate from sparingly soluble compounds added to soil and then only for soils very low in native phosphate. (Daft and Nicolson 1966; Gerdemann 1964, 1965.)

All the above results, whether in sand, soil or water culture and showing increases or decreases in the uptake of phosphate by plants in response to the presence of micro-organisms have one factor in common, namely, that effects are only apparent when phosphate is in short supply. This appears to be the major criterion which determines the relative importance of microbial effects on the phosphate nutrition of higher plants. Given this condition, in any particular situation, other factors which can influence the composition and metabolic activity of the microflora, such as the pH and substrate availability through exudation, may well assume importance and induce differences in the growth and nutrition of the plants. This is a field of work which as yet has scarcely been tapped.

In extreme situations many problems regarding the inorganic nutrition of plants may be better understood if the microbiology of the situation were to be examined. For example, it is conceivable that the ability of *Mercurialis* to grow successfully in very shaded conditions, on soil very low in available phosphate (Pigott 1966) may in part reflect a very low production of exudates by the roots resulting in an impoverished and thus less competitive rhizosphere flora. However, such interpretations can at the moment be little more than speculation. None the less, it is evident that any aspect of plant nutrition will be inadequately understood unless account is taken of the activities of the countless and very diverse micro-organisms present.

ACKNOWLEDGEMENTS

I thank Dr R. Scott Russell for his stimulating interest in this work and Miss U.C. Frankenburg and Miss C. Lambourne for their valuable assistance.

REFERENCES

AKHROMEIKO A.I. and SHESTAKOVA V.A. (1958). The influence of rhizosphere micro-organisms on the uptake and secretion of phosphorus and sulphur by the roots of arboreal seedlings. *Proc. 2nd Intern. Conf. Peaceful Uses Atomic Energy, Geneva*, **27**, 193–199.

BARBER D.A. (1967a). The effect of micro-organisms on the absorption of inorganic nutrients by intact plants. I. Apparatus and culture technique. *J. Exp. Bot.* **18**, 163–169.

BARBER D.A. (1967b). Influence of pH on the uptake of phosphate by barley plants in sterile and non-sterile conditions. *Nature*, **215**, 779–780.

BARBER D.A. and LOUGHMAN B.C. (1967). The effect of micro-organisms on the absorption of inorganic nutrients by plants. II. Uptake and utilization of phosphate by barley plants grown under sterile and non-sterile conditions. *J. Exp. Bot.* **18**, 170–176.

BARBER D.A., SANDERSON J. and RUSSELL R.S. (1968). The influence of micro-organisms on the distribution of ^{32}P labelled phosphate in roots. *Nature*, **217**, 644.

BEREZOVA E.F. (1965). Significance of the root system micro-organisms in plant life. *Plant microbe relationships*. (Ed. J. Macura and V. Vančura), pp. 171–177. Czechoslovakia Academy Science, Prague.

BOWEN G.D. and ROVIRA A.D. (1966). Microbial factor in short-term phosphate uptake studies with plant roots. *Nature*, **211**, 665–666.

CLARK F.E. (1967). Bacteria in soil. *Soil Biology*. (Ed. A. Burges and F. Raw), pp. 15–49. Academic Press, New York.

COOPER R. (1959). Bacterial fertilizers in the Soviet Union. *Soils and Fertilizers*, **22**, 327–333.

DAFT M.J. and NICHOLSON T.H. (1966). Effect of *Endogone* mycorrhiza on plant growth. *New Phytol.* **65**, 343–350.

GERDEMANN J.W. (1964). The effect of mycorrhiza on the growth of maize. *Mycologia*, **56**, 342–349.

GERDEMANN J.W. (1965). Vesicular-arbuscular mycorrhiza formed on maize and tulip tree by *Endogone fasciculata*. *Mycologia*, **57**, 562–575.

GERRETSEN F.C. (1948). The influence of micro-organisms on the phosphate intake by the plant. *Plant and Soil*, **1**, 51–85.

HAGEN C.E. and HOPKINS H.T. (1955). Ionic species in orthophosphate absorption by barley roots. *Pl. Physiol.* **30**, 193–199.

JENNY H. and GROSSENBACHER K. (1963). Root-soil boundary zones as seen in the electron microscope. *Proc. Soil Sci. Soc. Am.* **27**, 273–277.

LOUGHMAN B.C. and RUSSELL R.S. (1957). The absorption and utilization of phosphate by young barley plants. IV. The initial stages of phosphate metabolism in roots. *J. Exp. Bot.* **8**, 280–293.

MYŚKÓW, W. (1961). The influence of bacteria dissolving phosphates on the availability of phosphorus to plants. *Acta. Microbiol. Polon.* **10**, 195–201.

PARKINSON, D. (1965). The development of fungi in the root region of crop plants. *Plant Microbe Relationships.* (Ed. J. Macura and V. Vančura), pp. 69–75. Czechoslovakia Academy Science, Prague.

PIGOTT C.D. (1964). Nettles as indicators of soil conditions. *New Scientist*, **21**, 230–232.

ROVIRA A.D. (1956). Root excretions in relation to the rhizosphere effect. IV. Influence of plant species, age of plant, light, temperature and calcium nutrition on exudation. *Plant and Soil*, **11**, 53–64.

ROVIRA A.D. (1965). Plant root exudation and their influence upon soil micro-organisms. *Ecology of soil-borne plant pathogens.* (Ed. K.F. Baker and W. Snyder), pp. 170–184. University of California, Berkeley.

RUSSELL R.S. and MARTIN R.P. (1953). A study of the absorption and utilization of phosphate by young barley plants. I. The effect of external concentration on the distribution of absorbed phosphate between roots and shoots. *J. Exp. Bot.* **4**, 108–127.

SCOTT F.M. (1966). Cell wall surface of the higher plant. *Nature*, **210**, 1015–1017.

THE RELATION OF GROWTH TO THE CHIEF IONIC CONSTITUENTS OF THE PLANT

W. DIJKSHOORN
Institute for Biological and Chemical Research on Field Crops and Herbage, Wageningen, the Netherlands

Investigations into the requirement of plants for major nutrient ions have demonstrated three important points:

(1) essential ions are required in certain minimum amounts to permit growth, and this requirement is specific;

(2) when considering the ionic balance within the plant the metal cations have in common an ability to form salts both with the anions absorbed and with those produced by metabolism, and the cations may replace each other to some extent in this function without any apparent anomalies in ion accumulation and growth;

(3) the ability of cations to perform the functions of potassium in the tissues is restricted, in part, by their different rates of absorption and their rates of redistribution inside the plant.

Specific requirements for major ions

There is a minimum concentration of the essential elements within the plant such that the amount absorbed will entirely determine the rate of growth. For instance, when ryegrass was grown on a soil which released potassium at a sufficiently slow rate, the absolute amount of potassium in the foliage increased in strict proportion to its dry weight. There was a constant concentration of 0·2 me. per gram dried foliage and the growth could not continue unless sufficient potassium was absorbed to supply the new growth in this proportion (van Tuil 1965).

These plants had a greater percentage of dry matter compared with healthy plants. When sodium was supplied, the dry matter percentage was reduced to a value below the normal, the plants made some additional growth even though the potassium level in the leaves fell still lower to around 0·1 me. per gram dry matter.

Although the plants in both cases were deficient of potassium, they were different in appearance and composition. Grown in the presence of added sodium they showed excessive accumulation of this ion in the leaves and an increased water content and leaf expansion. Thus these plants were able to produce more dry matter for a given amount of absorbed potassium than those grown in media deficient in potassium and containing no additional sodium.

Although sodium can bring about additional growth in the presence of insufficient potassium, the different characteristics of ryegrass plants grown under these conditions compared with those grown in media with plenty of potassium, show that sodium cannot be considered as a substitute for potassium in a physiological sense.

When the growth is restricted to expanding, terminal leaves, as it is in the more advanced stages of growth in many dicotyledons, a minimum level of potassium can be maintained in the growing tissues by transfer of endogenous potassium. Since such potassium can be used repeatedly by the growing zones, the concentration in the whole leaf material may fall to very low values. Given this growth habit, the total yield of foliage may be less sensitive to shortage of the element, the real shortage being confined to the mature leaves which have ceased to grow.

Even though growth is continuing at the shoot apex, ions continue to move into the leaves lower down the shoot via the xylem system. When the leaves have reached their final size, and begin to export their assimilates and potassium to the other plant parts and growing zones via the phloem, calcium is retained, and its continued intake will mean that its concentration will rise very often to much higher values in the older leaves (Cooil 1948; Vickery 1961; Ward 1964; Kirkby and DeKock 1965). Thus for plants of this growth habit, the composition of leaves may vary considerably with position on the shoot.

There are sufficient data in the literature on yield and plant composition to evaluate the minimum concentration of the other cations required for the growth of the plants from the relation between the amounts absorbed and the dry weights produced. For calcium and magnesium it is in the order of 0·1 me. per gram dry matter.

Many plant species can absorb nitrate in excess of the amount required for metabolism, and this excess accumulates in the plant in the unchanged ionic form. Detection of nitrate in the tissues of these species can serve to indicate whether nitrate is in adequate supply (van Burg 1966).

This is similar to the situation found for sulphate. The presence of sulphate in the tissues in the unchanged ionic form is indicative of sufficient sulphate in the medium (Dijkshoorn and van Wijk 1967).

There is some evidence that the phosphate requirement for maximum growth is about twice the amount transferred into organic phosphates, and the organic phosphates are mainly in the TCA-insoluble fraction (Dijkshoorn and Lampe 1961).

Chloride is not metabolized. There is evidence that its presence is essential, but the amount needed for the growth seems to be very small, in the order of 0·01 me. per gram dry matter. Since chloride is readily absorbed by many plant species, its concentration in the plant may vary considerably with the amount in the medium, for instance, from 0·1–1·0 me. per gram dried plant material.

Cation-anion relationships

It has been pointed out that the effect of shortage of potassium may be greatly alleviated by other cations when the uptake systems permit their absorption in greater amounts. It is also known that the effectiveness of potassium as a plant nutrient may depend on anion absorption. Both aspects lead to the question as to the principle which governs the cation-anion balance in the plant. This will be discussed in relation to the following evidence.

The visual symptom of potassium shortage is leaf necrosis. It may appear soon after potassium has fallen to a value of 0·2 me. per gram dry matter. However, Boresch (1939) found that high chloride contents in a plant may bring about leaf necrosis at higher potassium concentration than 0·2 me. in the leaves. It would appear, therefore, that the potassium requirement depends, in part, on the absorbed anions.

Van Tuil (1965) grew poplar cuttings under conditions where there was a low cation availability. The total cation content in the leaves was 0·9 me. per gram dry matter, with 0·6 me. bound to carboxylates. The potassium content of the leaves varied from below 0·2–0·4 me. per gram dry matter, but all the plants had necrotic leaves. When the cuttings were defoliated and again supplied with the nutrient salts with varying amounts of potassium, the total cation content of the leaves, produced in a second growth period, was increased to 1·2 me. with 0·9 me. per gram dry matter linked to carboxylates. Although defoliation and the second supply of nutrient salts had not changed the range of the potassium content of the leaves the second leaf crop was necrotic only at the level of 0·2 me. of potassium per gram dry matter, necrosis disappearing as soon as potassium increased to a level of 0·3–0·5 me. per gram dried foliage. It would appear from this that the potassium requirement depends also on the total cation and carboxylate content of the leaves.

Richards (1956) reports that when nitrogen is supplied as ammonium, the effect of potassium deficiency may become much more severe than with nitrate.

Since chloride accumulates in the tissues at the expense of carboxylate anions (Böning and Böning-Seubert 1932), metallic cations accumulate, bound to carboxylates (Ulrich 1941), and metabolism of nitrates is the main source of carboxylates in the leaves (Dijkshoorn 1962), there is an underlying similarity between the above results: more potassium was required to cure the deficiency symptoms when the carboxylate content in the tissues was lower. It would appear therefore that information on the carboxylate content of plant is of considerable importance.

The ionic balance within plants

The ionic balance expresses the electroneutrality of the tissues as a whole, it being required that the total number of positive charges of whatever source must equal the negative charges.

Of the absorbed nutrient ions, nitrate, ammonium and sulphate are converted by metabolism into non-ionic organic nitrogen and sulphur. Some of the nitrate and sulphate anions are retained by the tissues in the unchanged ionic form, and thus contribute to the final inorganic anion content.

A great portion of the absorbed phosphate is in the form of organic phosphates, but there is no change in the ionic form $R_2PO_4^-$ when R denotes either H or some other constituent bound to the phosphate by esterification. Monovalent phosphate predominates in the common range of tissue pH 5–6. When the tissue pH is higher, divalent phosphate must be taken into consideration (Cooil 1948). In working out the ionic balance within a plant, it is convenient to group all the phosphate under the inorganic anions.

Since the pK of silicic acid is some four units higher than the pH of the plant tissues, there is no need to include silicates in the ionic balance sheet (Dijkshoorn 1963).

The sum: $NO_3^- + Cl^- + H_2PO_4^- + SO_4^{--}$ in milliequivalents is the inorganic anion content A. In ryegrass, it is in the order of 0·5 me. per gram dry matter.

Absorbed ammonium ions are almost all converted into organic nitrogen, but the metallic cations accumulate unchanged in the ionic form, irrespective of their dissolved or undissolved state, and their localization at the cellular level.

The sum: $K^+ + Na^+ + Mg^{++} + Ca^{++}$ in milliequivalents is the total cation content C. In ryegrass, C is in the order of 1·5 me. per gram dry matter.

It has been known for some considerable time that bicarbonates do not accumulate unchanged in plant tissues, but that the excess of cations over inorganic anions, (C–A), is balanced by carboxylate anions (Böning and Böning-Seubert 1932). All bicarbonates absorbed, or released by metabolism during nitrate assimilation, are converted into carboxylates; so that (C–A) represents the salts of the various carboxylic acids. The greater portion of these acids are the common organic acids, malic, citric, oxalic, succinic, etc., with a small fraction consisting of insoluble polyuronic acids (van Tuil 1965).

In most plants of economic importance, the tissue pH is between 5 and 6, and most of the organic acids exist in the form of their salts. The quantity (C–A) is often called the 'organic acid' content and has occasionally been mistaken for titratable acidity.

The kinetics of carboxylate formation

Detailed considerations have been advanced elsewhere that the metabolic conversion of absorbed nitrates is the main source of carboxylates in the growing plant (Dijkshoorn 1962). Potassium nitrate moves freely through the root tissues and xylem system to the assimilatory tissues in the leaves, where the nitrate anion is replaced via metabolism by its equivalent of carboxylate anion during the transfer of nitrogen to organic nitrogen compounds.

In the leaves of ryegrass, 2·5 me. of nitrate is converted into organic nitrogen per gram dry matter produced. The equivalent amount of carboxylate is produced *in situ*, but analysis reveals that the leaves retain only 1 me. per gram dry matter. The excess is removed via the phloem, which transfers the carboxylates of potassium with the other assimilates to the roots (Peel and Weatherley 1959). In the roots, carboxylate anions are decarboxylated, and the bicarbonate anions produced are released to the medium in exchange for nitrate anions. The nitrate then moves with the potassium ions which have remained in the roots into the xylem where these ions are transferred to the foliage, and the nitrate anions are metabolized. In this way, potassium circulates upward partnering nitrate, and downward partnering carboxylate, and organic nitrogen builds up in excess of those carboxylates retained by the leaves.

Experimental justification of this scheme has been produced in the experimental work of Dijkshoorn, Lathwell and de Wit 1968, who showed that increasing the concentration of nitrate ions in the tissues, brought about a reduction of the carboxylate content of the roots, but an increase in the carboxylate content of the shoot as a result of metabolism of the nitrate.

When nitrate was replaced by additional chloride, sulphate or phosphate in the mixture of nutrient salts, ryegrass continued to grow until the organic nitrogen content was diluted to around 1 me. per gram dried foliage. Analysis showed that in the absence of nitrate in the medium, the cations of the added salts are absorbed in excess of their anions, the excess being balanced by the uptake of bicarbonate anions arising from the dissociation of water and respiratory carbon dioxide.

Thus bicarbonate substitutes for nitrate amongst those ions absorbed when this latter ion is exhausted. The bicarbonate entering the roots is converted into carboxylate. But in contrast to nitrate, potassium bicarbonate is much more strongly retained by the root tissues (Wallace, Ashcroft and Lunt 1967) and there is only slow transfer of carboxylates or bicarbonates to the shoot. Owing to this slow transfer, it was observed that in ryegrass new growth made after the removal of nitrate contained only 0·5 me. of carboxylates, whereas with nitrate this value was 1 me. per gram dry matter.

When nitrate was removed and ammonium sulphate was added to meet the nitrogen requirement of the plant, the tissues remained high in organic nitrogen, but the carboxylate concentration in the plant fell to a still lower level. The sulphate added does not influence anion uptake since further increments in sulphate supply have little effect on the rate of sulphate uptake (Dijkshoorn 1959). But, with ryegrass, ammonium uptake increases the total uptake of cations by 2·5 me. per gram dry matter, which is equivalent to the amount required for growth. A study of the ionic balance reveals that this excess is completely balanced by bicarbonate uptake.

It can be assumed that all the ammonium which is absorbed enters the metabolic system in the form of ammonium bicarbonate. This is converted as follows:

$$NH_4^+ + HCO_3^- \rightarrow (NH_3) + H_2O + CO_2,$$

where (NH_3) denotes the various forms of organic nitrogen which are all substituted ammonia. The equation shows that the proton released by the ammonium ion during its conversion into non-ionic organic nitrogen, is neutralized by the partnering bicarbonate anion.

This means that the carboxylate content of plants fed with ammonium ions will depend on those bicarbonate anions absorbed along with the metallic cations. Mention has just been made of this process which, for ryegrass, is able to supply only 0·5 me. of carboxylates per gram dry matter produced.

Furthermore, the effects of ion competition during uptake will tend to reduce carboxylate production to a still lower level. Thus the removal of nitrate will enhance the uptake of other anions, such as chloride (de Wit, Dijkshoorn and Noggle 1963), and the introduction of ammonium ions may reduce the uptake of the other cations such as potassium (Tromp 1962). The uptake of bicarbonate in balance with potassium, sodium, magnesium and calcium will thus be reduced further. Indeed, it has been observed that if ryegrass is transferred to a solution containing ammonium chloride and other nutrient salts, but no potassium, there is a complete cessation of carboxylate production, the level dropping to the very low level of 0·25 me. per gram dry matter as the weight of the plant increases (Dijkshoorn, Lathwell and de Wit 1968).

If plants which contain a high amount of organic nitrogen and a low carboxylate content, as the result of growth in a medium containing ammonium salts, are transferred to a medium in which nitrate is the sole source of nitrogen, the nitrate cannot be metabolized by the mature tissues which are in any case already well supplied with organic nitrogen. It is only the growing tissues that develop the full capacity for nitrate assimilation. Hence, only the most recently formed tissue will contain carboxylates at a level of 1 me. per gram dry matter. Depending on the proportion of older tissue, the foliage as a whole will have a low carboxylate content for some considerable time after transference, the ionic balance being determined to a great extent by the ammonium nutrition.

If the nitrogen content of plants is reduced by starvation, added nitrate is rapidly absorbed and metabolized in all the leaves, the carboxylate concentration rising to the level of 1 me. per gram dry matter within 10 days after application of nitrate.

It can be seen therefore that information on both the nitrogen and the carboxylate content may render it possible to say something more about the history of the plant and its nutrient supply, such as for instance the course of nitrification in the soil during the growth of the plant.

Mention has already been made of the conversion of nitrate into organic nitrogen as a major route of synthesis for carboxylates in the green tissues. When absorbed and transferred to the foliage along with metallic cations, the utilization of nitrate in the formation of organic nitrogen compounds results in the production of bicarbonates:

$$K^+ + NO_3^- + 8H + CO_2 \rightarrow (NH_3) + 2H_2O + K^+ + HCO_3^-;$$

which are converted into carboxylates:

$$K^+ + HCO_3^- + RH \rightarrow K^+ + RCOO^- + H_2O,$$

where RH denotes a metabolic substance which is carboxylated, such as, for instance, pyruvate.

The reductive metabolization of sulphate proceeds in a similar way (Dijkshoorn 1962), but its small contribution to the synthesis of carboxylates can be neglected for the present purpose.

It is clear from the above that when hydrogen ions are taken up instead of metallic cations, the bicarbonate formed will be neutralized. In other words, when nitric acid is absorbed it is metabolized without the production of any carboxylates. The presence of metallic cations is essential for carboxylate production; the significance of nitrates in the synthesis of carboxylates is merely a result of the greater rate of upward transfer and higher rate of metabolism of nitrate anions in the shoot, when compared with bicarbonates absorbed in the absence of nitrate.

There may be an increased uptake of hydrogen ions by plants when grown on acid soils, or when a readily absorbed cation is replaced by a cation which is less readily taken up, as for instance, where calcium is substituted for potassium in the nutrient medium of grasses. In this latter case, carboxylate production is reduced and the carboxylate content in the plant may become lower than the normal.

In the dicotyledons, calcium can substitute for potassium in the uptake process much more effectively than in monocotyledons. The residual potassium is distributed throughout the plant such that there is a relatively uniform concentration; however calcium accumulates in the mature leaves so that the total cation and carboxylate content is raised to a much higher value than that for the young, expanding leaves (Cooil 1948). In the mature leaves, the major carboxylate anions is malate (Vickery 1961; Kirkby and DeKock 1965) and the same is true for potato tuber during the export of its reserves to a growing plant (Jolivot 1959).

Most soils can supply sufficient calcium. A potassium shortage will reduce leaf expansion and increase the proportion of mature leaves with high calcium and carboxylate content. Hence, unlike the situation for grasses, potassium shortage in dicotyledons tends to raise the carboxylate content of leaves to a value which is higher than when there is plenty of potassium.

The 'normal' carboxylate content

ıhas already been pointed out that the carboxylate content of the plant nay show wide variations depending on the nutrient supply. In an attempt ‹ determine the relation between yield and carboxylate content (C–A), de Wit, Dijkshoorn and Noggle (1963) have shown that when the nutrient supply is such that the highest possible yield is obtained, the carboxylate content per unit dry weight of foliage is maintained at a constant value which is characteristic of a plant species. This can be called the 'normal' carboxylate content.

The following details for ryegrass may be of interest in this respect. When potassium is replaced by calcium the carboxylate concentration declines in the subsequent growth period. On the other hand, with sodium in place of potassium, the carboxylate concentration may rise to a value higher than normal. When the former plants are transferred to a medium containing added sodium, the carboxylate content in the foliage returns to the normal level and may increase still further to the limit governed by the organic nitrogen content. Those plants growing in a medium containing sodium can have their carboxylate content reduced to subnormal levels when sodium is replaced by calcium. With all such treatments it is necessary to supply some potassium to support growth, but the amount should be small enough such that the plants are deficient of this element.

In this way it is possible to change the carboxylate concentration over a range of values on either side of the normal. But whatever way the carboxylate content changes, the plants will have low growth rates owing to the shortage of potassium.

In all cases the growth rate can be increased to a maximum by merely adding sufficient potassium to the medium. The plants will be healthy and show vigorous growth and their carboxylate content will rise or fall to the normal value (1 me. per gram dry matter), depending whether there was lower or higher than the normal content during the regime of potassium starvation.

It is this sort of evidence which has led to the conclusion that with subnormal growth the carboxylate content can vary quite considerably, the actual level depending on the cause of the growth restriction but that maximum growth response can be correlated solely with the normal carboxylate content.

From this it would seem that a given nutrient supply will support the maximum growth rate only when in addition to there being an adequate amount of each of the ions absorbed, the plant can both utilize the absorbed ions and maintain within its tissues the normal concentration of carboxylates (de Wit, Dijkshoorn and Noggle 1963).

The application of ion balance studies to ecology

The stability of the (C–A) level upon addition or removal of certain nutrient ions can differ considerably from one plant species to the other.

For instance, the substitution of calcium for potassium has little effect on the (C–A) value for sugar beet foliage of 3·5 me. per gram dry matter, since the drop in the potassium content is equalled by the rise in magnesium plus calcium content. In grasses, similar treatment will lower (C–A) from the normal value at 1 me. to a subnormal value of about 0·5 me. per gram dry matter.

It would seem that there is no general relation between nutrient supply and carboxylate content which is applicable to all plants. However, more information about this diversity in response of the ionic balance seems desirable if progress is to be made in our understanding the nutritional aspects of ecology.

Ideally, each plant species under consideration should first be submitted to a number of treatments involving systematic variations in the proportions of all nutrient ions. Much information can be gained about the relative rate of uptake of cations by examining the responses of plants which are grown in nutrient solutions of a given composition in which, for instance, potassium and calcium are supplied in different proportions with a constant sum of potassium plus calcium in milliequivalents of the supplied salts (de Wit, Dijkshoorn and Noggle 1963). Thus we may find out by plant analysis which of the two is absorbed more readily, and to what extent these ions compete in the uptake system. Similar tests for the anions can be made to complete the picture of the selectivity of the species.

Pot experiments with soil can be used to further study the relation between yield and the ionic balance. If only a vlaue for (C–A) is required, it can be obtained somewhat more simply by a determination of the ash-alkalinity and the nitrate content of the plant material. There is close agreement between the value for ash-alkalinity minus nitrate, and the (C–A) value calculated from analysis for the separate ions (van Tuil, Lampe and Dijkshoorn 1964). In addition, the ash can also be analysed for the individual cations and for phosphate.

For the appropriate information about nitrogen status, the plant material should be analysed with a modified Kjeldahl method to include all the nitrate. When nitrogen is expressed as its equivalent of nitrate, subtraction of the value for the nitrate content will give the organic nitrogen content in me. of the parent ion, which is the same as milligram atoms of nitrogen.

In cases where there are sufficient marginal treatments with reduced growth of plants, a graph representing yields as ordinates and the (C–A) values as abscissae may show that the points scatter in a region which may be represented as an optimum curve, the possible deviations in (C–A) being smaller the higher the yield. If the optimum is sharp, its position indicates the normal value for (C–A). If it is flat, either none of the treatments had the maximum possible growth and some other factor consequently has influenced the growth, or the normal (C–A) is less defined and conforms to a certain range of possible values.

It does not necessarily follow that the relation between (C–A) and yield will always be such that the meaning of higher and lower than normal will be firmly fixed. But the 'normal' carboxylate content, even when only approximately known, provides a convenient reference point in the investigation of unbalanced nutrition. In conclusion, an example is given.

All plants when grown with ammonium have lower values for (C–A) than when nitrate is the source of nitrogen. Most cultivated plants will make more growth when ammonium is replaced by nitrate, the latter form of nitrogen being required to shift (C–A) to the normal value. With nitrate, and alkaline media, (C–A) can sometimes be raised above the normal value, and this condition, known as 'lime-induced chlorosis', has been found to give rise to an increase in the carboxylate content (Rhoads and Wallace 1960).

In the Ericaceous plants the mere substitution of nitrate for ammonium evokes chlorosis. In blueberry, the total cation content was increased from 0·5–0·8 me. per gram dried leaves. Nitrate appeared less injurious when added as nitric acid (Cain 1952, 1954), or when ammonium was also given (Holmes 1960). At higher pH values the plants make poor growth even with ammonium ions present in the media (Oertli 1963). It is common experience that the Ericaceae make good growth when ammonium salts are present in the cultures and are ill adapted to all conditions known to increase the carboxylate content (Addoms and Mounce 1931; Bailey 1936; Stuart 1947; Cain and Holley 1955; Colgrove and Roberts 1956).

Thus substitution of nitrate for ammonium may raise the carboxylate content in one group of plants to the normal value required for maximum growth, in the other group to an excessive value, detrimental to the growth. In this and other aspects, the ionic balance seems capable of describing the nutrient habits in terms of the direction and extent of the shift of the carboxylate content (C–A) from its normal value, and of arranging the data in an orderly and logical fashion.

REFERENCES

Addoms R.M. and Mounce F.C. (1931). Notes on the nutrient requirements and the histology of the cranberry with special reference to mycorrhiza. *Pl. Physiol.* **6**, 653–666.

Bailey J.S. (1963). A chlorosis of cultivated blueberry. *Proc. Amer. Soc. hortic. Sci.* **34**, 395–396.

Böning K. and Böning-Seubert E. (1932). Wasserstoffionenkonzentration und Pufferung im Pressaft von Tabackblättern in ihrer Abhängigkeit von der Ernährung und Entwicklung der Pflanze. *Bioch. Z.* **247**, 35–67.

Boresch K. (1939). Weitere Untersuchungen der durch Chloride hervorgeruffenen Blattrandkrankheit der Johannisbeere. *Z. Pflanzenern. Düng. Bodenk.* **14**, 230–247.

Burg P.F.L. van (1966). Nitrate as an indicator of the nitrogen nutrition status of grass. *Proc. Xth Intern. Grassl. Congr. Helsinki*, 267–272.

Cain J.C. (1952). A comparison of ammonium and nitrate nitrogen for blueberry. *Proc. Amer. Soc. hortic. Sci.* **59**, 161–166.

Cain J.C. (1954). Blueberry chlorosis in relation to leaf pH and mineral composition. *Proc. Amer. Soc. hortic. Sci.* **64**, 61–70.

Cain J.C. and Holley R.W. (1955). A comparison of chlorotic and green blueberry leaf tissue with respect to free amino acids and basic cation contents. *Proc. Amer. Soc. hortic. Sci.* **65**, 49–53.

Colgrove M.S. and Roberts A.N. (1956). Growth of Azalea as influenced by ammonium and nitrate nitrogen. *Proc. Amer. Soc. hortic. Sci.* **68**, 522–536.

Cooil J.B. (1948). Effects of potassium deficiency and excess in guayule. II. Cation-anion balance in the leaves. *Pl. Physiol.* **23**, 403–424.

Dijkshoorn W. (1959). The rate of uptake of chloride, phosphate and sulphate in perennial ryegrass. *Neth. J. Agric. Sci.* **7**, 194–201.

Dijkshoorn W. (1962). Metabolic regulation of the alkaline effect of nitrate utilization in plants. *Nature*, **194**, 165–167.

Dijkshoorn W. (1963). The balance of uptake, utilization and accumulation of the major elements in grass. *Proc. 1st Regional Conference I.P.I. Wexford (Ireland)*, 43–62.

Dijkshoorn W. (1964). Le bilan ionique dans le diagnostic foliaire. *Jaarboek I.B.S. Wageningen* 1964, 133–144.

Dijkshoorn W. and Lampe J.E.M. (1961). Phosphorus fractions in perennial ryegrass. *Jaarboek I.B.S. Wageningen* 1961, 101–106.

Dijkshoorn W. and Wijk A.L. van (1967). The sulphur requirements of plants as evidenced by the sulphur: nitrogen ratio in the organic matter. A review of published data. *Plant and Soil*, **26**, 129–157.

Dijkshoorn W., Lathwell D.J. and Wit C.T. de (1968): Temporal changes in carboxylate content with stepwise change in nutrition. *Plant and Soil*. (In press.)

Holmes R.S. (1960). Effect of phosphorus and pH on ion chlorosis of the blueberry in water cultures. *Soil Sci.* **90**, 374–379.

Jolivot M.E. (1959). Variations des acides organiques dans le tubercule de semence de pomme de terre au cours de sa conservation hivernale et après plantation. *C.R. Acad. Sci. Paris*, **248**, 3208–3210.

Kirkby E.A. and DeKock P.C. (1965). Influence of age on the cation-anion balance in the leaves of Brussels sprouts. *Z. Pflanzenern. Düng. Bodenk.* **111**, 197–203.

Oertli J.J. (1963). Effect of form of nitrogen and pH on growth of blueberry plants. *Agron. J.* **55**, 305–307.

Peel A.J. and Weatherley P.E. (1959). Composition of sieve tube sap. *Nature*, **184**, 1955–1966.

RHOADS W.A. and WALLACE A. (1960). Possible involvement of dark-fixation of carbon dioxide in lime-induced chlorosis. *Soil. Sci.* **89**, 248–256.

RICHARDS F.J. (1956). Some aspects of potassium deficiency in plants. *Potassium Symposium I.P.I. Berne*, 59–73.

STUART N.W. (1947). Some studies on Azalea nutrition. *Nat. Hortic. Magazine*, **26**, 210–214.

TROMP J. (1962). Interaction in the absorption of ammonium, potassium and sodium ions by wheat roots. *Acta bot. neerl.* **11**, 147–192.

TUIL H.D.W. VAN (1965). Organic salts in plants in relation to nutrition and growth. *Agric. Res. Reports Wageningen*, **657**, pp. 1–83.

TUIL H.D.W. VAN, LAMPE J.E.M. and DIJKSHOORN W. (1964). The possibility of relating the ash-alkalinity to the organic salt content. *Jaarboek I.B.S. Wageningen*, 1964, 157–160.

ULRICH A. (1941). Metabolism of non-volatile organic acids in excised barley roots as related to cation-anion balance during salt accumulation. *Am. J. Bot.* **28**, 526–537.

VICKERY H.B. (1961). Chemical investigation of the tobacco plant. XI: Composition of the green leaf in relation to position on the stalk. *Connect. Agric. Res. Sta. Res. Bull.* 640.

WALLACE A., ASHCROFT R.S. and LUNT O.R. (1967). Day-night periodicity of exudation in detopped tobacco. *Pl. Physiol.* **42**, 238–242.

WARD G.M. (1964). Greenhouse tomato nutrition—a growth analysis study. *Plant and Soil*, **21**, 125–133.

WIT C.T. DE, DIJKSHOORN W. and NOGGLE J.C. (1963). Ionic balance and growth of plants. *Versl. Landbouwk. Onderz. Wageningen* 69. **15**, pp. 1–68.

ION UPTAKE AND IONIC BALANCE IN PLANTS IN RELATION TO THE FORM OF NITROGEN NUTRITION

E. A. KIRKBY
School of Agricultural Sciences,
University of Leeds

INTRODUCTION

It is axiomatic that during the process of ion uptake by plants electrochemical neutrality be maintained both in the plant and the nutrient medium in which the plant is growing. As plants absorb nutrient inorganic ions at different rates this equilibrium is largely controlled by an overall effect of H^+ or HCO_3^- excretion from the plant roots into the nutrient medium, and within the plant by the accumulation of organic anions. The acids of the tricarboxylic acid cycle accumulating as organic anions in cell vacuoles have been considered in this role by a number of authors. (Ulrich 1941; Jacobson and Ordin 1954; Chouteau 1960; Coic, Lesaint and Le Roux 1961; and others.) More recently the similar function of the uronic acids in cells walls has also been taken into account (Kirkby and Mengel 1967).

From short-term ion uptake studies by young root tissues from solutions of non-nitrogen containing simple salts, it has been concluded that when cations are taken up more readily than anions, stoichiometric quantities of organic acids are accumulated equivalent to the excess cations over anions absorbed. Similarly when anion uptake exceeds cation uptake much of the ion compensation results from a decrease in organic acid anions in the plant tissue and in particular the concentration of malate (Ulrich 1941; Jacobson and Ordin 1954).

In experiments carried out in this laboratory, however, in which the ionic balance of different tissues of the tomato plant were considered in relation to the form of N-nutrition of the plant, an apparently very different pattern emerged between ion uptake and organic anion accumulation. In the tissues of NO_3-fed plants where anion uptake exceeded cation uptake, very much greater concentrations of organic acid anions were present than in corresponding tissues of NH_4-plants where cation absorption was about 5 times greater than anion absorption (Kirkby and Mengel 1967). From these data it would appear that the presence of nitrogen containing ions in the nutrient medium considerably modifies the relationship between ion uptake and organic anion accumulation as found in the short term experiments discussed earlier. The present paper therefore considers ion uptake under the more normal situation in which whole plants are growing for longer periods in complete nutrient solutions and especially the relationship between ion uptake, the form of N-nutrition and organic anion accumulation in the plant. The relevance of this work is presented in relation to field situations in which plants are supplied with NO_3 or NH_4-nitrogen.

METHODS

Growth and harvesting of the plants

All plants discussed in this paper were grown in water culture for periods of 2–4 weeks. This was carried out using aerated nutrient solutions held in 7 black polythene containers, the plants being supported in 2 cm thick polystyrene sheets. Plants were usually grown in solutions where nitrogen was supplied either in the form of nitrate or ammonium. In one experiment urea was used as a N-source. The solutions employed were of the following type (me./l).

NO_3-solution		Urea-solution		NH_4-solution	
$Ca(NO_3)_2$	5·0	$CO(NH_2)_2$	5·0	$(NH_4)_2SO_4$	5·0
KH_2PO_4	2·0	KH_2PO_4	2·0	KH_2PO_4	2·0
$MgSO_4$	1·5	$MgSO_4$	1·5	$MgSO_4$	1·5
		$CaSO_4$	5·0	$CaSO_4$	5·0

The variable anion in the medium was thus sulphate, as this is most probably least likely to affect or be affected by the uptake of other ions (Mengel 1961). The micronutrients were added as described in an earlier publication (Kirkby and Mengel 1967).

At the beginning of this experiment the pH of all the solutions was increased to 5·5 with dilute $Ca(OH)_2$ solution. During the growth period the pH in all treatments was adjusted back to 5·5 every 2 or 3 days with $Ca(OH)_2$ solution or 0·02 N. H_2SO_4. Every 14 days the nutrient solutions were completely renewed. During the growth occasional qualitative checks were also made on the nitrate content in the solutions of the ammonium and urea series and the ammonium of the urea treatment. As these proved to be negative it is perhaps legitimate to assume that nitrogen was absorbed in the forms supplied.

In some experiments plants were grown using large quantities of circulating dilute solutions (0·4 me. N/l). This was done in order to simulate soil solution conditions and to give a better control of variation of pH in the nutrient medium. The apparatus used is shown in Fig. 1. Detailed technique is described elsewhere (Kirkby 1968).

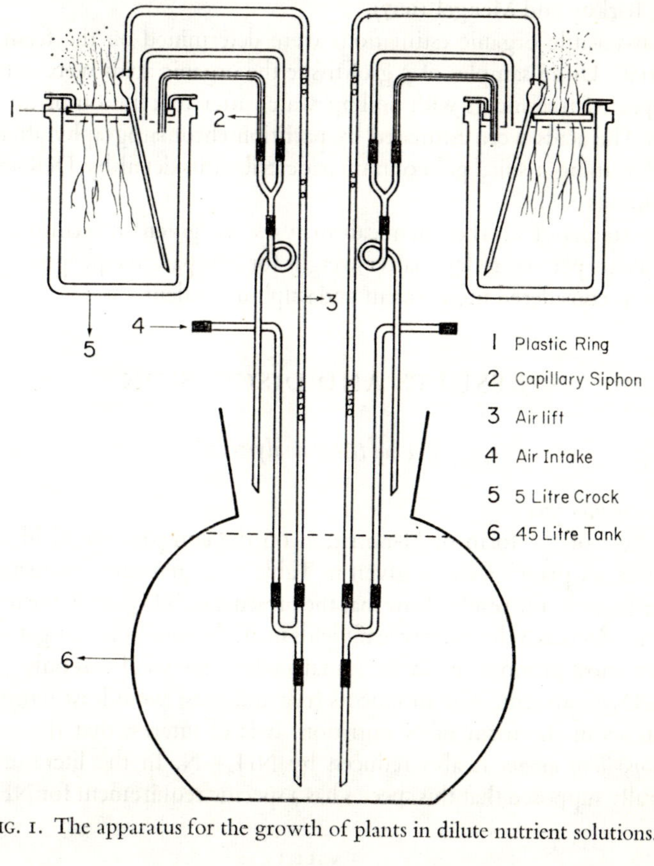

FIG. 1. The apparatus for the growth of plants in dilute nutrient solutions.

At harvesting plants were divided up into different tissues, replicates weighed and then bulked for chemical analysis. Weighed samples obtained as soon as possible after harvesting were stored in polythene bags at −15° C. The remaining weighed fresh material was dried at 85° C to constant weight and ground to a fine powder using a micro-hammer mill.

Chemical analysis

Chemical estimations were made on both the fresh and dried plant material. The dried material was used for the estimation of total nitrogen, total sulphur, chloride, total phosphorus, nitrate, inorganic sulphur, inorganic phosphorus, calcium, magnesium, potassium, sodium, oxalate and uronic acids. The methods employed are discussed elsewhere (Kirkby and DeKock 1965; Kirkby and Mengel 1967).

Non-volatile organic estimations were determined on the fresh plant material. Using samples of 5 g of tissue the organic acids were extracted by repeated treatments with boiling water after maceration of the plant tissue. The acids were estimated by partition chromatography after their isolation using a silica gel column as described in detail by DeKock and Morrison (1958).

The results of all the chemical analyses are given in terms of milli-equivalents per 100 g dry matter weight for comparative purposes. Phosphorus is considered monovalent and sulphur divalent.

RESULTS AND DISCUSSION

Leaf tissue analysis

1. *Dry matter yields*

The effect of the form of N-nutrition on the comparative yields of the leaves of six plant species is given in Table 1. In general agreement with earlier findings the results show that the presence of NH_4—N in the nutrient medium decreases dry matter leaf yields in all the species investigated. The effect is most pronounced in the tomato where the yield was only 32% of the NO_3-treatment. The monocots (rye and oats) were least sensitive to variations in the form of N-nutrition. It is of interest that the yield of *Chenopodium album* is also reduced by NH_4—N. In the literature it is generally supposed that this species has a specific requirement for NH_4—N

TABLE 1

The decrease in yield of the leaves of various plant species caused by NH_4-nutrition (results expressed as % of yield of leaves of NO_3-fed plants)

Tomato	Mustard	Rye	Oats	*Chenopodium album*	Buckwheat
32	51	84	81	50	51

and is unable to utilize NO_3—N. A more detailed discussion of this investigation is considered in another publication (Kirkby 1967). The general reduction in yield resulting from NH_4—N nutrition is shown clearly in Fig. 2 where the dry matter yields of leaves from buckwheat and mustard plants are plotted against the NH_4/NO_3 ratio in the nutrient medium. Such observations have also been made by Welte and Werner (1962) for maize, rape and oats growing in sand culture.

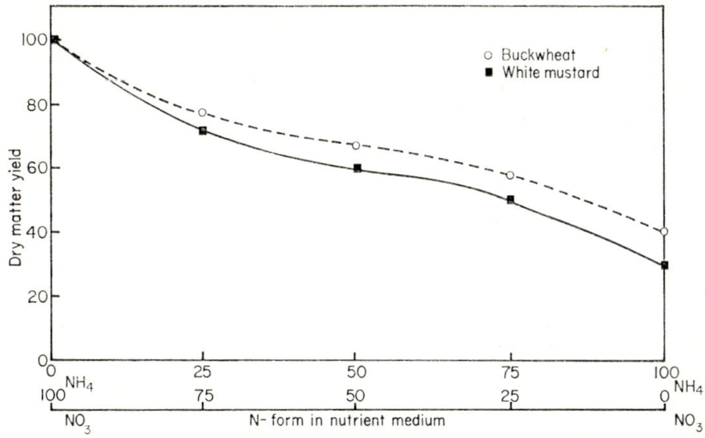

FIG. 2. The influence of the ratio of nitrate to ammonium N on the dry matter yields of White Mustard and Buckwheat (Sand culture experiment).

2. *Non-volatile organic acids*

Table 2 shows the influence of the source of N-nutrition on the non-volatile organic acids in the leaves of four of the plant species already considered. In all cases the concentration of total acids is much lower in the NH_4-plants. This is in agreement with earlier findings (Clark 1936; Chouteau 1960; Coic, Lesaint and Le Roux 1961; Kirkby and Mengel 1967 and others). The difference is particularly marked in tomato where the concentration of acids is about 8 times lower than in the leaves of the NO_3-plants. Rye is least sensitive to differences in the form of N-nutrition. In the case of tomato, malic, citric and oxalic acids are present in highest amounts whereas in white mustard and rye, oxalic is absent, and aconitic accounts for a small fraction of the total acids. In buckwheat leaves, oxalic is the predominant acid present.

The non-volatile organic acids are sufficiently strong (pK 1·5–5·5) to

TABLE 2

The influence of the form of N-nutrition on the non-volatile organic acids in the leaves of various plant species (me/kg fresh wt)

Plant Species	N-source	Fumaric	Succinic	Malonic	Aconitic	Malic	Citric	Oxalic	Total
Tomato	NO_3	1·2	1·2	1·8	—	130·2	47·4	60·6	242·4
	NH_4	0·5	0·8	1·1	—	5·9	10·6	11·8	30·7
Mustard*	NO_3	1·8	15·3	—	21·0	141·6	72·5	—	253·2
	NH_4	3·7	5·5	—	26·5	26·9	18·6	—	84·0
Rye	NO_3	1·3	2·4	—	42·7	20·2	11·3	—	77·9
	NH_4	0·8	1·3	—	21·8	8·4	7·6	—	39·9
Buckwheat	NO_3	1·2	1·0	—	—	20·2	8·6	132·4	163·4
	NH_4	0·3	0·6	—	—	3·5	2·3	32·3	39·0

* Dilute nutrient solution used. 25% of N as NO_3—N in NH_4 treatment

occur almost wholly as salts at the pH of the plant sap for most plant species. The accumulation of these acids can therefore be considered in relation to the ionic balance in the plant.

3. *Ionic balance*

As has already been described in detail, ionic balance in plant tissues is largely maintained by inorganic cations on the one hand and by inorganic and organic anions on the other. The influence of the form of N-nutrition on the inorganic cation content of five of the plant species already considered in relation to dry matter yields are given in Table 3. In agreement with

TABLE 3

The influence of the form of N-nutrition on the inorganic cations in the leaves of various plant species (me./100 g dry wt)

Plant Species	N-source	Ca	Mg	K	Na	Total
Tomato	NO_3	161	30	58	19	268
	NH_4	62	25	29	15	131
Mustard	NO_3	107	28	81	5	221
	NH_4	72	22	59	7	160
Rye	NO_3	50	18	86	1	155
	NH_4	31	14	80	1	126
Chenopodium album	NO_3	112	76	62	18	268
	NH_4	42	22	40	14	118

earlier workers (Arnon 1939; Chouteau 1960; Coic, Lesaint and Le Roux 1961; Werner and Welte 1962) the content of inorganic cations is lower for NH_4—N in all the species, the levels each of the cations Ca, Mg and K being affected. Again as for the non-volatile organic acids, rye is less sensitive to differences in the form of N-nutrition.

The effect on the ionic balance of the different species is demonstrated in Fig. 3. Here organic anion fractions have been determined by alkalinity of the ash estimations. As is shown in both figures irrespective of the form of N-nutrition, a fairly close balance is obtained for cation and anion totals for each treatment.

A more detailed examination of the ionic balance in tomato leaves in relation to NO_3, urea or NH_4-nutrition is given in Table 4. The cation and organic anion results follow the same pattern as already discussed (Tables 2, 3,

TABLE 4

The influence of the form of N-nutrition on the cation-anion balance in tomato leaves (me./100 g dry wt)

N-source	CATIONS	ANIONS							
	Total	NO_3^-	$H_2PO_4^-$	SO_4^-	Cl^-	TCA acids	Oxalic	Uronic	Total
Nitrate	268	4	13	22	12	117	41	44	253
Urea	210	—	15	31	23	41	25	40	175
Ammonium	131	—	15	35	14	11	8	46	129

Fig. 3). NH_4-nutrition reduces the content of total organic anions considerably. Examining the organic anion constituents more closely it can be seen that both the non-volatile TCA cycle acids and oxalate are affected. Uronic acid values, however, show little difference between treatments. This is understandable as it is less likely that the cell wall constituents are influenced by the form of N-nutrition. The inorganic anion concentrations vary little between treatments which infers that a higher proportion of inorganic cations in the NH_4-treatment are bound in inorganic form. In all cases urea is intermediate between the other two forms of N-nutrition.

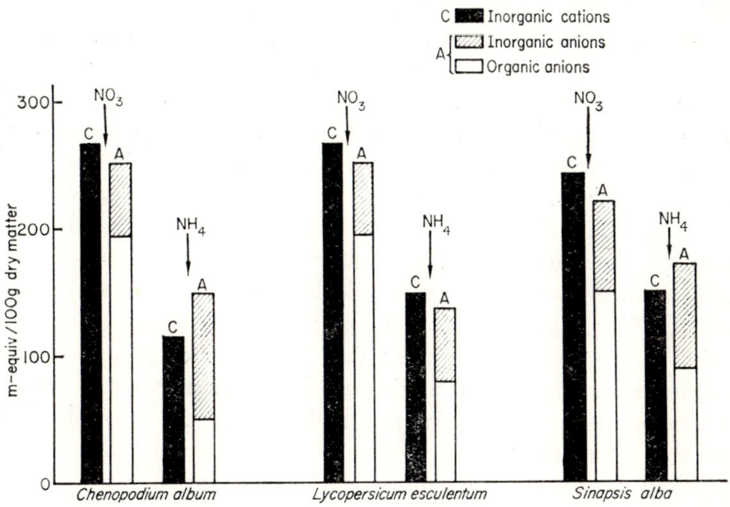

FIG. 3. Cation-anion balance in three plant species in relation to the form of N-nutrition.

The results for the leaf tissues considered show that quite apart from the form of N-nutrition, total cations are fairly well balanced by total anions. Discrepancies between cation and anion totals do occur but may be accounted for either by chemical errors or the omission of charges such as proteins, amino acids, phosphorylated compounds, organic bases and hydrogen ions from the balance. The result, however, gives very good support to the concept of a cation-anion balance in the leaf maintained largely by the diffusible the indiffusible ions which were determined.

The form of N-nutrition exerts a considerable effect on dry matter yields and levels of ionic constituents in the leaf tissues in all the species examined. In comparison with NO_3-nutrition, the yields of leaves from plants supplied with NH_4—N are smaller and contain lower contents of cations and organic acid anions. From the results presented in this paper and the findings of a number of other workers it would appear that the effects of the different N-forms are general for most plant species. As there is a very important interaction between growth and ion uptake it may be supposed that the differences observed could have resulted from concentration or dilution effects caused by variations in growth between treatments. Such effects, however, should be similar for cation and anion constituents. As the results showed, this was not the case so that the effects of the two N-forms must be considered in relation to ion uptake by the plant.

In quantitative studies in ion uptake in which N-assimilation is to be investigated in relation to organic ion accumulation, total ion uptake must be considered, so that a complete ionic balance sheet can be drawn up for the whole plant. For this purpose the results of ion uptake and assimilation by whole tomato plants (leaves, petioles, stems and roots) are discussed for plants grown with NO_3, urea or NH_4 as a sole N-source. This has enabled comparative balance sheets to be considered for plants supplied with nitrogen, the nutrient required in greatest quantities by the plant as an anion, molecule or cation. These balance sheets which are discussed for each treatment are similar to those calculated by Van Tuil (1965) from the results of leaf analysis by Cooil (1948) for guayule, and Coic (1961) for maize. As was pointed out by Van Tuil, however, such balance sheets for leaf tissues are complicated particularly by the uncertainty of the form in which some nutrients move into the leaf. Data for the whole plant is probably presented for the first time in the present paper as is certainly the comparison between the three forms of N-nutrition.

Ion uptake and ionic balance by whole plants

In considering the effects of the form of N-nutrition on both the ionic balance of the whole plant and in the nutrient medium it must be remembered that in comparison with other nutrients the plant has a very high inherent requirement for nitrogen. This means that for plants supplied with NO_3—N high amounts of anions must be taken up by the plant whereas with NH_4—N nutrition high quantities of cations must be absorbed. Moreover as neither NO_3^- nor NH_4^+ remain in ionic form to any extent in the plant and are largely assimilated to non-charged organic-N compounds their charges must be transferred to other metabolites. These effects are now considered in detail in relation to ion uptake by three N-forms.

1. *Nitrate nutrition*

A balance sheet for ion uptake and assimilation and organic anion accumulation for whole tomato plants grown in complete nutrient solution with NO_3 as the N-source is given in Table 4. Inorganic anion uptake (A_u) exceeds inorganic cation uptake (C_u) by 10 me. per 10 plants. This implies either that 10 me. H^+ are taken up or that 10 me. $OH^-(HCO_3^-)$ are excreted by the plant root into nutrient medium. This is in accordance with the increase in alkalinity of the pH of the nutrient medium over the growing period (Fig. 4) which is generally observed for NO_3-nutrition.

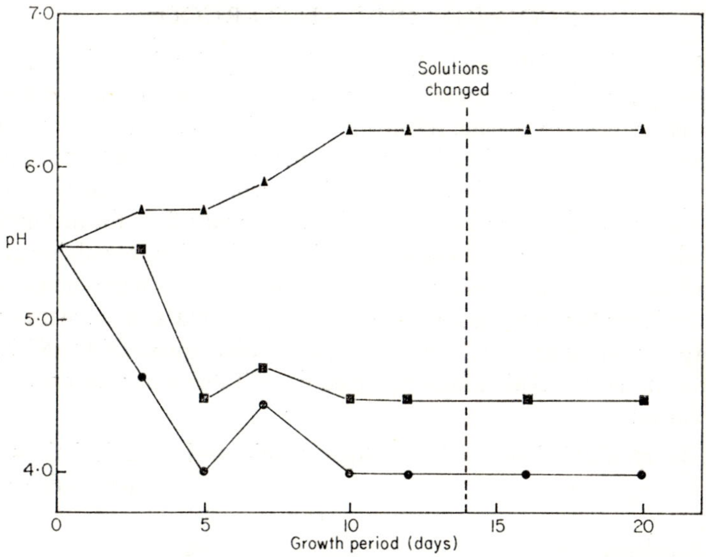

FIG. 4. pH changes in the nutrient media during the growth of tomato plants (pH adjusted back to 5·5 after every determination). (▲) NO_3, (■) Urea, (●) NH_4.

As has been considered by Dijkshoorn (1962), the process of NO_3^--reduction and the production of large quantities of uncharged organic N compounds may lead to an alkalization effect within the plant with the production of strongly OH^- anions

$$NO_3^- + 8H \rightarrow NH_3 + 2H_2O + OH^-$$

A similar but quantitatively less important process must also hold for sulphate reduction

$$SO_4^{--} + 8H \rightarrow SH_2 + 2H_2O + 2OH^-$$

A fraction of the OH^- produced in this way may be returned to the nutrient medium as already mentioned in exchange for excess anion uptake by the plant. Charges remaining within the plant must be transferred to stable metabolites. As already described this role is filled by organic acid anions and includes anions of the tricarboxylic cycle acids, oxalic acid and uronic acids. In the normal functioning of the tricarboxylic acid cycle there is no reason why any one member of the cycle should accumulate and as has been shown, the concentrations of these acid anions in this plant vary between species. Organic anion accumulation may take place either by the neutralizing effect of OH^-

$$RCOO^-[H^+ + OH^-] \rightarrow H_2O + RCOO^-$$

or via the effect of OH^- on CO_2 fixation into organic acids as for example by phosphoenolpyruvate carboxylase to form oxalacetate which could then be reduced to malate. Whatever mechanism is involved, however, electrons should be transferred from metabolized NO_3^- and SO_4^{--} to either $OH^-(HCO_3^-)$ and passed into the nutrient medium or to organic anions and accumulated as such within the plant. Organic anions formed in this way should be balanced along with the inorganic anions remaining after assimilation by the inorganic cation uptake. Also the sum of the equivalents of assimilated N and S (organic N and organic S) should be equal to the sum of organic anions accumulated and the excess anion uptake $OH^-(HCO_3^-)$ excretion. These relationships extracted from Table 5 may be expressed as follows:

1. Inorganic Cation Uptake C_u = Organic Anions + Inorganic Anions Accumulated

		130	+	42
me./10 plants	$\underline{\underline{174}}$		$\underline{\underline{172}}$	

2. Organic N + Organic S = Organic Anions + $OH^-(HCO_3^-)$ Excretion (excess anion uptake)

me./10 plants				
	138 + 4	130	+	19
	$\underline{\underline{142}}$		$\underline{\underline{140}}$	

The results give striking support to Dijkshoorn's concept of NO_3-assimilation in relation to ion uptake and organic anion accumulation. It seems likely therefore that this type of mechanism is operative in the tomato plant and probably in other higher plants. This would account for excess anion uptake in association with high contents of cations and organic

TABLE 5

The ionic balance sheet of tomato plants grown with nitrate as the souurce of nitrogen (me./10 plants)

Uptake				Utilization			Accumulation					
Anions		Cations		Anions		Cations	Anions		Cations		Organic anions found	
N	147	K	65	N_o	138		NO_3	9			TCA and oxalic acids	97
Cl	11	Na	11				Cl	11				
P	13	Mg	21				H_2PO_4	13			Uronic acids	33
S	13	Ca	77	S_o	4		SO_4	9				
A_u	184	C_u	174	OH^-	142		A	42	C	174		
			10	OH^-	10		Inorg. salts			42		
					132	⟶	Org. salts			132		130

A_u=anion uptake N_o=organic N A=inorganic anions
C_u=cation uptake S_o=organic S C=inorganic cations

acid anions in plant tissues. In some plant species with acid saps and with high H^+ where free acids are present the situation will be modified in that inorganic cation uptake will be lower than in the situation previously considered.

The question of whether the charge from NO_3^- is passed to an organic anion or out into the nutrient medium is as HCO_3^- is of particular interest in relation to the cation/anion uptake ratio and hence the whole mineral nutrition of the plant. From the work of Bear (1950) it was often held that for a given plant species the cation/anion uptake ratio

$$\frac{Ca+Mg+K+Na}{N+S+P+Cl} = \text{constant}$$

This would imply that the ratio of the OH^- directed towards organic anion accumulation to that of HCO_3^- excretion is constant for a given species. More recent evidence of Dijkshoorn (1958) suggests however that this is not the case, younger plants and plants supplied with high levels of nitrate showing lower cation/anion uptake ratios and hence more predominant excretion of HCO_3^-. The evidence that the cation/anion uptake ratio of plants supplied with NO_3—N is invariably less than unity (see Walker 1960) along with the usual observations of increases in pH in the nutrient medium with NO_3-nutrition in water culture (Fig. 4) would also tend to contradict the widely held view that the root surfaces of plants are acid.

The level of NO_3^- reduction, the site at which the reduction takes place as well as the factors influencing the reduction must all play a part controlling ion uptake. If reduction takes place in the root as is supposedly the case in apple trees and asparagus (Bonner 1950), then there should be a greater possibility for exchange of OH^- for NO_3^- uptake to predominate. In this case the cation/anion uptake ratio should be low. This explanation may possibly account for the low cation/anion uptake ratios in the grasses and cereals in the order of 0·5 and the fact that relatively low concentrations of organic acid anions are accumulated as was found in rye. For tomato plants where the ratio is higher 0·95, HCO_3^- excretion must play a relatively minor role, organic anion accumulation and cation uptake being much more a predominant. In this case reduction should probably be more prominent in the upper parts of the plant (Bonner 1950).

The site at which NO_3^- reduction takes place may also influence cation translocation, for if NO_3^- is reduced in the leaf then it must be translocated to the site of reduction with an inorganic cation. Here calcium may play an important role for it moves in the transpiration stream and is accumulated in leaf tissues. On reduction of NO_3^- the charge could pass directly to an organic anion. This could well account for the often almost stoichiometric relationship between calcium and oxalate (Olsen 1939) or calcium and malate (Kirkby and DeKock 1965) found in leaf tissues. Experiments are at present in progress in our laboratory to investigate the relationships between NO_3^- and SO_4^{--} reduction and cation translocation.

2. *Urea nutrition*

The balance sheet for urea nutrition is shown in Table 6. In this case nitrogen is taken up by the plant as a molecule and does not contribute to the ionic balance. Cation uptake (C_u) exceeds anion uptake (A_u) by 41 me. per 10 plants. Again this is in accordance with the decrease in pH in the nutrient medium (Fig. 4). This acidity results from H^+ exchange of OH^- uptake for excess cation removal by the plant. As there is no charge transfer in urea assimilation the only alkalization effect is caused by sulphate reduction. Organic anions are accumulated largely in response to excess cation uptake. For urea nutrition the following equations extracted from Table 6 should hold:

1.	Inorganic Cation Uptake	=	Organic Anions Accumulation	+	Inorganic Anions Accumulation
	C_u		43	+	30
me./10 plants	78			73	

2.	Organic S + H^+ excretion Excess Cation Uptake	=	Organic Anion Accumulation
me./10 plants	7 + 41		
	48		43

The closeness of these experimental results to the theoretical values again supports the concepts discussed.

3. *Ammonium nutrition*

Table 7 gives the balance sheet data for NH_4^+-nutrition. In this case cation uptake exceeds anion uptake by 75 me. per 10 plants. This again is in agreement with the increase in acidity in the nutrient medium as H^+ exchange into the medium for excess cation uptake (Fig. 4).

The assimilation of NH_4^+ results in an acidification within the plant.

$$NH_4^+ \rightarrow NH_3 + H^+$$

H^+ produced in this way must to a larger extent be returned to the nutrient medium to balance the higher uptake of cations by the plant. Here the

TABLE 6

The ionic balance sheet of tomato plants grown with urea as the source of nitrogen (me./10 plants)

Uptake		Utilization		Accumulation		
Anions	Cations	Anions	Cations	Anions	Cations	Organic anions found
Cl 13	K 29			Cl 13		TCA and oxalic acids 26
P 9	Na 6			H_2PO_4 9		
S 15	Mg 10	S_o 7		SO_4 8		Uronic acids 17
	Ca 33					
A_u 37	C_u 78	OH^- 7		A 30	C 78	
OH^- 41 →		OH^- 41		Inorg. salts	30	
		48 →		Org. salts	48	43

A_u=anion uptake N_o=organic N A=inorganic anions
C_u=cation uptake S_o=organic S C=inorganic cations

TABLE 7

The ionic balance sheet of tomato plants grown with ammonium as the source of nitrogen (me./10 plants)

Uptake				Utilization				Accumulation				
Anions		Cations		Anions		Cations		Anions		Cations	Organic anions found	
Cl	5	N	61			N_0	61	Cl	5		TCA and oxalic acids	4
P	5	K	10					H_2PO_4	5			
S	6	Na	4	S_0	1			SO_4	5			
		Mg	4									
		Ca	12									
A_u	16	C_u	91	OH^-	1	H^+	61	A	15	C 30	Uronic acids	11
						OH^-	1	Inorg. salts		15		
						H^+	60					
OH^- ——→						OH^-	75					
							15 →	Org. salts		15		15

A_u = anion uptake N_0 = organic N A = inorganic anions
C_u = cation uptake S_0 = organic S C = inorganic cations

tendency to accumulate organic anions should be much lower because of a general reduction in pH in the plant tissue which should tend to depress the dissociation of the organic acids, and a largely non metabolically controlled supply of NH_3 should result in a general depletion of carbohydrates. This could account for the much lower organic acid anion and cation contents in plants supplied with NH_4—N and the higher percentage of cations associated with inorganic anions. Differences in the cation contents of leaf tissues supplied with NH_4—N as compared with NO_3—N may possibly be accentuated where NO_3^- reduction is taking place in the leaf. If NO_3^- reduction occurs in the root as has been discussed already lower concentrations of organic anions and cations are more likely to be accumulated in the leaf and hence differences between the two N-sources will be less marked. This may possibly account for the results for rye where the differences between NO_3 and NH_4-nutrition are not so great as the other species investigated.

It is also possible that the lower content of inorganic cations in NH_4-fed plants may result from a restricted uptake because of their competition with NH_4^+ ions or H^+ ions resulting as a consequence of NH_4^+ assimilation. There appears, however, to be no evidence for competition between NH_4^+ and any of the inorganic cations for specific carrier sites.

In the case of NH_4-nutrition the following equations extracted from Table 7 should hold:

1. Inorganic Cation Uptake = Organic Anion + Inorganic Cations Accumulation

$$\underline{\underline{30}} \qquad = \qquad 15 + 15 = \underline{\underline{30}}$$

2. Organic S + H^+ excretion − Organic N = Organic Anion Accumulation
   (Excess cation uptake)

$$1 + 75 - 61 = 15$$
$$\underline{\underline{15}} \qquad \underline{\underline{15}}$$

The results confirm the general concepts discussed.

The data discussed for whole tomato plants grown in NO_3 urea or NH_4-nutrition shows that it is possible to demonstrate direct relationships between N and S assimilation, organic anion accumulation and ion uptake. The significance of this should be important in studies concerning ion uptake by whole plants. The concept of a cation-anion balance is also important in understanding competitive effects between cations or between anions.

Such a balance implies that different cations compete for the bulk of anions taken up and vice versa. This probably explains why the increase of one ion species in the nutrient medium decreases the uptake of a similarly charged ion species where there is no specific competition for a carrier site.

Agronomic and ecological aspects

Ammonium sulphate, ammonium nitrate, sodium nitrate and urea are commonly used solid nitrogen fertilizers in this country so that the three forms discussed may all be introduced to agricultural soils. Both urea and NH_4—N, however, are normally rapidly transformed to NO_3—N and under most conditions this is the predominant source available to the plant. Where nitrification is inhibited, however, as for example by low soil temperatures or acid soil conditions where the nitrifying bacteria are not active then NH_4—N may be taken up by plants in large quantities. Some time ago the Dow Chemical Company manufactured a pyridine derivative 'N serve' which inhibits nitrification, and if applied with NH_4^+ fertilizers could be expected to result in a more efficient utilization of nitrogen by preventing the loss of the easily leached NO_3^- ion. Under conditions where plants take up NH_4—N, however, reduction in growth may be expected particularly by dicotyledenous species as well as a reduction in the uptake of inorganic cations. This may be of particular significance in magnesium nutrition where high applications of ammonium and potassium salts may induce magnesium deficiency (Van Itallie 1937, Mulder 1956). With the present tendency of increasing crop yields, the use of highly purified fertilizers and the omission of magnesium salts as a regular fertilizer dressing, an increase in the occurrence of magnesium deficiency may be expected particularly on light acid soils which are heavily fertilized with potassium and ammonium salts.

For soil conditions where fertilizer additions are not made then the extreme cases of nitrate and ammonium nutrition may be considered in relation to soils with a mull or mor humus. For the mull humus soil, usually of higher pH and well aerated, the abundant nitrogen form available is NO_3—N. In the mor humus soil, low pH conditions and poor aeration must lead to a predominance of NH_4—N in the soil profile. This difference between 'mull' and 'mor' may give greater variations in cation uptake than would be expected from the estimation of available mineral cations in the soil, for not only is a less abundant supply of cations available for growth in the mor soil but also inorganic cation availability to the plant will be reduced by NH_4—N. Moreover the predominant supply of NH_4—N in the soil profile will accentuate the already high leaching of mineral cations and in particular that of calcium. The form of nitrogen nutrition may therefore play a very important role in the selection of plant species growing in a given soil and may be of greater significance than is generally recognized.

REFERENCES

ARNON D.I. (1939). Effect of ammonium and nitrate nitrogen on the mineral and sap characteristics of barley. *Soil Sci.* **48**, 295–307.

BEAR F.E. (1950). Cation and anion relationships in plants and their bearing on crop quality. *J. Amer. Soc. Agron.* **42**, 176–178.

BONNER J. (1950). *Plant Biochemistry*, New York.

CHOUTEAU J. (1960). Les equilibres acides-bases dans le tabac Paraguay. *Thesis, Bordeaux.*

CLARK H.E. (1936). Effect of ammonium and nitrate nitrogen on the composition of the tomato plant. *Plant Physiol.* **11**, 5–24.

COIC Y., LESAINT C. and LE ROUX F. (1961). Comparaison de l'influence de la nutrition nitrique et ammoniacale combinée ou non avec une deficience en acide phosphorique, sur l'absorption et le metabolisme des anions-cations et plus particulièrement des acides organiques chez le mais. Comparaison du mais et de la tomate quant a l'effet de la nature de l'alimentation azotée. *Ann. Physiol. Veg.* **3**, 141–163.

COOIL J.B. (1948). Effects of potassium deficiency or excess in guayule. II. Cation-anion balance in the leaves. *Plant Physiol.* **23**, 403–424.

DEKOCK P.C. and MORRISON R.I. (1958). The metabolism of chlorotic leaves. 2. Organic Acids. *Biochem. J.* **70**, 272–277.

DIJKSHOORN W. (1958). Nitrate accumulation, nitrogen balance and cation-anion ratio during the regrowth of perennial rye-grass. *Neth. J. Agric. Sci.* **6**, 211–221.

DIJKSHOORN W. (1962). Metabolic regulation of the alkaline effect of nitrate utilization in plants. *Nature*, **194**, 165–167.

JACOBSON L. and ORDIN L. (1954). Organic acid metabolism and ion absorption in roots. *Plant Physiol.* **29**, 70–75.

KIRKBY E.A. and DEKOCK P.C. (1965). The influence of age on the cation-anion balance in the leaves of Brussels sprouts (*Brassica oleracea* var. *Geminifera*). *Zt. Pflanzenernähr Düng Bodenk.* **111**, 197–203.

KIRKBY E.A. and MENGEL K. (1967). Ionic balance in different tissues of the tomato plant in relation to nitrate, urea or ammonium nutrition. *Plant Physiol.* **42**, 6–14.

KIRKBY E.A. (1967). A note on the utilization of nitrate, urea and ammonium nitrogen by *Chenopodium album*. *Zt. für Pflanzenernähr und Bodenk.* **117**, 204–209.

KIRKBY E.A. (1968). Influence of ammonium and nitrate nutrition on the cation-anion balance and nitrogen and carbohydrate metabolism of white mustard plants grown in dilute nutrient solutions. *Soil Sci.* **105**, 133–141.

MENGEL K. (1961). *Ernährung und Stoffwechsel der Pflanze*, Gustav Fischer Verlag. Jena, p. 209.

MULDER E.G. (1956). Nitrogen-magnesium relationships in crop plants. *Plant and Soil*, **7**, 341–376.

OLSEN C. (1939). Absorption of calcium and formation of oxalic acid in higher green plants. *Compt. rend. Lab. Carlsberg. Ser. Chim.* **23**, No. 8, 101–123.

ULRICH A. (1941). Metabolism of non-volatile organic acids in excised barley roots as related to cation-anion balance during salt accumulation. *Amer. J. Bot.* **28**, 526–537.

VAN ITALLIE TH.B. (1937). Magnesiummangel und Ionverhältnisse in Getriedepflanzen. *Bodenk. u Pflanzenernähr*, **5**, 303–334.

VAN TUIL H.D.W. (1965). Organic salts in plants in relation to nutrition and growth. *Thesis, Wageningen.*

WALKER T.W. (1960). Uptake of ions by plants growing in soil. *Soil Sci.* **89**, 328–332.

WELTE E. and WERNER W. (1962). Ionenaustauscherversuche über die Beeinflussung der Kationenaufnahme der Pflanzen durch die Stickstoff—Form. *Agrochemica*, **6**, 337–348.

INTRA-SPECIFIC VARIATION IN MINERAL NUTRITION OF PLANTS FROM DIFFERENT HABITATS

P. J. GOODMAN
Welsh Plant Breeding Station, Plas Gogerddan,
Nr. Aberystwyth, Cards.

INTRODUCTION

Herbage plants in different places vary in size, morphology and sometimes in physiological characters. If these characters are retained when plants from the various habitats are grown together, the plants are called ecotypes (Turesson 1922). Such ecotypes have become adapted to their specialized habitats by natural selection. Ecotypes are not usually abruptly separated in space, but grade into one another, in a succession, or cline in response to some continuously variable factor of the environment, such as altitude, climate or soil (Simpson *et al.* 1959).

Edaphic ecotypes varying in their tolerance of heavy metals have been studied extensively by Bradshaw *et al.* (1965). Jowett (1964) found that such tolerance in *Agrostis tenuis* populations around mine workings was rather specific to the metal of the immediate substratum. The practicability of breeding for tolerance, in this case to salt, was suggested by Dewey (1962), working on *Agropyron desertorum*. As far as I am aware, this suggestion has seldom been implemented.

Antonovics *et al.* (1967) have more recently shown that ecotypes varying in their response to mineral nutrients are often found on soils differing in fertility. If this phenomenon is a general one, it holds considerable promise for agriculture. Thus it may be possible to select among edaphic ecotypes those which yield most for each unit of fertilizer applied. Such variations in fertilizer response are already known among cultivated varieties of many crop species (Vose 1963), and the ability to respond strongly to fertilizers is an important factor in yield.

Another character which may vary between ecotypes is their ability to take up nutrients. Snaydon and Bradshaw (1961) showed that the ability of *Festuca ovina* races to take up calcium from low concentrations explained their success on acid soils. Mineral content also affects the nutritional value of herbage plants to the farm animal. Thus Butler and Johns (1961) found differences in the sodium, calcium, manganese, aluminium, copper, iron, zinc, nitrate, sulphate and acid soluble phosphorus content of seven plants of ryegrass. This species, *Lolium perenne* L., was the one chosen for my experiments, as it is important in agriculture. The wide habitat range and variability of *L. perenne* (Beddows 1967) also make it very suitable as a model for studying the mechanism and speed of adaptation in ecological conditions.

Natural selection may occur at any stage of a plant's growth. The seedlings of each generation show many types of recombination, but the extremes are eliminated by soil conditions. This means that mature plants are more adapted to their habitat than seedlings. For this reason, most workers have preferred to use clonal material rather than seedlings, but Crossley (1963) using seedlings and tillers showed that the two sources of material gave similar results, differing only in the range of variation. In screening material for breeding programmes, however, the restriction of the amount of genetic variation caused by using clones would give an unrealistic assessment of the agricultural value of the material. For this reason, I have restricted my work to studying seedling populations.

NUTRIENT ADAPTATION IN NATIVE POPULATIONS

Populations of *L. perenne* were collected, a few tillers at a time, from soils with the widest range that could be found for nitrogen, phosphorus, potassium and sodium content (Table 1). Collections in natural habitats were supplemented by six samples taken from the Rothamsted Park Grass experiment in which the plots, on what was originally an old pasture, have been treated with different fertilizers consistently since 1856 (Warren and Johnston 1964). To prevent any 'carry over' effects from the parent plants, the tillers were grown for several months in John Innes No. 1 compost. The flowering heads were bagged in pairs and their seed was collected in summer 1966.

After ripening, the seed populations were tested for their responses to a range of nutrient levels, by sowing them in pots of low nutrient status soil (Table 1), with factorial nutrient additions. Two experiments were carried out. In the first experiment, with first generation seed of twenty populations, statistical confounding had to be used as the material was scarce. The second experiment used second generation seed of twelve populations. Both experiments had a $4\times2\times2$ ($N\times P\times K$) design, with two replications.

TABLE 1

Sources of material

Population	Treatment	Soil analysis		
Park Grass plots		Kjeldahl nitrogen (%)	Soluble phosphorus (ppm)	Soluble potassium (ppm)
7D*	P, K, Na, Mg	0·23	100	670
8D*	P, Na, Mg	0·24	140	80
12D*	Unmanured	0·28	5	80
17B*	Nitrate N, lime	0·26	10	60
17D*	Nitrate N	0·26	10	60
20(2)*	N, P, K, manure	0·26	5	70
Location	**Habitat**			
Aber Falls, Caerns.*	Mountain stream	0·17	0·7	39
Afon Dulyn, Caerns.*	Mountainside	0·24	0·8	69
Barton, Beds.	Chalk downland	0·32	19	230
Borth, Cards.	Raised bog edge	1·19	4	223
Bridgeham, Norfolk	Sandy gravel heath	0·08	160	80
Catsholme, Norfolk	New black fen	1·17	16	342
Ely, Cambs.†	Old black fen	1·71	67	38
Gedney, Lincs.	New silt fen	0·14	6	234
Holme, Hunts.*	Old black fen	1·63	19	736
Puffin Island*	Bird island	Not available	Not available	Not available
Rothamsted, (Fosters)†	Flinty clay field	0·19	190	220
St. Neots, Hunts.*	Water meadow	0·73	21	428
Thetford, Norfolk†	Sandy gravel heath	0·09	22	7
Wansford, Hunts.‡	Limestone meadow	0·08	180	150
Wigtoft, Lincs.	Old silt fen	0·15	31	163
Ynyslas, Cards.*	Sand dunes	0·03	5	0·03
Aberystwyth, Cards.	Cultivar S.24	Not available	Not available	Not available
Potting soil (clay loam + 20% sand)		0·18	1·3	69

* Seeds used in two ecotype experiments.
† Soils used in microplot experiments.
‡ Seed and soil used.

Nutrient solutions, modified from Shive and Robbins (1942), were mixed into the soil at the beginning of each experiment. Nitrogen was added as sodium nitrate. Ionic concentrations were kept constant by substitution of other ions to balance the varying nutrient concentrations. Sodium was used to balance potassium, so that antagonistic effects of these elements could not be detected. The first experiment used 1·5 kg soil, with the nutrient additions listed in Table 2. The second experiment used 1 kg soil, so that the effective concentrations were proportionally stronger.

TABLE 2

Nutrient treatments in the first experiment (ppm in soil)

	Nutrient		
Level	Nitrogen	Phosphorus	Potassium
0	0	0	0
1	40	120	160
2	100	—	—
3	200	—	—

Total leaf length was measured for each pot at fortnightly intervals. This variate was used as a non-destructive measure of yield instead of dry weight as plants were scarce. Total leaf length was very highly correlated with leaf area (correlation coefficient $r=0{\cdot}93^{***}$) and with total plant dry weight (correlation coefficient $r=0{\cdot}90^{***}$). The first and second experiments were harvested for total plant dry matter after 5 weeks and 11 weeks respectively.

Since the results of the two experiments were closely similar, they will be considered together. The populations differed in yield consistently throughout the experiment. The leaf lengths of the three highest-yielding, the two lowest-yielding, and one intermediate-yielding population in the second experiment, are shown in Fig. 1. The highest yielding populations were Aberystwyth cultivar S.24, and two populations from the Rothamsted Park Grass experiment, all of which received annual applications of phosphorus and potassium. The lowest-yielding populations were the unmanured plot of the Park Grass experiment, the Afon Dulyn mountain population, and the sand dune population. The intermediate-yielding population from the bird island apparently rich in nutrients, is

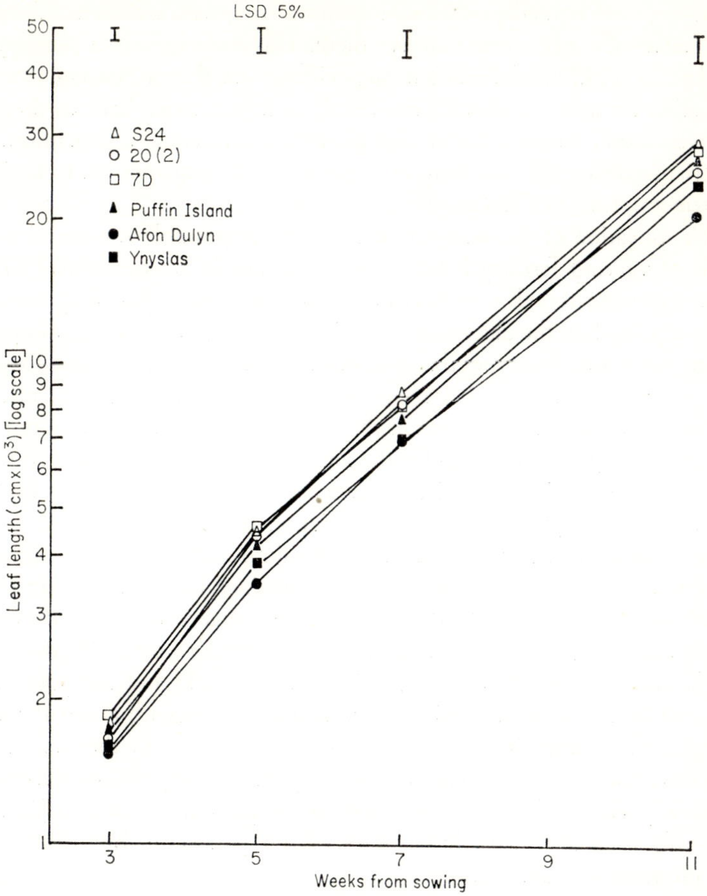

FIG. 1. Leaf length growth of ecotypes (160 per ecotype) in the second ecotype experiment.

interesting as it exceeded the yield of S.24 when it was tested in the experiments of Antonovics *et al.* (1967), and in our first, though not in the second experiment. Thus the high-yielding and intermediate-yielding populations came from habitats well supplied with phosphorus and potassium, while the low-yielding populations came from soils of low nutrient status.

A general relationship thus exists, that nutrient-rich habitats are populated by high-yielding ecotypes. This agrees with Turesson's (1922) observation that unadapted plants introduced to specialized habitats assume the habits of the ecotypes that live there. This convergence of phenotype and genotype in each habitat is strong evidence that there is constant natural selection for adapted plants. Presumably in nutrient-rich habitats there is active natural selection for strong growth to overcome competition. In nutrient-starved habitats there may equally well be adaptation to limited nutrient uptake, and less growth.

Identification of the nutrient or nutrients responsible for selection in soils of varying fertility is not easy. For example, nutrients tend to be plentiful or scarce all together, rather than separately, and they interact strongly, both in the native soils, and in the testing pots. The first experiment, with more ecotypes than the second experiment, was found to give the stronger correlations between ecotype responses and soil nutrients.

Soil nutrient concentrations in the ecotype habitats were measured by Warren and Johnston (1964) for the Park Grass soils, and by the National Agricultural Advisory Service, Trawscoed, using standard methods, for two other soils (Table 1). For convenience, the nutrient concentrations were transformed to the N.A.A.S. 'potassium and phosphorus indices'. The largest correlation coefficients were obtained between yield and soil nutrient status by using leaf length as the measure of yield. Between yield and soil nitrogen, the correlation coefficient $r=0{\cdot}35$ NS. Between yield and soil phosphorus, $r=0{\cdot}50^{*}$, while for yield and soil potassium $r=0{\cdot}33$ NS. The most significant correlation coefficient, $r=0{\cdot}58^{**}$ was obtained by comparing the yield with the least value of either soil phosphorus or soil potassium. This strongly suggests that populations have been naturally selected for yield in response to soil phosphorus content (Fig. 2), possibly associated with low values of soil potassium, but having no relationship with soil nitrogen.

The effects of nutrients on leaf length during the experiments were consistent with those on shoot dry weight at the end of the experiments. Phosphorus application increased yield significantly ($P<0{\cdot}001$), while the effect of nitrogen was small and non-significant, and that of potassium was even smaller.

The absence of a significant nitrogen effect was largely explained by the significant interaction of phosphorus with nitrogen. At the higher level of phosphorus, nitrogen increased yield, but at the lower level, yield was unaffected by nitrogen (Table 3). Phosphorus was thus a limiting factor in these experiments, suppressing the response to nitrogen to a variable extent. Clearly the experiment would have given more information on nitrogen effects if a third and higher level of phosphorus had been included.

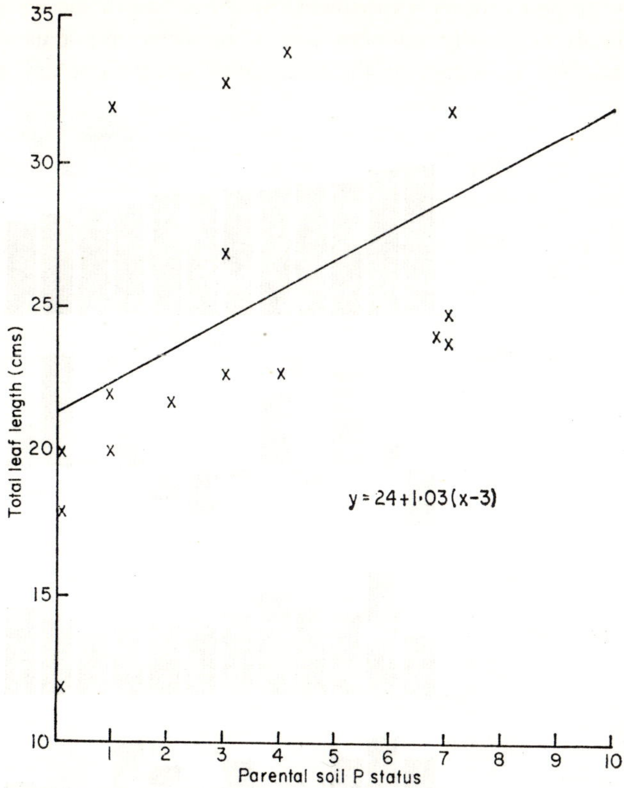

FIG. 2. Dependence of the total leaf length of F_1 seedlings in the first ecotype experiment (16 per ecotype) on the parental soil phosphorus index.

TABLE 3

Effect of nitrogen and phosphorus on Mean Shoot dry weight (grams). Mean of all populations

	N_0	N_1	N_2	N_3	Mean
P_0	343	323	344	332	335
P_1	455	466	459	499	470
Mean	398	394	401	416	403

S.E.(N) = 7·6. S.E.(P) = 5·4. S.E.(N × P) = 10·7.

In the second ecotype experiment, total leaf length during growth showed only marginally significant interactions between populations and nutrient levels. However, at the final harvests for shoot dry weight,

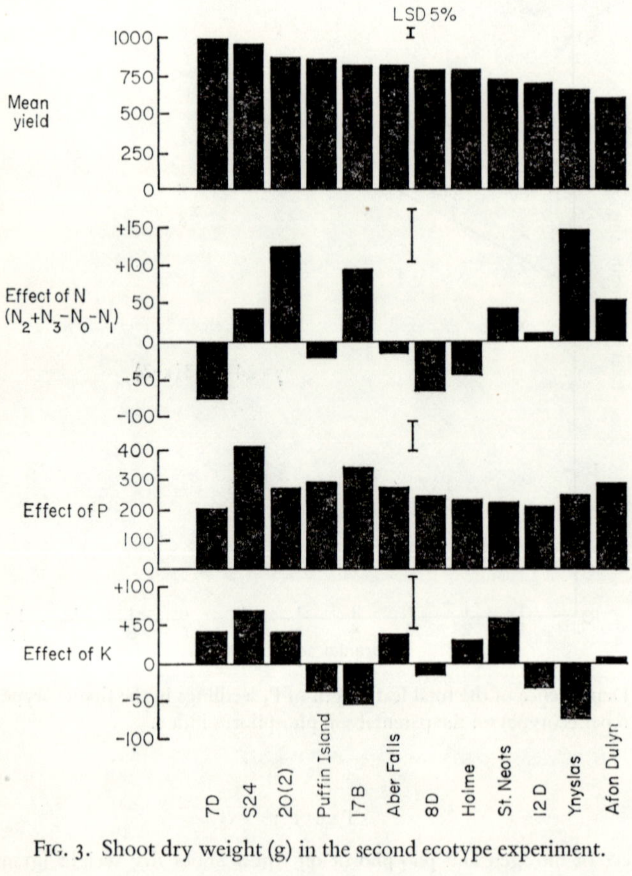

FIG. 3. Shoot dry weight (g) in the second ecotype experiment.

significant ($P<0{\cdot}001$) effects were found for population interactions with phosphorus, nitrogen and potassium (Fig. 3).

Phosphorus had such a large positive effect on all populations that the interactions never became negative. When nitrogen and potassium interacted with populations, however, both positive and negative effects were found (Fig. 3). The negative nitrogen effects are attributed to the limiting effect of phosphorus, while the negative potassium effects may be caused by the antagonism of sodium in this experimental design. It is interesting to note that the two obviously maritime populations, from Puffin Island and Ynyslas, responded to sodium. In general, however, there were no significant correlations between the responses of the populations to nutrients, and the soils from which the populations came. This lack of correlation may have resulted from interactions between nutrients either in their selection effects, or in the pots of test soils. Correlated responses have been found by other workers, for example between calcium and phosphorus and lead (Jowett 1964), and a similar situation seems likely to apply to the present ecotypes.

In summary, the material from natural habitats showed a wide range of yields and of responses to several nutrients. This range resulted from natural selection by habitat factors, which we cannot define precisely. Although the populations used in the experiments were chosen from habitats of long standing, their precise history was not known and thus it was of interest to determine how many generations were needed to effect a detectable change in the ecotype characters. Also the distance required to separate ecotypes is unknown, and this too, needed investigation.

RATE OF CHANGE IN POPULATIONS

Using a novel technique, Crossley (1963) investigated the speed of change in populations of cultivated varieties of *Dactylis glomerata* and *Lolium perenne*. Cultivars of these grasses are not homozygous, which exactly reproduces the ecological situation, in which unadapted seed arrives in a habitat. Selection occurs during seedling growth. Crossley (1963) reported a very rapid rate of adaptation, such that differences were found between the performance of progenies after only one generation of seed multiplication on chalk soil, compared with the performance of the original seed stock.

Crossley's technique has been extended to a group of soils, in a 4×4 Latin square of microplots, sown with perennial ryegrass, Aberystwyth S.24. The plots were open-ended upright boxes, 0·8 m^2 and 0·6 m deep, of paving slabs set on edge. The four soils used were a black fen (Ely), a flinty clay (Rothamsted), a sandy gravel (Thetford), and a loam-over-Oolite (Wansford) (Table 1). They were fertilized with 10·5 g K_2HPO_4, 5 g K_2SO_4 and 25 g $(NH_4)_2SO_4$. The seed sowing rate was deliberately heavy (71 g/m^2), to encourage competition between plants. The plots were sown in October 1966, and 0·4 m^2 in the centre of each plot was harvested for seed in July 1967. Progenies from the different plots were then compared with the original seed stock using the same methods as in the second ecotype experiment (i.e. a 4 × 2 × 2 (N × P × K) factorial in exhausted soil). The pot experiment began in August, 1967, and lasted for 6 weeks.

Progeny yields were consistently smaller than parent stock yields (Fig. 4). This large difference probably resulted from the small amount of fertilizer added, and from the short growing period given to the plants before seeds were collected. Of more importance than the difference in size between the generations was the fact that the progenies differed among themselves. Ely and Thetford progenies yielded significantly more than Rothamsted and Wansford progenies (Fig. 4). The reason for this particular grouping is obscure, since Thetford soil was nutrient-starved and Ely rich in nitrogen only (Table 1).

In this type of experiment, there is always the possibility that nutrients may be 'carried over' from the parent plants to the progeny seeds. According to Austin (1966), phosphorus supply can affect the growth of progeny by 'carry over'. For this reason, the seeds were analysed for phosphorus, but there was no obvious correlation between seed phosphorus contents and the seedling yields. Germination tests on the progeny seeds also failed to show differences in vigour or viability. Similarly, seed size, though it differed among the progenies, was not consistently related to performance (Table 4).

Despite the uncertainty as to the cause of the differences in progeny yields, they had an important feature in common with the ecotype yield differences. The different yields were associated with significantly different population × phosphorus interactions. This considerably increases the interest of the experiment. It is remarkable that these population differences were induced in an out-breeding species after only 1 year's growth while separated from other populations by less than 1 yard (0·9 m).

The rapidity of the change in population character seen in this experiment needs further investigation. In particular, we need to repeat the experiment over a number of generations to find whether natural selection of populations continues equally strongly after the first year. Further field investigations will then be needed in as many different habitats as possible, for example, on heavy metal, saline, acid and calcareous soils. The effect may also differ between species depending, for example, on whether they are annual or perennial. Presumably unadapted genotypes would be more quickly eliminated from an annual than from a perennial species.

As remarkable as the rate of change of populations is the small distance needed between them, since pollen must have spread over all plots, and the

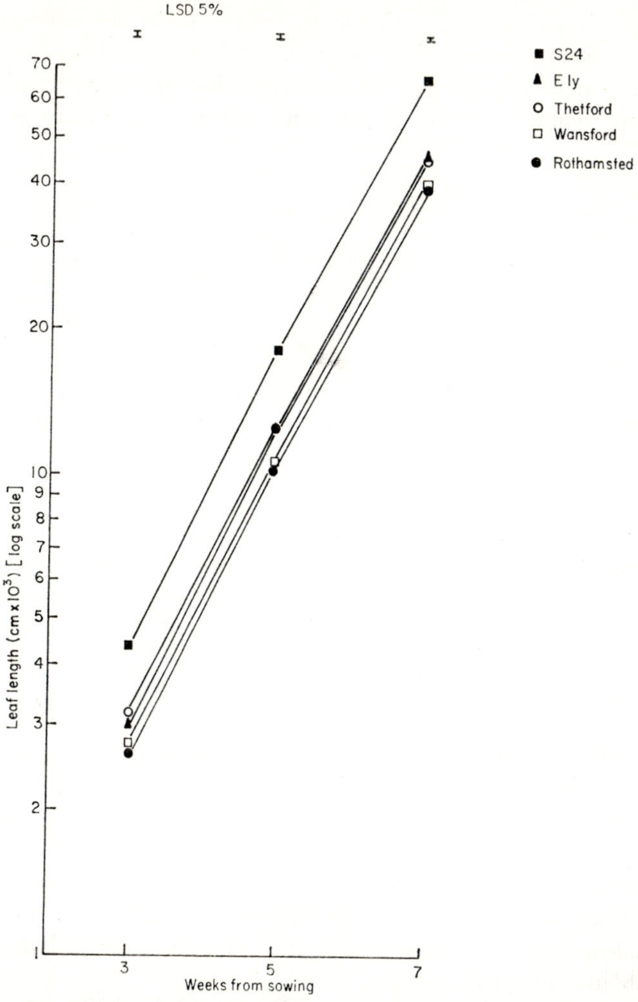

FIG. 4. Leaf length growth of F_1 Seedlings (640 per plot) in the microplot experiment.

effect is therefore likely to be largely due to the maternal genotype. Such effects are not unknown, since Jain and Bradshaw (1966) and Snaydon (1963) have also found different genotypes less than 50 cm apart. Those workers used tiller material, however, which would have been further selected by the soil after germination. The present results, using seed, are both more dramatic and agriculturally more important.

The evidence of this experiment, in support of Crossley's (1963) results, suggests that natural selection must act very rapidly and over very small distances, in herbage species. They must therefore have considerable ability to adapt to new environments, and they must have very many ecological races.

TABLE 4

Characters of seeds raised on different soils

Soil	Progeny yield (g)	Parent yield (g)	Seed characters: Mean weight (g)	%P	%Germination
Thetford	3586	274	1·61	0·175	93
Ely	3542	526	1·26	0·182	96
Wansford	3131	151	1·29	0·159	89
Rothamsted	2826	511	1·39	0·187	92

UPTAKE DIFFERENCES

Although the most important differences between populations are those directly affecting yield, these differences are correlated with, and actually caused by, physiological and biochemical differences within the plant. For example, Snaydon (1962) showed that different populations of *Trifolium repens* not only yielded differently, but differed in the expression of 'lime-induced chlorosis'. Their ability to take up calcium at low levels in the soil was all-important for growth, thus establishing a causative link between metabolism and yield.

The small amount of ecotype material available has limited the number of physiological investigations we have been able to do. The percentage nitrogen content of the twenty populations of the first ecotype experiment showed no significant differences. Nitrogen uptake was proportional to dry matter yield, and consequently showed significant differences, the largest yielding ecotypes containing most nitrogen.

Uptake of phosphorus has been studied in a separate experiment, using tillers of S.24, Ynyslas and Puffin Island first generation progenies grown in low-nutrient-status soil, with additions of nutrients (modified Shive and

Robbins 1942), including either 0·163 g KH_2PO_4 per 300 g soil, or no phosphorus. Eight weeks later, 4 μci ^{32}P and 0·026 NaH_2PO_4 solution were introduced through a glass tube below the roots. Leaves were harvested 20 hr and 53 hr later, dried, ignited, and precipitated as phosphomolybdate (Piper 1950). After filtration, the precipitates were dried, weighed and counted with a Panax end-window Geiger counter.

The populations did not differ significantly in dry matter yield, percentage phosphorus content or phosphorus uptake as measured by chemical analysis (Table 5), though phosphorus treatment had a significant effect on

TABLE 5

Phosphorus uptake by different populations.

Population	Treatment	Mean leaf dry weight (g)	% P	P yield (mg)	Specific activity 10^2 dpm/g (dw)
S.24	P_0	1·12	0·73	8·4	47·8
	P_1	2·33	1·14	26·6	29·6
	Mean	1·74	0·94	17·5	38·7
Puffin Island	P_0	1·78	1·40	24·9	34·4
	P_1	2·40	1·00	24·0	13·7
	Mean	2·09	1·20	24·5	24·0
Ynyslas	P_0	1·60	1·52	24·3	29·2
	P_1	3·03	1·13	34·2	7·6
	Mean	2·32	1·33	29·3	18·4
	S.E. (pops)	0·85 NS	0·31 NS	12·6 NS	4·4**

all these variates. By contrast, the specific radioactivity of the leaves differed significantly between populations, between exposure times and between phosphorus treatments.

This experiment showed that small numbers of plants from different populations could be indistinguishable by conventional growth analysis and nutrient testing methods. Differences could still be found, however, in the uptake of a radioisotope. These differences obviously arose in the plants at maturity since the total previous unlabelled phosphorus uptake of plants showed a different pattern from that of the radioisotope. The isotope differences could have been caused by variations in the size of the root system which would have led to the interception of more isotopically labelled phosphate or by differences in the rate of phosphorus uptake.

There is the further possibility that the mechanism of uptake in the different genotypes is particularly efficient at certain phosphorus concentrations, as was found by Snaydon and Bradshaw (1962) for phosphate in *T. repens*. Whichever explanation of phosphorus uptake differences is correct, the largest yielding ecotypes of *L. perenne* have the greatest capacity to absorb phosphate, and this must be an important factor in their success.

Furthermore, preliminary experiments suggest that the large yield of the Puffin Island ecotype is related to a faster rate of photophosphorylation than that of the Afon Dulyn mountain population. In addition to differences in uptake of phosphorus, there are thus variations between populations in phosphorus use within the plant (Treharne, K.J. unpublished data).

In contrast to photophosphorylation it appears that nitrate reduction is more efficient in the Afon Dulyn population than in the Puffin Island population. Thus in a number of biochemical characters there are variations between populations, comparable to the variations in yield.

The ecological success and agricultural value of herbage grasses must therefore depend on many different morphological, physiological and biochemical characters which need measurement. As more systems are investigated, so more biochemical reasons for large yield will be found. While these are individually important, as in maize (Hageman *et al.* (1967)) it seems likely that the factor of overriding significance will be the integration of the many enzyme systems.

DISCUSSION

A wide diversity of response to mineral nutrition has been found in *L. perenne*. Ecotypes exist which show virtually every combination of responses to nitrogen, phosphorus and potassium. Previous workers have similarly found in *L. perenne* differential response to many other nutrients (Butler and Johns 1961; Crossley 1963; Vose 1963; Antonovics *et al.* 1967).

Nutrient response has often been found by other workers to have strong correlations with the soil analysis (Antonovics *et al.* 1966; Snaydon 1963), but in such cases it is apparent that soil factors such as heavy metal and salt contamination or pH were overwhelming in the environment. Obviously the populations had been highly stressed, and there was a strong unidirectional selection for tolerance. Even in heavy metal tolerance, however, there were indications that *Agrostis* was adapted to low calcium as well

(Jowett 1964), so that the selection is not entirely specific. In my experiments there was no strong unidirectional selection, so that responses have been selected to a wide range of nutrients. These may or may not be significantly adaptive. In such more tolerable environments a balance will exist between the production of new genetic variation either by segregation, cross pollination or entry of seeds, and its elimination by soil factors.

A remarkable study by Snaydon (1963) has shown that populations of *Anthoxanthum odoratum*, which is widespread in the Park Grass experiment, differed in many characters. These varied from morphological and physiological characters, such as panicle height and posture and pH response, to disease susceptibility. This suggests that even extremes of soil pH can cause a wide range of plant characters to be selected simultaneously. If a single habitat factor can be identified as important in the present experiments it is probably shown by the correlation between leaf length of seedlings and the soil phosphorus concentration in the parental habitat. Since many other workers have reported similarly, it seems that phosphorus is the most important major nutrient in ecological situations. Whitehead (1966) reported that the phosphorus content of herbage from many Welsh grasslands was very small. Thus the range of variation in soil phosphorus may account for the wide range of ecotypes found in Welsh habitats (Antonovics *et al.* 1967; Crossley 1963). Just as a range of plant characters may be selected by one habitat factor, so in natural habitats many selection pressures may be applied simultaneously to populations. For example, the pressure of grazing, exposure, drainage and many other factors as well as soil type, may vary. Charles (1964) has shown that artificial mixtures of genotypes of a single species can be unbalanced, or single genotypes eliminated by pressure of management or nitrogen fertilization within 3 years of sowing. Such an effect is caused by an extreme environment altering the conditions of competition greatly in favour of one or another component of a mixture. Thus the natural habitats from which ecotypes were taken may have had strong selective pressures due to factors other than soil type. In particular, the small yielding ecotypes of the mountains may have been selected for their persistence under hard grazing.

It is possible that factors other than soil nutrient status can select plants for nutrient response. This may have happened in the microplot experiment. In that experiment, use was made of the heterogeneity that exists in herbage varieties. Natural selection favoured the strongest competitors on each soil type. In spring, before flowering, the populations varied in density on the different soil types in a way apparently unrelated to the soil nutrient status. It may be that the soil physical condition during winter had a greater selective effect than the soil nutrient status. Nevertheless, after one generation, not only the yield, but also the phosphorus response varied. It seems likely that the differential phosphorus response was a secondary effect of natural selection rather than a primary one. Nevertheless, the populations had acquired different characters in the remarkably short period of one generation.

In selecting for yield, whether in new environments or old ones, plants have to be well adapted to their conditions. To fit the complexity of ecological situations, selection for several characters simultaneously, for example in mineral nutrition, may be more effective than a single selection criterion. This approach is complicated by the physiological interaction between nutrients, such as the suppression of nitrogen response by relative shortage of phosphorus which was found in the present experiments. Obviously, selection is needed at several levels of each of the interacting factors. This is essential, in any case, in selecting for the uptake of nutrients which are important for animal nutrition, but which give no competitive advantage to the plant. By using factorial methods, it should be possible to select desirable crop plants to suit widely differing agricultural situations. The agricultural process of selection closely parallels the events which take place in ecological situations, fitting plants to their habitats. Use of agricultural model systems may thus help in solving ecological problems.

ACKNOWLEDGEMENTS

Professor A.D. Bradshaw and Mrs J. Lovett helped in finding ecotypes in North Wales; Mr G.V. Dyke, Miss J.M. Thurston and Miss M. Ford helped to find material in the Rothamsted Park Grass experiment; Drs R. Hull and K. Scott helped me in the microplot experiment at Broom's Barn Experimental Station, near Bury St. Edmunds; the National Agricultural Advisory Service, Trawscoed, analysed the soils. Professor Bunting and Dr R. Snaydon allowed me to see the latter's manuscript. Mr D.M. Hughes gave much help in the seedling experiments. To all these, and to Professor P.T. Thomas, who suggested the work, and provided the facilities, I am extremely grateful.

REFERENCES

Antonovics J., Lovett J. and Bradshaw A.D. (1966). The evolution of adaptation to nutritional factors in populations of herbage plants. *The use of isotopes in plant nutrition and physiology*. IAEA, Vienna.

AUSTIN R.B. (1966). The influence of the phosphorus and nitrogen nutrition of pea plants on the growth of their progeny. *Plant and Soil*, **24**, 359–368.

BEDDOWS A.R. (1967). *Lolium perenne* L. *J. Ecol.* **55**, 567–587.

BRADSHAW A.D., MCNEILLY J.S. and GREGORY R.P.G. (1965). *Industrialization, evolution and the development of heavy metal tolerance in plants. Ecology and the Industrial Society.* (Ed. G.T. Goodman, R.W. Edwards and J.M. Lambert). Blackwell, Oxford.

BUTLER G.W. and JOHNS A.T. (1961). Some aspects of the chemical composition of pasture herbage in relation to animal production in New Zealand. *J. Austr. Inst. of Agric.* **27**, 123–133.

CHARLES A.H. (1964). Differential survival of plant types in swards. *J. Br. Grassld Soc.* **19**, 198–204.

CROSSLEY G.K. (1963). Variation in the nutritional requirements of natural and bred strains of some British pasture grasses. Ph.D. Thesis, Univ. of Wales.

DEWEY D.R. (1962). Breeding crested wheatgrass for salt tolerance. *Crop Sci.* **2**, 403–407.

HAGEMAN R.H., LENG E.R. and DUDLEY J.W. (1967). A biochemical approach to corn breeding. *Adv. Agron.* **19**, 45–86.

JAIN S.K. and BRADSHAW A.D. (1966). Evolutionary divergence among adjacent plant populations. I. The evidence and its theoretical analysis. *Heredity*, **21**, 407–441.

JOWETT D. (1964). Population studies on lead tolerant *Agrostis tenuis*. *Evolution*, **18**, 70–80.

PIPER C.S. (1950). *Soil and Plant Analysis.* University Press, Adelaide.

SHIVE J.W. and ROBBINS W.R. (1942). Methods of growing plants in solution and sand cultures. *Bull. New Jersey Agric. Exp. Sta.* 636.

SIMPSON G.G., PITTENDRICH C.S. and TIFFANY L.H. (1959). Life. *An introduction to Biology.* Routledge & Paul, London.

SNAYDON R. W. (1963). Final report on investigations of physiological adaptation in ecotypes of *Trifolium repens, Festuca ovina* and other herbage species. *Cyclostyled document, Dept. of Agric. Bot. Univ. of Reading.*

SNAYDON R.W. and BRADSHAW A.D. (1961). Differential response to calcium within the species *Festuca ovina* L. *New Phyt.* **60**, 219–234.

SNAYDON R.W. and BRADSHAW A.D. (1962). Differences between natural populations of *Trifolium repens* in response to mineral nutrients. I. Phosphate. *J. Exp. Bot.* **13**, 422–434.

TURESSON G. (1922). The genotypical response of the plant species to the habitat. *Hereditas, Lund*, **3**, 211.

VOSE P.B. (1963). Varietal differences in plant nutrition. *Herb. Abstr.* **33**, 1–13.

WARREN R.G. and JOHNSTON A.E. (1964). The Park Grass experiment. *Rep. Rothamsted exp. Sta. for* 1963, 240–262.

WHITEHEAD D.C. (1966). *Nutrient minerals in grassland herbage.* Mim. Public. No. 1/1966 C.A.B. Hurley, Berks.

DISCUSSION ON MINERAL NUTRITION OF THE WHOLE PLANT SYSTEM

Recorded by DR R.W. SNAYDON

Dr Rorison considered some of the complicating factors which impede the interpretation of field situations in terms of laboratory findings. These factors can be broadly classified into environmental variability and interaction, and biotic variability and interaction. He drew attention to environmental variability, both in time and space, and touched upon the complicating effects of environmental factor interaction. Biotic variability includes the changing response of the individual plant with time and developmental stage, and intergenotypic variation considered later by Dr Goodman. Biotic interactions include microbial effects (later considered by Dr Barber) as well as competition.

In discussion Dr Rorison stressed that it would be dangerous to extrapolate from studies of seedling response over relatively short time spans to other stages of the life cycle and longer time spans. Dr Woolhouse drew attention to the limitations of single harvest analyses and total analyses in defining nutrient status; such single figures are, as Professor Epstein pointed out, the integration of a whole series of processes occurring both in time and space.

Environmental variability in time and space was discussed with reference to soil phosphorus. Dr Rorison pointed out that the seasonal variation in plant requirement might be important in relation to observed seasonal variation in nutrient availability, and referred to the example of *Scabiosa columbaria* in his paper.

The importance of environmental factor interactions was stressed by Professor Pigott, who pointed out that in *Urtica* response to phosphorus was highly dependent upon the nitrogen level present.

Dr Rorison's paper brings into focus some of the problems in extrapolating from controlled nutrient conditions to field situations and, as he mentioned, L.T. Evans (1963) has already considered this problem from the point of view of climatic factors. The numerous complicating factors probably explain our ignorance of the factors limiting the distribution of many species and it is perhaps surprising, but also encouraging, that Dr Rorison and others have found such good correlations between the response of some species to controlled nutrient conditions and the field response and distribution of these species.

Dr Barber's paper, and the subsequent discussion, highlight our previous disregard of the possibly large effects of microflora upon plant mineral nutrition.

The effects of the microflora upon nutrient uptake is probably dependent upon the species of micro-organisms present. Dr Rovira stated that inoculation with soil suspension led to increased phosphorus uptake in tomato, clover, wheat and pine. This contrasts with the reductions found by Dr Barber using laboratory contaminants and indicates possible adaptive specificity of the rhizosphere microflora in nutrient uptake, but, as Dr Barber states in his paper, the literature is equivocal on this point. Professor Epstein and others questioned the validity of the low levels of phosphorus (0·01 ppm P) used by Dr Barber. Dr Barber agreed that the lowest levels would be transient, but similar effects had been observed at higher phosphorus levels and in short term uptake studies.

Dr Barber stated that the number of micro-organisms in his solutions was low (1×10^5/g) in comparison with that in soil (2×10^9/g). Although the biomass and nutrient content of the microflora population at any one time was small, the turnover rate for nutrients was high in comparison with root tissue. Dr Rovira found that only approximately 10% of the phosphorus in or on plant roots was in microbial form at any one time.

Mr Barley pointed out that the microflora might influence root development and morphology, as well as directly influencing uptake processes; these two effects might be studied by inoculation at various times before the experimental uptake period. Dr Rovira agreed that there was now evidence that the microflora influenced root growth, and even flower development, as well as the uptake, translocation and incorporation of nutrients. Dr Rovira concluded that the observed effects of the microflora upon the plant nutrition were sufficient to warn physiologists, and perhaps to call for reinterpretation of nutrient uptake data obtained in non-sterile conditions. From an ecological point of view it would obviously be interesting to know to what extent species with contrasting edaphic tolerances have different and specific rhizosphere microfloras, and whether such differences contribute to differences in mineral nutrition.

Dr Dijkshoorn was asked by Dr Grubb whether the absence of antagonism between Ca and K was confined to grasses since Olsen (1942) had found none in *Tussilago farfara*. He replied that although there was a general distinction between grasses and dicotyledons, it was not absolute; e.g. antagonism between these two cations was found in ryegrass, when K supply was deficient. Mr Kirkby asked whether cation uptake ever exceeded anion uptake under normal soil conditions, as opposed to ammonium nutrition. Dr Dijkshoorn replied that cation uptake exceeded anion uptake in buckwheat (*Fagopyrum esculentum*) under field conditions, and in other species when N supply was low.

In reply to a question by Professor Weatherley, Dr Dijkshoorn stated that K was apparently involved in the translocation of carboxylate from shoot to root via the phloem, but that its function could, to some extent, be replaced by Na, though the latter appeared to be less efficient. Professor Harley asked whether carboxylate had been detected in the phloem, and which organic anions were associated with K in translocation. Dr Dijkshoorn gave no analytical data but felt that the detection and form of carboxylate groups was less important than whether translocation was rapid enough to satisfy demand. He concluded that, since the requirement for carboxylate was likely to be less than the requirement for these compounds for growth of roots, the translocating system was capable of supplying the demand.

The discussion following Mr Kirkby's paper epitomizes the physiological rather than the ecological approach; most of the discussion emphasized the unifying concepts of the mechanisms involved. Little attention was paid to the obvious differences between species in response to the form of nitrogen supply; it is these differences which are important ecologically.

Dr Grubb asked whether the reduction in cation uptake under ammonium nutrition could be ascribed to the increasing acidity of the culture solutions; and he quoted Olsen who found no reduction in plant yield with NH_4^+ in a solution of pH 5. Both Mr Kirkby and Professor Mengel cited experiments where dry weight yields were depressed by NH_4^+ even when solutions were kept at as constant a pH as possible, and concluded that NH_4^+ had a direct effect; Professor Mengel added that NH_4^+ might be a source of H^+ within the plant, and so influence both yield and cation content. Dr Grubb was not convinced that nitrate reduction occurred solely in the leaves. Mr Kirkby agreed that there was no doubt that some NO_3^- reduction occurred in roots, and tissues of the plant other than the leaves, to a degree which varied with the species. As he had already mentioned, the site of NO_3^- reduction might be closely related to ion uptake and organic acid accumulation within the plant. A discussion centred around whether organic acid production was a byproduct of metabolism or induced by the ionic environment within the plant. Mr Kirkby stressed the close equivalence of total anions and cations in a wide range of plants under contrasting conditions; this implied a causal relationship. Dr Jennings suggested that the related question of whether growth and ion uptake was controlled by organic acid production or vice versa might perhaps be answered by constructing a mathematical model; the limiting reactions might thus be defined. Although the possibility of TCA exports from root to soil was mooted, no mention was made of the widely recorded export and cyclic exchange of inorganic cations by the roots; this would obviously have a bearing upon cation/anion balance in the plant. Dr Rovira doubted that the results obtained in solution culture could be extrapolated to soil conditions; he had found that ammonium was used as efficiently as nitrate in sterile soil, and that nitrification occurred in non-sterile soils at pH 4·5. Dr Woolhouse wondered whether this utilization of NH_4^+ in sterile soil might parallel Hewitt's observation that tomato utilized NH_4^+ if succinate was provided. Ammonium nutrition might impose a strain upon the production of 4-carbon units, which might be relieved from soil sources. Mr Nye felt that the widely held view that roots excreted H^+ should be questioned; Cunningham found that anion uptake exceeded cation uptake in 50 species. This indicated that hydroxyl or bicarbonate ions rather than H^+ would be exported. Dr Scott Russell struck an ecological note by condemning generalization and stressing the wide differences between species, for example, in response to NH_4^+.

It is perhaps indicative of the nascent state of studies of intraspecific variation in mineral nutrition, that technique attracted more attention in the discussion of Dr Goodman's paper than did the nature, mechanisms and ecological significance of intraspecific variation. The relative merits of seed and tiller material were discussed. Dr Rorison expressed concern that tillers previously grown in high nutrient conditions might be buffered against subsequent response to low nutrient conditions. Dr Goodman drew attention to another potential problem in the large cumulative effects of somatic selection of tillers (Hayward and Breese, 1965). On the other hand seed populations might not adequately represent the genetic structure of natural populations, and would exaggerate the genotypic variability present in the populations. Dr Kenworthy asked whether the differences in final plant weight of populations might be due to differences in time of germination, since relative growth rates were identical. Dr Goodman agreed that the differences between populations appeared within three weeks, but the differences were not correlated with seed size, P content or percentage germination. Professor Pigott observed that a range of N or P treatments produced large initial differences in leaf area of *Urtica* which were then maintained. Professor Bradshaw had also observed large initial differences in tiller material; he interpreted this in terms of differences between populations in the rapidity of phenotypic adjustment to contrasting nutrient conditions. Dr Kenworthy asked whether specific differences in rhizosphere microflora might not account for some of the intraspecific differences in mineral nutrition. Dr Goodman replied that intraspecific differences had been observed in isolated leaf systems, but that this did not eliminate the possibility that some differences might also be due to the microflora. Dr Goodman's paper draws attention to the correlation between intraspecific differences in plant size, under uniform conditions, and the nutrient status of the natural habitat; Rorison (this volume) and previous workers have observed similar correlations at the interspecific level. These observations indicate the possible adaptive significance of plant size in relation to soil nutrient status. The absence of correlation in Dr Goodman's data between the response of each population to N and P and the nutrient status of its natural habitat contrasts sharply with the fairly high correlations previously recorded and may, he points out, reflect the effects of nutrient interaction; on the other hand it may be the result of limited population sampling followed by wide genetic recombination within the resulting seedling populations. The ecological significance of intraspecific variation in mineral nutrition has yet to be determined. Dr Rorison (this volume) indicated that this variation may account for the wide ecological amplitude of such species as *Festuca ovina*. Dr Goodman's data on *Lolium perenne* indicates that rapid, apparently adaptive, changes of populations over short distances may allow close adaptation of populations to their immediate environment in time and space.

REFERENCES

Breese E.L., Hayward M.D. and Thomas A.C., (1965). Somatic selection in perennial ryegrass. *Heredity* **20**, 367–379.

Evans L.T. (1963). Extrapolation from controlled environments in the field. *Environmental Control of Plant Growth* (Ed. L.T. Evans) pp. 421–435. Canberra.

Olsen C. (1942). Water culture experiments with higher green plants in nutrient solutions having different concentrations of Ca. *C.R. Lab. Carlsberg* **24**, 69–97.

SECTION FOUR

MECHANISMS OF MINERAL NUTRITION

THE PHYSIOLOGY OF THE UPTAKE OF IONS BY THE GROWING PLANT CELL

D. H. JENNINGS
Department of Botany, Leeds University†

It has been known for some considerable time, especially as the result of the now classical work of Hoagland and his colleagues in the late 1920s that plant cells are capable of accumulating inorganic salts against a concentration gradient. Since these salts which are accumulated were found to be for the most part in the vacuolar sap and thus fully ionized, it was concluded that the plant cell must be doing work in the process. It will be realized that, in view of the ionic nature of inorganic salts, the exact point at which any driving force exerts its effect cannot be determined, for the reason that it is the movement of the constituent ions which must be considered. Movement of an ion of one charge will lead to the movement in the reverse direction of an ion of the same charge or the movement of an ion of opposite charge in the same direction.

It is appropriate to digress a little on the difficulty of achieving any substantial separation of ions of opposite sign. This can be demonstrated by a simple calculation. Supposing there is a 10-litre spherical vessel containing an 0·1 M solution of potassium chloride and there is available some mechanism whereby all the chloride ions can be removed, leaving behind only potassium ions in the vessel. Since there would be 1 g ion of potassium in the vessel, the charge would be 1 Faraday or approximately 10^5 coulombs. Taking the radius of the vessel as 13 cm, its capacity is therefore 14×10^{-12} farads. The potential on it would therefore be 7×10^{15} volts. This can be looked at another way. If there is a charge separation to the extent of a potential of one volt being generated, the difference in the number of ions of one sign over those of opposite sign will be no more than about 1 part in 10^{15}. This means of course that, even if a plant cell were to generate a potential of 1 volt between the cell interior and the external solution, the charge separation which results would not be detectable by chemical analysis. Of course it follows from this that even if potentials of this order of magnitude are generated it can be taken that for most purposes there is an almost exact balance of positive and negative charges within plant cells.

So in tracking down the mechanism of salt accumulation, the investigator must focus his attention upon the movement of individual ions into the cell. The procedure used by both animal and plant physiologists has been

† Now at Botany Department, Liverpool University.

to consider in the first instance those physical forces which might be acting on an ion, to see how far its distribution between the inside of the cell and the external medium might be governed by such forces. For the most part, it is only by eliminating the possibility that the observed distribution is not explicable in terms of the observed physical forces that investigators have been able to demonstrate that the cell is itself doing work on the ion.

An ion will always be acted upon by at least two physical forces: these are the chemical potential gradient and the electrical potential gradient. These forces are $d\mu/dx$ and $zFd\psi/dx$ respectively, both in joule. mole^{-1}. cm^{-1}, where μ is the chemical potential, z the algebraic valency, F the Faraday and ψ the electrical potential in volts. A third force may also be involved, namely the frictional drag brought about as a consequence of the movement of water in the system. For the moment, this latter force can be justifiably neglected. If this is done, the total force acting on an ion will then be the sum of the chemical and electrical potential gradients which can be denoted by $d\bar{\mu}/dx$, where $\bar{\mu}$ is the electrochemical potential of the ion under consideration.

Given a system of two solutions separated by a membrane, as might be the case in a living cell suspended in an aqueous solution, the situation where no physical force is acting on the ion will be that in which its electrochemical potential on both sides of the membrane is the same. Thus

$$\bar{\mu}_o = \bar{\mu}_i,$$

therefore

$$\mu_o + zF\psi_o = \mu_i + zF\psi_i,$$

and

$$RT \ln a_o + zF\psi_o = RT \ln a_i + zF\psi_i,$$

for which

$$\psi_i - \psi_o = \Delta\psi = \frac{RT}{zF} \ln \frac{a_o}{a_i},$$

where the subscripts o and i refer to the two sides of the membrane, a is the chemical activity of the ion and R and T have their usual meanings. The final derivation is the Nernst equation for the potential difference when the ion is in passive equilibrium. This equilibrium is dynamic, in that, while there is no net change in concentration of the ion on either side of the membrane, there will be movements of fluxes of the ion in either direction across the membrane.

If a cell is in flux equilibrium with the ions in its external environment and if it should prove possible to measure the electrical potential difference between the cell interior and the external medium, it should be possible to obtain a clear idea about those ions passively distributed as the result of the cell doing work. The jargon has it that these latter ions are being 'pumped' or 'actively transported' across the cell membranes.

The only really satisfactory micro-electrode experiments with plant cells have been those using algae with large or relatively large cells. The virtues of using such cells lie in the fact that there is little doubt about the location of the electrode once it has been inserted inside the cell, that it is possible to measure the cytoplasmic and vacuolar concentrations of most of the appropriate ions and that there appears to be little difficulty in achieving flux equilibrium. Furthermore the evidence which is now available indicates that the effect of solvent drag can be ignored, since of course when there is ion flux equilibrium there is no net movement of water across the cells and furthermore that there is only one major pathway in the cell membranes along which the water can move (Dainty 1963).

Once information has been obtained about the appropriate electrical potential differences and ion concentrations in the various phases, the procedure has been to calculate the electrical potential difference from the observed concentrations of the ion in the two phases for which the electrical potential difference applies. It should be noted that the assumption has been made that for biological systems concentrations can be substituted for activities, but it should be realized that this need not necessarily hold. If the two values for the electrical potential difference are the same, i.e., the Nernst equation applies to the distribution of the ion in question, it can be taken that the ion is passively distributed across the membrane. If the Nernst equation does not hold, it has been taken that the cell is doing work on that ion—the ion is being actively transported across the membrane. Some characteristic data for the equilibrium distribution of ions across the membranes of large internodal cells *Nitella translucens* is summarized in Fig. 1.

There have been attempts to make similar measurements in higher plant cells. There is some question about the validity of the exercise, since there is for the most part no certainty about the location of an electrode once it has been inserted inside the cell. Usually it has been assumed that the tip of the electrode resides in the vacuole, but there is good evidence that when electrodes are inserted into cells of potato tuber (Macklon and MacDonald 1966) the electrode is still in the cytoplasm (Jennings 1968a). This could be the case for those other micro-electrode investigations of higher plant cells, so that caution must be used in interpreting the results, since the total electrical force acting on ions moving between the outside and the inside of the cell may not have been measured. There are also the further difficulties that with higher plant cells the concentrations of an ion in the cytoplasm and the vacuole cannot be readily determined directly, although an indirect estimate can be made (Pitman 1963), and that it is not always easy to obtain higher plant cells in flux equilibrium. In spite of these problems the results of microelectrode experiments with higher plant cells have been presented in Table 1, on the assumption that all the physical forces acting on the ions have been taken into account.

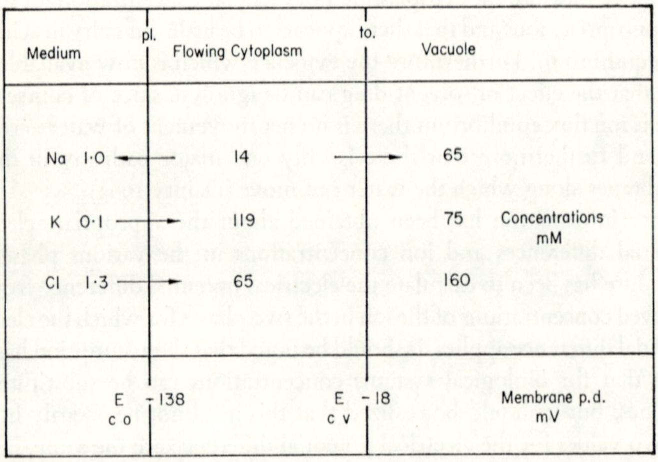

FIG. 1. A schematic presentation of the ionic state of *Nitella translucens*, giving the ionic concentrations and membrane potential differences. The arrows indicate the direction of active transport of a particular ion. Thus at the plasmalemma (*pl.*) Na is pumped out of the cytoplasm while K and Cl are pumped into the cytoplasm. At the tonoplast (*to.*) K and Cl are close to passive equilibrium while Na is pumped into the vacuole. (Redrawn from Spanswick and Williams 1964).

Most of the information about the biochemical nature of the pumps in algal cells has come from the study of the active transport of ions in *Nitella translucens* (MacRobbie 1962, 1964, 1965, 1966; Smith 1966). The evidence indicates that there is a coupled pumping in of potassium and pumping out of sodium ions by an ATPase similar to that found in animal cells (Baker 1966) and that the ATP for the pump is generated by photophosphorylation. The inwardly directed active transport of chloride and phosphate is independent of ATP, and appears to be coupled to the electron transfer reaction associated with the second photosystem of the photo-chemical reaction of photosynthesis. The same situation appears to hold for *Hydrodictyon africanum* (Raven 1967a, b) and probably for those members of the Characeae other than *Nitella*. However, in *Chaetomorpha darwinii*

TABLE I

Summary of results of microelectrode experiments on higher plant cells

Tissue	K^+	Na^+	Ca^{++}	Mg^{++}	SO_4''	Cl'	HPO_4'	NO_3'	HCO_3'	
Pea root	Out	Out	Out	Out	In	In		In		Etherton (1963); Higinbotham, Etherton and Foster (1967)
Pea epicotyl		Out	Out			In				
Oat root		Out	Out	Out	In	In	In			
Oat coleoptile		Out	Out		In					
Red beet	In					In			In	Poole (1966)
Hookeria lucens		Out				In				Sinclair (1967)

In = Into the cell. Out = Out of the cell. Blank means no information available.

(Dodd *et al.* 1967) the biochemical basis of inwardly directed potassium pump and the outwardly directed sodium pump appears to be different.

Potassium might be pumped into and sodium ions might be pumped out of higher plant cells through the mediation of an ATPase (Guener and Neumann 1966; Dodds and Ellis 1966). There is certainly good circumstantial evidence that ATP is involved in ion uptake in higher plants (Jennings 1963).

Part of any discussion about the function of ion-pumps in the plant cell must be a consideration of the relationship of ion-pumps to the electrical potentials across the cell membranes. There are three ways in which a potential can develop between two phases. First there may be some mechanism whereby an asymmetrical distribution of ions can be produced across the membrane. It is possible to think of a number of ways in which this might happen. Thus if one of the phases is equated with the cell interior, ions such as organic acid anions or hydrogen ions may be produced as a consequence of the metabolic breakdown of a non-ionic substrate. Asymmetry may also be produced by the active transport of ions across the membrane. Once such an asymmetrical distribution of ions has been achieved and if the membrane is permeable to the passive movement of ions, there will be a tendency for the above ions to establish in the first instance or re-establish in the second instance a uniform concentration on both sides of the membrane. The electrical potential difference across the membrane will be a function of this movement of ions towards a uniform concentration. That the potential difference is brought about by the diffusion of ions has led to the potential being described as a *diffusion potential.*

On the other hand, an ion-pump itself may contribute directly to an electrical potential difference through the transfer of charge as well as ions across the membrane. Such a pump is called *electrogenic.* In contrast, in that situation where the potential is solely a diffusion potential the pumps are described as *non-electrogenic,* since they are not transferring charge and are thus electrically neutral. As will be clear later both pumps can be present in the same cell.

Less is known about the third instance, which is that situation where it is possible to have a current flowing across a membrane without such a movement of ions. In this instance the current can be carried by electrons. This can be brought about by an oxidation-reduction system, such that the reaction producing electrons on one side of the membrane has a more negative redox potential than that receiving electrons on the other side of the membrane. Lundegårdh (1954) and Conway (1953) have both proposed theories of ion uptake based on such a mechanism as this. The presumed active uptake of chloride in algal cells might be brought about by this sort of mechanism which has hitherto been undetectable since micro-electrode measurements do not take into account a redox potential driving force across the chloroplast membrane.

Thus it can be seen that the ion pumps will play an important role in the accumulation of salts within plant cells. The pumps will move ions in against an electro-chemical potential gradient and bring about the movement of other ions by the generation of an electrical potential difference either directly or indirectly. While metabolism can bring about ion uptake by a cell through driving the pump, it should also be clear from what has already been said that metabolism will also play a part in ion uptake either by the production of ions from non-ionic metabolic substrates or by the production of a suitable membrane redox potential. The metabolic uptake of ions without the intervention of pumps may be important ecologically in the specific instance where there is movement into the plant of cations as the result of the production of organic acid anions, since this process can be very dependent on the carbon dioxide concentration in the external medium (Briggs, Hope and Robertson 1961; Hiatt and Hendricks 1967).

There is of course great biological significance in the presence of ion pumps in plant cells. Perhaps the most important consequence is that the plant cell is provided with a mechanism for concentrating salts within its interior. Indeed the process could have considerable specificity with regard to the ion being transported into the cell. Given a potassium-sodium pump as found in the outer membrane of many animal cells, the cell will select in favour of potassium from mixtures of potassium and sodium. One instance for which there is now good circumstantial evidence for the selective effect of an ion-pump within a plant has been indicated by Jennings (1967). The instance concerns the ability of roots to have a selective action on the cations reaching the shoot, particularly potassium in favour of sodium. Part of this selective action is via an inwardly directed sodium-pump at the tonoplast.

Dainty (1962) has pointed out that, while anion-pumps seem to be relatively rare in animal cells, these pumps are a logical evolutionary consequence of the development of a cell wall and the consequent ability to withstand high hydrostatic pressures. The anion-pump is the chosen mechanism for producing the high internal osmotic pressure in the vacuole. The anion-pump or pumps involved need not necessarily transport inorganic ions. The pumping of organic acid anions into the vacuole may be very important in higher plants. Of course it is important to remember that ion-pumps may affect the water relations of a plant cell not only by

19

changing the osmotic potential of the cell but by affecting the rate of water flow into the cell as a result of frictional drag (Kedem 1965). This latter effect could be particularly important in the movement of water across tissues, although in the one instance—exuding isolated maize roots (House and Findlay 1966a, b)—in which water movement across a plant tissue has been analysed from the correct physico-chemical standpoint, the bulk of the water moving across the tissue appears to be doing so under the action of simple osmotic forces.

The reader must constantly keep in mind the fact that almost everything which has been written above about ion-pumps in higher plants is almost all speculation. Since this is so it would seem unwise to speculate further. However, there is some virtue in such speculation in view of the considerable amount of recent information about the behaviour of the sodium-potassium pump in red blood cells under somewhat abnormal physiological conditions (Garrahan and Glynn 1967a, b, c, d, e) which can be added to what is already known about the behaviour under more normal conditions (Whittam 1964; Baker 1966). In the latter instance the pump in the membrane expels sodium ions and takes up potassium ions and the energy for these movements comes from the adenosine triphosphatase-mediated hydrolysis of adenosine triphosphate on the inner surface of the cell membrane. The crucial observations made by Garrahan and Glynn are that, when there is a high concentration of sodium and a very low concentration of potassium on the outside and a high concentration of potassium and in particular a very low concentration of sodium on the inside of the membrane, there is a net synthesis of ATP within the cells. The general point would seem to be this: if conditions are such that the overall reaction associated with transport ATPase activity leads to an increase in free energy, the transport system will run backward at a measurable rate and ATP will be synthesized from the energy derived from the ionic concentration gradients.

These observations are of considerable significance, because they offer an explanation of the ability of sodium ions to bring about increased growth both in terms of fresh weight and dry weight of a number of plants (Jennings 1968b). The most striking response of a number of plants to an increased uptake of sodium is increased succulence (Richards and Shih 1940a, b; Black 1958). This response is certainly not a simple cation effect, because increased uptake of potassium is associated with a decreased cell water content. The interesting aspect of the response to sodium ions is that both increased light and increased aridity also bring about increased succulence. The effect of light can be interpreted in terms of ATP synthesis via photophosphorylation, whilst the effect of increased aridity can be interpreted in terms of an increase of sodium over potassium reaching the shoots. Sodium itself brings about ATP synthesis by the reversal of a transport ATPase. However, the ability of the ATPase to act in this manner depends on other properties of the plant cell, two of which are the presence of a sodium pump at the tonoplast and the presence of suitable permeability characteristics associated with both the plasmalemma and the tonoplast. The increased ATP synthesis might allow increased wall synthesis or increased wall extensibility. The final outcome however is that the increased water content of the cell has a diluting effect on sodium ions which might otherwise be toxic at higher concentrations. The plant cell is thus responding in a homeostatic manner to the presence of high concentrations of an unfavourable ion.

Virtually all the above discussion about the ionic relations of plant cells has centred upon the presence of ion pumps in these cells. It is worth reiterating that the identification of these pumps has been obtained from cells in equilibrium conditions, where there is no net flux of ions across the membranes. It is because there has been considerable interest in ion pumps *per se* and because in discussing the means by which they can be identified there has been no need to talk in any great detail about the passive movement of ions that the above discussion has for the most part ignored these fluxes. As will be indicated later, when the growing plant cell is considered, all ion movements between the cell and the external medium, whether these movements be active or passive, must be taken into consideration; the growing cell is never in flux equilibrium. But even considering non-growing cells—and virtually all the work on the ion relations of plant cells has been carried out with cells which are non-growing—comparisons between the ion content of the cells in media of differing ionic composition demand some knowledge of the fluxes of the individual ions. It is specially important to know to what extent individual fluxes are active or passive. The effectiveness of any ion pump will depend to a large measure on the passive permeability of the cell membranes. Much of what has been said about the function of ion-pumps will remain as surmise until there is quantitative data about individual fluxes both active and passive. The essential point to be made is that the effectiveness of any pump will depend on the passive permeability of the cell, to the extent that with an increasing passive permeability the more likely that any pumping activity will be 'short-circuited' by the passive movement of ions.

A quantitative expression of this can be seen in the results of Spanswick, Stolareck and Williams (1967) obtained with *Nitella translucens*. When cells are pretreated in 5 mM sodium chloride and transferred to calcium-free solutions containing different ratios of sodium to potassium, the electrical potential across the plasmalemma behaves as a diffusion potential. In the presence of calcium, the effects of sodium and potassium on the plasmalemma potential are very much smaller, while the effect of calcium ions is relatively large. A likely explanation is that under these latter conditions the passive movement of ions across the plasmalemma is reduced and an electrogenic pump plays a much larger role in the production of the potential difference across the outer membrane of the cell. Thus it would appear that the pump is only able to have a direct effect on the membrane potential when there is a considerable reduction in the passive movement of ions. Similarly it must be clear that the functions of pumps outlined above will only be expressed given certain levels of passive fluxes.

It is appropriate at this point to discuss a little further the effects of bivalent cations, especially calcium ions, on cell permeability. There is now increasing evidence that calcium can reduce the leakage of a wide variety of compounds from plant cells (Table 2). The process of leakage investigated has been that produced by ethylene diamine tetra acetic acid, high concentrations of monovalent cations or that into distilled water. The evidence available indicates that such leakage is a passive process. This fact and the fact that high rates of leakage can be induced by monovalent cations led Jennings (1963, 1964) to postulate that there are pores in the outer membrane of plant cells which can be filled by calcium ions and that such calcium-filled pores are opened in the presence of suitable concentrations of monovalent cations. Other bivalent cations, particularly those of strontium, barium, magnesium and manganese can replace calcium ions but for the most part not so effectively. The observations of Jones and Jennings (1965) indicate that the effectiveness of the various types of bivalent cations is likely to differ from species to species of plant, although calcium in many cases appears to be the most effective.

There is evidence that calcium ions can affect the movement of ions between the cell and the external medium in a more indirect manner. It has been generally found that even small amounts of calcium can stimulate the uptake of uni/univalent salts when the salt concentration is low, though at higher concentrations the stimulation is less marked. This is the so-called 'Viet's effect' (see Jennings 1963 for a summary of the experimental data). A number of explanations have been put forward based on specific interactions of calcium with metabolism, with carriers or with the membrane structure of the cell. The effect of calcium however seems relatively unspecific, for other bivalent cations can have the same quantitative effect

TABLE 2

A summary of observations on the effect of calcium on the loss of solutes by plant cells

Tissue	Solutes lost	Conditions in which loss was observed	Concentration of calcium salt used to reduce loss	Reference
Beech mycorrhizal roots	Inorganic phosphate and phosphate in combined form	Induced at 2° C by high concentrations (10^{-2} M and above) of potassium, sodium and ammonium chlorides	1 mM chloride	Jennings (1964)
Potato tuber discs	Potassium ions	Predicted to happen from theoretical analysis of the effect of the counter-ion on the uptake of chloride at 0°C.		Laties, MacDonald and Dainty (1964)
Imperfect fungus: *Dendryphiella salina*	Potassium ions	Loss induced by 0·4 M sodium chloride and is relatively temperature insensitive	50 mM chloride	Jones and Jennings (1965)
Red beet discs	Potassium, sodium and chloride ions	Loss induced by 0·1 ethylene diamine tetra-acetic acid (EDTA) at 24° C	0·38 mM chloride	van Steveninck (1965a, b)
Corn scutellum slices	Sucrose	Distilled water and in the presence of 20 mM EDTA 30° C	20 mM chloride	Garrard and Humphreys (1967)
Germinating *Lilium longiflorum* pollen	Glucose	,, ,,	1·27 mM nitrate	Dickinson (1967)
Imperfect fungus: *Dendryphiella salina*	Mannitol and Arabitol	Loss into distilled water at 22° C	10 mM chloride	Ashford and Jennings (1968)

and cations of higher valency can produce a similar stimulation. The most satisfactory explanation of the effect has been put forward by Pitman (1964), who has suggested that uptake of anions is limited by diffusion through a negatively charged surface or membrane, which however will be more permeable to the anions when divalent cations rather than univalent cations are the counterions.

These observations on the effect of calcium on cell permeability have an importance which is not confined to studies on ion movements in plant cells. Thus from the ecological point of view, it is of interest that the reduced permeability brought about by calcium ions can in part explain the ameliorating action of bivalent ions, in particular calcium ions, against the toxic effect of high concentrations of monovalent cations—of which sodium ion is the most important under natural conditions—on the growth of plants (Hyder and Greenway 1965; Jones and Jennings 1965). However, the essential point which comes out of all these observations is the necessity to pay close attention to the passive permeability of plant cells. From the information above it is clear that high rates of leakage of metabolites like phosphorylated compounds might be very important in determining rates of metabolism (Ashford and Jennings 1968). Similarly, the loss of soluble compounds from root cells by passive leakage will be an important determinant of the rhizosphere microflora which itself will affect the growth of the plant (Garrett 1956, 1963; Harley 1959). The ecological significance of this sort of speculation requires examination.

One final point needs to be made about passive fluxes. It is that membranes may have as selective an action on the passive movement of ions into a cell as they may have on active movement (Briggs 1963; Jennings 1967). This fact is too often ignored.

So far much of what has been said has been of a qualitative nature and has served to illustrate how active and passive transport might affect the ion balance of non-growing plant cells. The extrapolation from this information to the cell which is growing presents very real difficulties. The magnitude of the difficulties can be seen by considering the simple analogy where the physiological experiment can be likened to the 'still' of a ciné film, the running time of which is equivalent to the life-span of the cell, organ or whole plant. Thus if an extrapolation is to be made, it will be from what is happening at one point in time to what might be happening as time varies. At the very least, therefore, information is required about what is happening at several points in time during a life-span before growth can be interpreted in terms of ion movements.

Before embarking on a discussion about the approach which the investigator requires to use, it is necessary to establish the precise question or questions which must be asked, if the information about ion movements between a cell or indeed a whole plant and its external environment is to be of value in the interpretation of the growth process. What the investigator should be searching for are those conditions in which the rate of growth is limited by the rate of movement of solutes between the cell and its external environment. Ideally, the best approach is to set up the appropriate differential equations and solve them. Such equations are designed to define the rate of flow of a solute across a boundary, i.e. in this instance a cell membrane, where the boundary conditions are themselves changing, i.e. the cell membrane is increasing in area as a result of growth. The equations will only be soluble if the change in boundary conditions can be defined in some precise way.

The type of analysis required can be illustrated by considering the simplest model, in which there is a spherical cell of radius r, surface area A and volume V_i, with an internal concentration of solute C_i, suspended in a suitable medium in which the same solute is at a concentration C_0 which can be maintained at this level indefinitely.

The total number of atoms of solute in the cell, N_i, is given by

$$N_i = C_i V_i \tag{1}$$

Suppose that the net flux of atoms from the medium into the cell is $J_{o \to i}$, then the net flow from $o \to i$ is $J_{o \to i} . A$. Consequently

$$\frac{dN_i}{dt} = J_{o \to i} . A. \tag{2}$$

Furthermore, the volume of the cell and the area are related to the radius as follows:

$$V_i = \frac{4}{3} \pi r^3, \tag{3}$$

and

$$A = 4\pi r^2 \tag{4}$$

The rate of change in the volume of the cell is given by

$$\frac{dV_i}{dt} = 4\pi r^2 \cdot \frac{dr}{dt} \tag{5}$$

$$= A . \frac{dr}{dt} \tag{6}$$

From equation 1, the rate of change of the internal concentration of solute with time is given by

$$\frac{dC_i}{dt} = \frac{V_i dN_i/dt - N_i dV_i/dt}{V_i^2} \tag{7}$$

$$= \frac{V_i \cdot {}_iJ_{o\to i} . A - N_i A . dr/dt}{V_i^2}$$

Using equations 2 and 6

$$\frac{dC_i}{dt} = \frac{A}{V_i}\left(J_{o\to i} - \frac{N_i}{V_i} . \frac{dr}{dt}\right)$$

$$= \frac{A}{V_i}\left(J_{o\to i} - C_i \cdot \frac{dr}{dt}\right)$$

dr/dt and V_i can be put in terms of A, but there is also a need to define A in terms of C_i—that is define the boundary conditions for the movement of solute across the cell membrane. This can be done by assuming that the solute, once it has entered the cell, is involved in an enzyme reaction the rate of which governs the rate of growth of the cell. If this is so and it is assumed that there is at any moment a constant concentration of enzyme in the cell, the rate of production of cell membrane is given by

$$\frac{dA}{dt} = \frac{R_{max} C_i}{K_m + C_i},$$

where R_{max} is the maximum rate of membrane formation and K_m is the Michaelis constant for the enzyme.

The differential equation which gives the change of internal concentration of the solute with the flux of the solute across the cell membrane is complicated, and does not appear to be readily soluble. The practical approach is by numerical analysis. If the above equations were in fact applicable to a growing cell the most satisfactory procedure is to grow the cell under conditions where the concentration of solute in the external solution remains constant, or substantially so, and determine the area and volume, and the flux across the membrane of the solute at a number of points in time during the growth crycle. From the values obtained, the internal concentration of the solute for the same points in time can be calculated. Agreement between the measured internal concentration and calculated values would indicate the model to be correct. Given that the model is substantially correct, it should then be possible to find out those conditions in which the net flux is limiting growth and also confirm predictions about the primary site of action of the solute within the cell.

Any prediction of the primary site of action of any solute is bound to be fraught with difficulties, particularly when ions are under consideration. Nevertheless possibilities do suggest themselves. For instance, in the case of potassium a not unreasonable possibility might be the conversion of phosphoenolpyruvate by pyruvate kinase (2.7.1.40), an enzyme known to be activated by potassium ions. The point to note here is that the primary site of action of the ion is not necessarily a reaction directly responsible for membrane synthesis, but can be and indeed is more likely to be a reaction of more general metabolic significance, influencing the synthesis of all protoplasmic components.

Even if the above model were applicable to plant cells—one of the few possibilities where the model could be applicable might be young fungal hyphae—there is little doubt that it is based on some fundamental assumptions which may not be tenable. Two assumptions spring quickly to mind; first that once the solute enters the cell it immediately assumes a uniform distribution within the cell and second that the concentration of enzyme remains constant throughout the growth cycle. The production of a suitable model for an actual plant cell will be based on a greater number of assumptions in so far as the number of boundaries will increase; at the very least the presence of the tonoplast will have to be taken into account.† With increasing sophistication of any model there will be an added problem in that with the increase in number of differential equations only a small amount of numerical juggling will be required to make the model fit the observed facts. Nevertheless this approach ought to be tested; if it were to be successful there would be a considerable advance in our understanding of the growing plant cell.

This sort of approach will depend very much on the presence of suitable experimental material. Indeed theoretical considerations apart there is a need for hard facts about the ion relations of growing plant cells. The most suitable material for initial studies might be synchronous cultures of cells—the study of Jung and Rothstein (1966) using synchronized mouse leukemic lymphoblast cells indicates the possibilities of this approach. Unicellular algae and yeasts can be obtained in synchronous culture (Cameron and Padilla 1967), but it would be of the greatest value to be able to use higher plant tissue cultures whose cell divisions had been synchronized. Even the study of non-synchronized plant tissue cultures has yielded valuable data about the change in potassium sodium selectivity as cells move from the

† It should be pointed out that it is not too difficult to produce differential equations taking into account more than one boundary; it is not appropriate to give details of such equations here.

non-growing to the growing state (Sutcliffe and Counter 1959). Another candidate for experimental study would be the giant cells of *Valonia* for which there is already some interesting information. Aikman and Dainty (1966) have indicated that, as these cells grow in size, the rate of growth could well be limited by the net flux of potassium into the cell. There are obvious virtues in using Valonia for studying the ionic relations of growing cells, since the total driving force on an ion can be measured, it is possible to obtain information about cytoplasmic ion concentrations and there appear to be no solute-solvent interactions as ions move across the cell membranes (Gutknecht 1967).

The real point which needs to be made is that before the plant physiologist can be any value to the ecologist and indeed also to agriculturalists with regard to providing some understanding of the role of transport processes in the mineral nutrition of plants, he must devote more attention to systems which are more relevant to the growing plant. A cautionary note has been struck by Pitman (1967) in this respect. He has pointed out that many of the experiments with excised roots have been using tissues in a state of mineral depletion and the behaviour of the roots in this state with regard to potassium-sodium selectivity differs markedly from roots with a much higher salt content. Thus, even where the more traditional approach is concerned, forethought is needed if the data to be obtained are to be of value to those other biologists concerned about physiology of the growing plant. However there is little doubt that something more positive than this is required. There is a need for a *quantitative* picture of the permeability properties of growing plant cells, such that a clear idea is presented of how various processes interact. At the moment it is only possible to make a few rather tentative and very qualitative suggestions about such interactions. In view of this, it is not surprising that it is the extreme ecological situations where a specific factor is exerting its effect—the effect of aluminium on calcicole species being a noteworthy example—which have proved to be the ones in which most progress has been made towards an understanding of the underlying physiological mechanism. There is a need to interpret those situations which are less extreme. Jennings (1968b) has pointed out that mesophytes and halophytes appear to differ in degree and not in kind with respect to their response to sodium. At the very least, therefore, to attempt to explain this sort of ecological situation, it will be necessary to get some sort of quantitative information about some of the physiological parameters of the plants and make an assessment of their relative importance. However, it is to be hoped that more than just this will be attempted. Plant nutrition has suffered from the lack of an integrated outlook. This is what all of us, whether physiologists or ecologists, should be striving for. Without such an outlook, real progress in our understanding of the subject will not be made.

REFERENCES

AIKMAN D.P. and DAINTY J. (1966). Ionic relations of *Valonia ventricosa*. Some contemporary studies in marine science (Ed. H.Barnes), 37–43. George Allen and Unwin, London.

ASHFORD A.E. and JENNINGS D.H. (1968). Unpublished observations

BAKER P.F. (1966). The sodium pump. *Endeavour* **25**, 166–171.

BLACK R.F. (1958). Effect of sodium chloride on leaf succulence and area of *Atriplex hastata* L. *Aust. J. Bot.* **6**, 306–321.

BRIGGS G.E. (1963). Rate of uptake of salts by plant cells in relation to an anion pump. *J. exp. Bot.* **14**, 191–197.

† BRIGGS G.E., HOPE A.B. and ROBERTSON R.N. (1961). *Electrolytes and plant cells.* 217 pp. Blackwell. Oxford.

CAMERON I.L. and PADILLA G.M. (1966). *Cell synchrony: studies in biosynthetic regulation.* 392 pp. New York.

CONWAY E.J. (1953). A redox pump for the biological performance of osmotic work and its relation to the kinetics of free ion diffusion across membranes. *Int. Rev. Cytol.* **2**, 419–445.

† DAINTY J. (1962). Ion transport and electrical potentials in plant cells. *Ann. Rev. Pl. Physiol.* **13**, 379–402.

DAINTY J. (1963). Water relations of plant cells. *Adv. Bot. Res.* **1**, 279–326.

DICKINSON D.B. (1967). Permeability and respiratory properties of germinating pollen. *Pl. Physiol.* **20**, 118–127.

DODD W.A., PITMAN W.G. and WEST K.R. (1966). Sodium and potassium transport in the marine alga, *Chaetomorpha darwinii. Aust. J. biol. Sci.* **19**, 341–354.

DODDS J.J.A. and ELLIS R.J. (1966). Cation stimulated adensine triphosphatase activity in plant cell walls. *Biochem. J.* **101**, 31P.

ETHERTON B. (1963). Relationship of cell transmembrane potentials to potassium and sodium accumulation ratios in oat and pea seedlings. *Plant Physiol.* **38**, 581–585.

GARRAHAN P.J. and GLYNN I.M. (1967a). The behaviour of the sodium pump in red cells in the absence of external potassium. *J. Physiol.* **192**, 159–174.

GARRAHAN P.J. and GLYNN I.M. (1967b). The sensitivity of the sodium pump to external sodium. *J. Physiol.* **192**, 175–188.

GARRAHAN P.J. and GLYNN I.M. (1967c). Factors affecting the relative magnitudes of the sodium: potassium and the sodium: sodium exchanges catalysed by the sodium. *J. Physiol.* **192**, 189–216.

GARRAHAN P.J. and GLYNN I.M. (1967d). The stoicheiometry of the sodium pump. *J. Physiol.* **192**, 217–235.

GARRAHAN P.J. and GLYNN I.M. (1967e). The incorporation of inorganic phosphate into adenosine triphosphate by reversal of the sodium pump. *J. Physiol.* **192**, 237–256.

† These references are valuable surveys of the subject of the ion relations of plant cells.

GARRARD L.A. and HUMPHREYS T.E. (1967). The effect of divalent cations on the leakage of sucrose from corn scutellum slices. *Phytochem.* **6**, 1085–1095.

GARRETT, S.D. (1956). *Biology of root-infecting fungi.* 293 pp. Cambridge.

GARRETT, S. D. (1963). *Soil fungi and soil fertility.* 165 pp. Oxford.

GRUENER N. and NEUMANN J. (1966). An ion stimulated adenosine triphosphatase from bean roots. *Physiologia Pl.* **19**, 678–682.

GUTKNECHT J. (1967). Membranes of *Valonia ventricosa*: apparent absence of water-filled pores. *Science*, **158**, 787–788.

HARLEY J.L. (1959). *The biology of mycorrhiza.* 233 pp. London.

HIATT A.J. and HENDRICKS S.B. (1967). The role of CO_2 fixation in accumulation of ions by barley roots. *Z. Pflanzenphysiol.* **56**, 220–232.

HIGINBOTHAM N., ETHERTON B. and FOSTER R.J. (1967). Mineral ion contents and cell transmembrane electropotentials of pea and oat seedling tissue. *Pl. Physiol.* **42**, 37–46.

HOUSE C.R. and FINDLAY N. (1966a). Water transport in isolated maize roots. *J. exp. Bot.* **17**, 344–354.

HOUSE C.R. and FINDLAY N. (1966b). Analysis of transient changes in fluid exudations from isolated maize roots. *J. exp. Bot.* **17**, 627–640.

HYDER S.Z. and GREENWAY H. (1965). Effects of Ca^{++} on plant sensitivity to high NaCl concentrations. *Plant and Soil* **23**, 258–260.

† JENNINGS D.H. (1963). *The absorption of solutes by plant cells.* 204 pp. Edinburgh.

JENNINGS D.H. (1964). The effect of cations on the absorption of phosphate by beech mycorrhizal roots. *New Phytol.* **63**, 348–357.

JENNINGS D.H. (1967). Electrical potential measurements, ion pumps and root exudation—a comment and a model explaining cation selectivity by the root. *New Phytol.* **66**, 357–369.

JENNINGS D.H. (1968a). Microelectrode experiments with potato cells: a reinterpretation of the experimental findings. *J. exp. Bot.* **19**, 13–18.

JENNINGS D.H. (1968b). Halophytes, succulence and sodium as a plant nutrient—a unified theory. *New Phytol.* **67**, 899–911.

JONES E.B.G. and JENNINGS D.H. (1965). The effect of cations on the growth of fungi. *New Phytol.* **64**, 86–100.

JUNG C. and ROTHSTEIN A. (1967). Cation metabolism in relation to cell size in synchronously grown tissue. *J. gen. Physiol.* **50**, 917–932.

KEDEM O. (1965). Water flow in the presence of active transport. *Symp. Soc. exp. Biol.* **19**, 61–73.

LATIES G.G., MACDONALD I.R., and DAINTY J. (1964). Influence of the counter-ion on the absorption isotherm for chloride at low temperatures. *Pl. Physiol.* **39**, 254–262.

LUNDEGÅRDH H. (1954). Anion respiration: the experimental basis of a theory of absorption, transport and exudation of electrolytes by living cells and tissues. *Symp. Soc. exp. Biol.* **8**, 262–296.

MACROBBIE E.A.C. (1962). Ionic relations of *Nitella translucens. J. gen. Physiol.* **45**, 861–878.

MACROBBIE E.A.C. (1964). Factors affecting the fluxes of potassium and chloride ions in *Nitella translucens. J. gen. Physiol.* **47**, 859–877.

MACROBBIE E.A.C. (1965). The nature of the coupling between light energy and active ion transport in *Nitella translucens. Biochim. biophys. Acta* **94**, 64–73.

MacRobbie E.A.C. (1966). Metabolic effects on ion fluxes in *Nitella translucens*. 1. Active influxes. *Aust. J. biol. Sci.* **19**, 363–370.

Macklon A.E.S. and MacDonald I.R. (1966). The role of transmembrane electrical potential in determining the absorption isotherm for chloride in potato. *J. exp. Bot.* **17**, 703–717.

Pitman M.G. (1963). The determination of the salt relations of the cytoplasmic phase in cells of beetroot tissue. *Aust. J. biol. Sci.* **16**, 647–668.

Pitman M.G. (1964). The effect of divalent cations on the uptake of salt by beetroot tissue. *J. exp. Bot.* **15**, 444–456.

Pitman M.G. (1967). Conflicting measurements of sodium and potassium uptake by barley roots. *Nature, Lond.* **216**, 1343–1344.

Poole R.J. (1966). The influence of the intracellular potential on potassium uptake by beetroot tissue. *J. gen. Physiol.* **49**, 551–563.

Raven J.A. (1967a). Ion transport in *Hydrodictyon africanum*. *J. gen. Physiol.* **50**, 1607–1626.

Raven J.A. (1967b). Light stimulation of active transport in *Hydrodictyon africanum*. *J. gen. Physiol.* **50**, 1627–1640.

Richards F.J. and Shih S.-H. (1940a). Physiological studies in plant nutrition. X. Water content of barley leaves as determined by the interaction of potassium with certain other nutrient elements. Part 1. The relationship between water content and nutrient composition. *Ann. Bot. N.S.* **13**, 164–175.

Richards F.J. and Shih S.-H. (1940b). Physiological studies in plant nutrition. X. Water content of barley leaves as determined by the interaction of potassium with certain other nutrient elements. Part II. The relationship between water content and composition of leaves. *Ann. Bot. N.S.* **13**, 403–425.

Sinclair J. (1967). Nernst potential measurements on the leaf cells of the moss *Hookeria lucens*. *J. exp. Bot.* **18**, 594–599.

Smith F.A. (1966). Active phosphate uptake by *Nitella translucens*. *Biochim. biophys Acta* **126**, 94–99.

Spanswick R.M., Stolarek J. and Williams E.J. (1967). The membrane potential of *Nitella translucens*. *J. exp. Bot.* **18**, 1–16.

van Steveninck R.F.M. (1965a). The effects of calcium and tris(hydroxy-methyl) aminomethane on potassium uptake during and after the lag phase in red beet tissue. *Aust. J. biol. Sci.* **18**, 227–233.

van Steveninck R.F.M. (1965b). The significance of calcium on the apparent permeability of cell membranes and the effects of substitution with other divalent ions. *Physiologia Pl.* **18**, 54–69.

Sutcliffe J.F. and Counter E.R. (1959). Absorption of alkali cations by plant tissue cultures. *Nature, Lond.* **153**, 1513–1514.

Whittam R. (1964). Transport and diffusion in red blood cells. 228 pp. London.

THE PROPERTIES OF MECHANISMS INVOLVED IN THE UPTAKE AND UTILIZATION OF CALCIUM AND POTASSIUM BY PLANTS IN RELATION TO AN UNDERSTANDING OF PLANT DISTRIBUTION

R. L. JEFFERIES AND D. LAYCOCK,
School of Biological Sciences, University of East Anglia

G. R. STEWART AND A. P. SIMS†
Botany Department, University of Bristol

INTRODUCTION

In studies of the ecological aspects of the mineral nutrition of plants it is well known that the presence or absence of particular ions in the external environment can affect the growth of a plant thereby profoundly influencing their distribution in natural habitats. For example calcicole species like *Scabiosa columbaria* (L.), and *Leiocolea turbinata* (Raddi) Buch, are generally restricted to soils rich in calcium whereas calcifuge species such as *Deschampsia flexuosa* (L.) Trin and *Cephalozia connivens* (Dicks) Lindb. tend to be confined, in the British Isles, to soils containing very low levels of calcium. The presence of high levels of sodium and/or chloride ions can also profoundly influence the type of vegetation found in a habitat; halophytic species like *Spartina* × *townesendii* H. and J. Groves, and *Triglochin maritima* L., are rarely found outside saline or brackish environments whereas glycophitic species are normally absent from such localities. In contrast to these examples, however, some plant species are distributed over a wide range of ionic habitats. *Festuca ovina* L., is found in both acidic and calcareous soils and it is worthwhile considering possible reasons why particular species may enjoy a widespread distribution.

In some cases the apparent wide range of distribution of a species reflects nothing more than the persistence of similar microenvironments in different regions. It is well known, for example, that certain calcifuge plants are able to invade calcareous grassland where incipient leaching of the surface soil has occurred (Grime 1963). In many other instances widespread distribution of a species is achieved as a consequence of the existence of

† Present address: School of Biological Sciences, University of East Anglia.

genetically distinct populations, each of which is adapted to a relatively narrow range of conditions. For example Snaydon and Bradshaw (1961) have demonstrated that the species *F. ovina* is composed of quite distinct physiological populations which are adapted to grow over a characteristic range of calcium concentrations. However some plants show extreme phenotypic plasticity and their biochemical mechanisms are capable of considerable adjustment in response to changes in the external ionic environment thereby enabling the plants to grow over a wide range of ionic conditions. At this point it is worthwhile considering some of the ways in which ions may affect the metabolism and growth of plants (Bollard and Butler 1966). Thus:

(a) The concentration of an ion in a habitat may be insufficient to meet the metabolic requirements of plants of a particular species.

(b) The assimilation and/or utilization of an ion by plants may be impaired as a result of antagonism between competing ions in the external environment.

(c) The presence of low concentrations of toxic ions in the soil may profoundly affect the growth of some plants as a result of the inhibition of key enzymic processes. Lead and other heavy metals are especially potent inhibitors of enzymes and their mode of action is often quite specific.

(d) The presence of elevated levels of certain ions in the soil, notably calcium and sodium may result in non-specific ionic effects resulting in the inhibition of a very large number of enzymes in the cells of plants.

It is useful to consider some of the regulatory mechanisms which may enable plants to overcome the deleterious effects of ions; we have placed these mechanisms in two distinct categories. The first group includes those mechanisms which are involved in the control and maintenance of the internal ionic environment of the cell, mechanisms that frequently involve selective ion pumps. These ion pumps may be either inwardly directed, that is they are normally concerned in the maintenance of the supply of essential nutrients from the external environment, or they may be outwardly directed thereby controlling the internal concentration of an ion which otherwise might reach toxic levels. The second category of mechanisms includes those which ensure the maintenance of an appropriate level of enzymic activity in spite of considerable fluctuations in the levels of different ions which occur within the cell. This enzymic adjustment can be achieved in several distinct ways. Mechanisms controlling the level of synthesis of an enzyme, can operate through induction, repression or de-repression of synthesis, thereby compensating for a specific or general loss of enzyme activity caused by elevated levels of inhibitory ions (Monod and Jacob 1961; Umbarger 1961; Orgel 1964). Another possible solution to loss of enzyme activity involves changes of the configuration of enzymes whereby their tolerance limits are modified to fit different internal ionic conditions.

It is clear that any adaptive process in plants has to be considered in relation to both genotypic and phenotypic plasticity. Thus whilst it is pertinent to look for differences in the properties of regulatory mechanisms in plants of populations which are adapted to markedly contrasting habitats it is also important to recognise that even within a habitat large changes of ionic conditions may occur. Plants with limited phenotypic plasticity may thus be unable to adjust to fluctuating ionic conditions and consequently be at a competitive disadvantage when compared with other species which can make the adjustment. In the present paper we have examined mechanisms which are involved in both the genotypic and phenotypic adaptation of plants to different ionic environments.

DIFFERENCES IN THE PROPERTIES OF ION ABSORPTION MECHANISMS IN SPECIES OF LIVERWORTS

General considerations

Selective ion transport may be achieved in plants as a result of the presence of specific ion-pumps, or as a consequence of differential permeability of the cellular membranes to ions. In order to distinguish active and passive ion fluxes it is necessary to estimate the magnitude of the physical 'driving forces'—the algebraic sum of the gradient of chemical activity and of electric potential which are acting on an ion that is diffusing across membranes between two compartments (Dainty 1962). If the physical forces are insufficient to account for the asymmetric distribution of an ion between the two phases then the ion may be actively transported across the membrane.

Additional evidence for the existence of energy-driven pumps can be obtained from studies of the effect of specific inhibitors on ion transport processes. Although the chemical nature and mode of action of these ion-pumps are for the most part poorly understood it appears that they are under genetic control (Epstein and Jefferies 1964). Since differential permeability of the membranes to ions may result from the presence of specific proteins which line pores within these structures it is evident that membrane characteristics can also be modified in the event of genetic alteration. However short term changes in the rate of assimilation of an ion into a plant may occur as a result of a modification of the binding sites on pumps or as a consequence of alterations of the structure of the cellular membranes brought about by the presence of other ions in soil solution. A knowledge therefore of the nature of antagonistic and synergistic effects between ions is of the utmost importance in any understanding of the factors which influence the movements of ions between an individual plant and its environment.

Material and methods

In an attempt to determine to what extent the influx of cations into plants of two liverwort species *C. connivens* and *L. turbinata* resulted from active or passive transport processes the distribution of ions between the external solutions and tissues of these plants has been measured together with the electrical potential differences which exist between the leaf cells and solutions. The standard micro-electrode techniques were used in these studies and measurements were made after the plants had been immersed in the appropriate solutions for 60 hr. The concentration of all ions, except calcium, was similar in the various solutions which were used in the experiments (Table 2). Measurements of influx of radioactive potassium (^{42}K) were made under different conditions using the continuous culture method developed by Scott (1961) and Sinclair (1965).

TABLE 1

Concentrations of cations (mM) in different soil solutions

Species	Site	Grid references	Potassium	Sodium	Calcium	pH
Cephalozia connivens	Buxton Heath, Norfolk.	TG 176215	0·05	0·2	0·3	4·0
Leiocolea turbinata	Cherry Hinton, Cambridgeshire.	TL 483577	50·0	10·5	6·5	8·1

Results

There are considerable differences in the concentration of ions present in the two soil solutions where these liverworts are found (Table 1). Industrial effluent drains into the chalk pit at Cherry Hinton and probably accounts for the high level of sodium and potassium which occurs in this water. The levels of calcium in the soil solutions at the two sites reflect the normal differences found between calcareous and non-calcareous soils.

There are differences in electrical potential between the cells of young and mature leaves and the experimental solutions which suggests that there may be considerable changes in the permeability of cells to electrolytes as the leaves mature. At the end of the period of immersion of the plants there was little change in the concentration of the different ions in the tissues and it would appear therefore that as equilibrium conditions prevail the Nernst equation can be applied (see p. 262). The calculated potentials based on the distributions of sodium and calcium between the solutions and plants of both species nearly all fall within the range of the measured potentials which indicates that these ions may be passively distributed across the cell membranes (Table 2). However, the concentration of potassium in tissues of both these liverworts is far higher than that predicted from the Nernst equation. This suggests that at least some of the potassium which is entering the tissues of these plants is actively pumped into the cells which implies that the internal concentration of potassium is under metabolic control.

In order to determine any differences in the ability of plants of these liverwort species to accumulate potassium, the influx of this ion into the tissues at different concentrations of potassium has been measured. The rate of uptake of an ion can be described by an equation which is of the same form as the Michaelis–Menten equation of enzyme kinetics (Epstein and Hagen 1952), in which

$$v = \frac{V_{\max} S}{K_m + S}$$

where v is the rate of uptake by a plant at an external concentration S of the ion and $V_{\max}$ is the maximum rate of uptake. If Michaelis–Menten postulates apply, the K_m can be considered to be a measure of the affinity of a specific carrier protein for potassium. The relationships between the rates of uptake of potassium into the tissues of each of these liverworts and the external concentrations of this ion are shown in Fig. 1. It can be seen from the data that an appreciable difference in the affinity of the transport systems for potassium exists between the two species. The Michaelis constant of the 'calcifuge' plant (0·087) mM is considerably lower than that of the 'calcicole' plant (0·36 mM). Moreover these values can be correlated with the differences in the concentration of potassium in the soil solutions where these plants grow. Although the ionic composition

TABLE 2

Ion concentration (mM) and equilibrium potentials (mV) of *Cephalozia connivens* and *Leiocolea turbinata*

Plants	Ion	C_o	C_i	Equilibrium potential (relative to medium)	Range of measured potential difference of young and mature leaf cells	Mean potential of mature leaf cells.
Cephalozia connivens	K	0·1	97·8	−173	−61 to −140	−81 ± 12†
	Na	0·1	16.3	−123		
	Ca	0·1	14·7	−63		
	K	0·1	91·8	−172	−44 to −131	−64 ± 10
	Na	0·1	14·5	−125		
	Ca	3·0	21·3	−25		
Leiocolea turbinata	K	0·1	150	−184	−54 to −138	−67 ± 9
	Na	0·1	13·2	−123		
	Ca	0·1	24	−69		
	K	0·1	157	−185	−35 to −133	−37 ± 7
	Na	0·1	1·4	−90		
	Ca	3·0	17·9	−51		

† Standard Error of the Mean

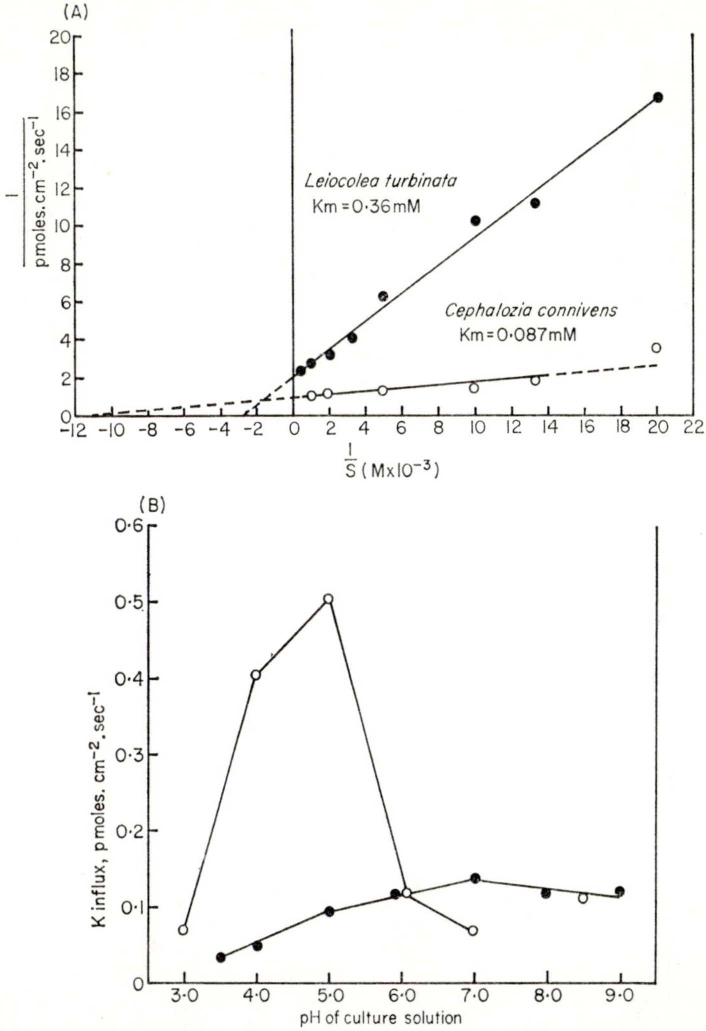

FIG. 1. (a) Relationship between the external concentration of potassium and the rate of uptake of this ion into *Cephalozia connivens* (○) and *Leiocolea turbinata* (●) plotted as a double reciprocal plot.

Concentration of sodium chloride in all solutions is $1{\cdot}0 \times 10^{-4}$ M. The levels of calcium chloride are $1{\cdot}0 \times 10^{-4}$ M and $3{\cdot}0 \times 10^{-3}$ M respectively in solutions which contain plants of *Cephalozia* and *Leiocolea*.

(b) Effect of hydrogen ion concentration on the potassium influx into plants of *Cephalozia connivens* and *Leiocolea turbinata* is shown.

The concentration of potassium chloride in all the solutions is 1×10^{-4} M, and that of sodium chloride $4{\cdot}5 \times 10^{-4}$ M. The levels of calcium chloride are 1×10^{-4} M and 3×10^{-3} M respectively, in solutions which contain plants of *Cephalozia* (●) and *Leiocolea* (○).

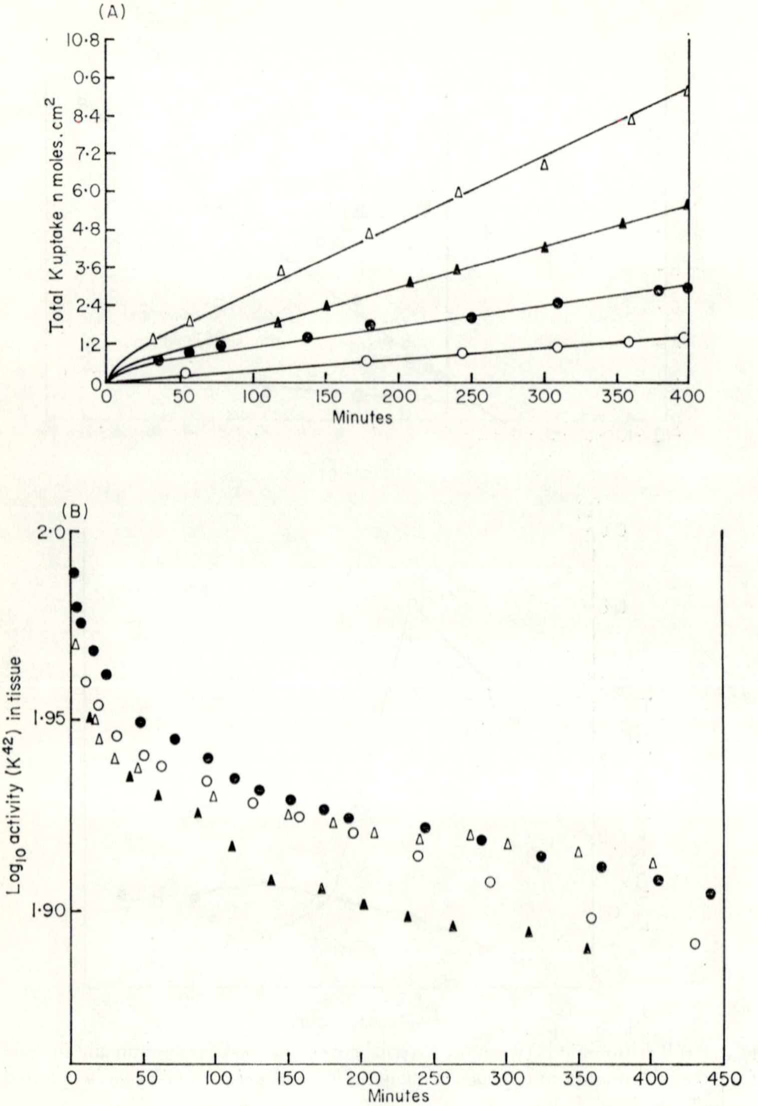

FIG. 2. Total uptake and efflux of potassium in plants of *Cephalozia connivens* and *Leiocolea turbinata*.

Plants are bathed in solutions containing 3×10^{-3} M calcium chloride and $0{\cdot}1 \times 10^{-3}$ M calcium chloride.

Sodium chloride and potassium chloride are also present in the solutions at a concentration of $1{\cdot}0 \times 10^{-4}$ M.

of experimental solutions is considerably different from the soil solutions the results do provide an important insight into one aspect of the physiological basis of edaphic adaptation in plants. However, it must be remembered that since the values of the Michaelis constant and also the maximum rates of accumulation may be affected by the prevailing ionic environment in which the tissue is immersed, much more information is required as to the extent to which changes in these parameters reflect fluctuations in the ionic environments in which the plants grow.

We have examined some ways in which the presence of other ions may influence the transport of potassium in these liverworts. The effects of the hydrogen ion concentration in the external solutions on the uptake of potassium by the two plants have been examined, since there is a marked difference in the pH of the soils where these liverworts were collected (Table 1). It can be seen in Fig. 1 that the pH of the external solutions markedly influences the net uptake of potassium into plants of both species. In both cases the highest net accumulation rate of this ion into the tissues of both liverworts occurs when the hydrogen ion concentration of the solutions correspond to the pH values of the soils in which these plants grow. Calcium also influences the uptake of potassium into the two liverworts (Fig. 2). In the presence of 3×10^{-3} M calcium chloride or calcium sulphate the potassium influx into *Cephalozia connivens* shows a reduction of about 30% compared with the corresponding rate into plants immersed in a solution which contains 1×10^{-4} M calcium. On the other hand, the potassium influx from the 3×10^{-3} M calcium solution into plants of *Leiocolea turbinata* shows an increase of approximately 60% compared with the corresponding rate from the solution containing 1×10^{-4} M calcium chloride. These results demonstrate that the level of calcium chloride in the external solution can influence the magnitude of influx of potassium into the tissues of these species.

FIG. 2—*continued*

(A) uptake data.
(B) efflux data.

△——△ Total uptake of potassium of efflux of radioactive potassium when plants of *Cephalozia* are immersed in a solution containing $1{\cdot}0 \times 10^{-4}$ M calcium.

▲——▲ Total uptake of potassium or efflux of radioactive potassium when plants of *Cephalozia* are immersed in a solution containing $3{\cdot}0 \times 10^{-3}$ M calcium

○——○ Total uptake of potassium or efflux of radioactive potassium when plants of *Leiocolea* are immersed in a solution containing $1{\cdot}0 \times 10^{-4}$ M calcium.

●——● Total uptake of potassium or efflux of radioactive potassium when plants of *Leiocolea* are immersed in a solution containing $3{\cdot}0 \times 10^{-3}$ M calcium.

In Fig. 2 the efflux of radioactive potassium from tissues of both species are shown when the plants are immersed in solutions which contain two different concentrations of calcium ions. With the exception of the efflux of potassium during the first 100 min, the rates of loss from the plants in the different solutions are essentially similar. The small loss of activity during the course of the experiment indicates that in contrast to the effect of calcium on the uptake of potassium, different concentrations of the divalent ion in the external solution have relatively little influence on the efflux of the cation from the two liverworts. Although it is well established that calcium can have a marked effect on the differential permeability of the tissues to ions its role in influencing the movement of other ions in plant tissues is complex. In these experiments as the effect of different concentrations of calcium chloride on potassium uptake occurs immediately on addition of this salt to the solutions, it suggests that the action of calcium chloride is on the surface region of the cell probably at the plasmalemma (Pitman 1964). The enhanced uptake of potassium on addition of high levels of calcium observed with *Leiocolea turbinata* may be a further example of the phenomena first reported by Viets (1944). Pitman (1964) has suggested that enhanced potassium uptake may be associated with an increased chloride influx due to presence of high concentration of calcium chloride. However, a hypothesis which must be considered is that a high concentration of calcium in the external solution is necessary in order to maintain the integrity and efficiency of the potassium pump in this liverwort. Calcium may influence the activity of the specific protein which is involved in potassium transport in a manner analagous with the effect of this divalent ion on the activity of malic dehydrogenase enzyme as described in section II. High concentrations of calcium sulphate or chloride (3×10^{-3} M) partially inhibit the transport of potassium into the tissues of *Cephalozia connivens*; once again it seems likely that the divalent ion may affect the energy-dependent potassium transport in this liverwort tissue. These results indicate that the efficiency of the uptake of potassium in both liverworts is markedly influenced by the concentration of other ions present in the experimental solution.

If measurements of influx are to have greater ecological significance it is desirable to obtain information on the magnitude of the ion fluxes between plants and external solutions over much longer periods of time. One such investigation has involved an examination of the ionic relations of *Wolffia arrhiza* (L.) This plant was chosen because of its structural simplicity and rapid growth under closely defined nutritional conditions. Clones were grown over long periods of time under axenic conditions in nutrient solutions in which the concentrations of calcium differed.

One experiment involved the transfer of plants grown in a culture medium containing a low concentration of calcium to solutions where the calcium concentrations varied between 0·5 and $4{\cdot}0 \times 10^{-3}$ M. It was only in solutions which contained the high concentrations of calcium that changes in growth rate occurred. In these cultures the restoration to a new steady state condition took up to 5 days and concurrent with changes of growth rate were alterations in the influx of potassium (Table 3). It would

TABLE 3

Potassium influxes and internal concentrations of potassium and sodium in plants of *Wolffia arrhiza* at various time intervals, after the plants have been transferred from a solution containing 5×10^{-4} M calcium to one which contains 3×10^{-3} M calcium

Days	Potassium influx moles $\times 10^{-4}$ g.f.w.$^{-1}$ hr^{-1}	Concentration of potassium in plants M $\times 10^{-2}$/l	Concentration of sodium in plants M $\times 10^{-3}$/l
0	0·74	4·22 ± 0·17†	16·60 ± 1·30†
1	1·32	2·32 ± 0·35	10·35 ± 2·32
2	1·08	5·65 ± 0·2	7·65 ± 0·55
3	0·91	6·92 ± 1·07	9·35 ± 1·25
5	—	5·85 ± 0·275	7·67 ± 1·37
8	0·96	—	—
9	—	6·40 ± 0·275	5·62 ± 0·70
10	—	5·92 ± 0·35	5·97 ± 0·97
11	0·86	—	—
15	0·73	—	—

† Standard Error of the Mean

thus seem that adjustments in the rate of uptake of this ion appear to be strictly regulated and controlled in such a way as to maintain a consistently high internal concentration of potassium in spite of considerable changes of growth rate (Fig. 3). However, it is also evident that an increase in the external level of calcium ions can also effect the internal concentrations of potassium and calcium especially during the first 2 days of transfer. (See also Fig. 8 with *Lemna minor*.) As will become apparent in the next section the growth lag observed may reflect the time necessary for the modification of enzyme systems to occur resulting from enhanced levels of calcium in the tissues.

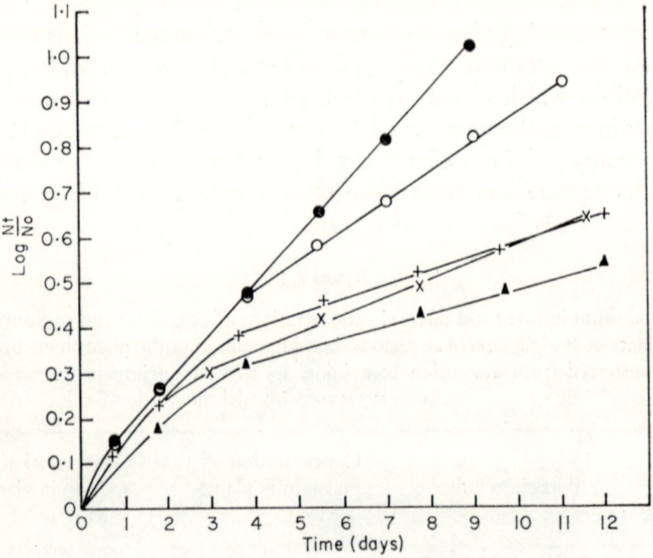

FIG. 3. Growth rates of plants of *Wolffia arrhiza* in nutrient solutions containing different concentrations of calcium.

Plants were grown in a solution containing a calcium concentration of $5{\cdot}0 \times 10^{-4}$ M calcium chloride and transferred to different solutions.

The different calcium concentrations in these solutions are shown by the following symbols.

(●——●) 0·5 and $1{\cdot}0 \times 10^{-3}$ M; (+——+) $3{\cdot}0 \times 10^{-3}$ M;
(○——○) $2{\cdot}0 \times 10^{-3}$ M; (▲——▲) $4{\cdot}0 \times 10^{-3}$ M
(×——×) $2{\cdot}5 \times 10^{-3}$ M;

THE PHENOTYPIC MODIFICATIONS OF THE PROPERTIES OF ENZYMES IN RESPONSE TO CHANGING IONIC ENVIRONMENTS

General considerations

In instances where the entry of an ion into a tissue is not strictly controlled a possibility exists that enzymes themselves may undergo modifications which enables them to function equally well in the presence of different concentrations of the ion. Clear evidence that variations in the response of plant enzymes to calcium can occur has already been presented (Sims, Stewart and Folkes 1969). An examination of plants of a single species growing in different calcium regimes revealed that the response of the enzyme malic dehydrogenase to calcium ions was not a fixed characteristic but a property that could vary depending upon the concentration of calcium in the soil in which the plant was growing. The large number of plant species in which this enzyme adjustment has been demonstrated indicates that it may represent a fairly general mechanism associated with phenotypic plasticity whereby plants can adapt to fluctuating ionic conditions. We propose to outline some of the experimental evidence which indicates that changes of the properties of malic dehydrogenase can result from a modification of the sub-unit structure of the enzyme and to consider some of the biological implications of this type of regulatory mechanism. *Lemna minor*, a species of widespread distribution, was chosen for the present study since cloned material can be used throughout and so ensure that any observed physiological changes are phenotypic in origin. Plants were collected from a wide range of ecological habitats in an attempt to discover whether qualitative and quantitative differences in regulatory mechanisms occur within this species.

Materials and methods

Contrasting clones of *L. minor* were obtained from the following localities

- S_1 from a stream near the brickworks at Gillingham, Dorset (clay).
- S_2 from a pond at Milton Abbas, Dorset (chalk).
- S_4 from a stream near Ringwood, Hampshire (alluvium).
- S_6 from a drainage culvert near Shapwick, Somerset (peat).
- S_8 from a roadside ditch near Minehead, Somerset (sandstone).
- S_{22} from a moorland stream near North Bovey, Devon (granite).
- S_{29} from a stream at Gomshall, Surrey (greensand).

Methods used throughout have been described in an earlier paper (Sims, Stewart and Folkes 1969).

Properties of Clones S_1

Plants of a single clone of *L. minor* were grown under axenic conditions in culture solutions which contained different concentrations of calcium chloride but otherwise were of similar ionic composition. After approximately two weeks of growth the plants were harvested and the enzyme and protein content of the frond analysed. The effect of different concentrations of calcium chloride on the activity of malic dehydrogenase was examined in desalted enzyme preparations; it appears that as a result of differences in the level of calcium in the culture medium a progressive modification takes place in the response characteristics of the enzyme towards calcium as exemplified by its degree of activation, the concentration of calcium giving optimal activation and sensitivity to inhibition. (Fig. 4).

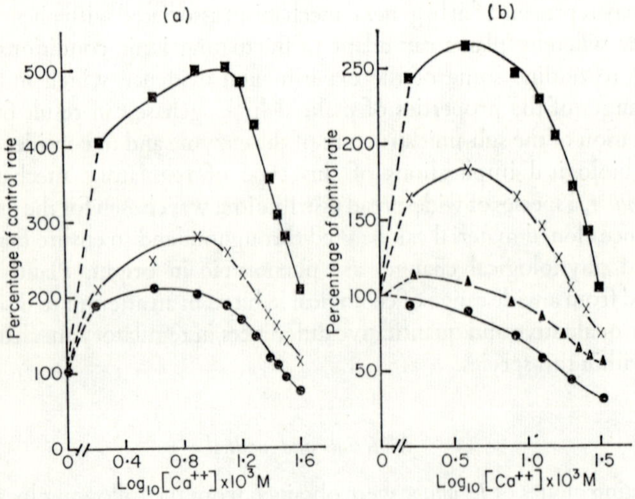

FIG. 4. The modification of the response of malic dehydrogenase to calcium ions under different conditions of growth.

(a) The modification of the response of malic dehydrogenase from clone S_1 grown on different levels of calcium.

Determination of enzyme activity were carried out at 25° C using a recording spectrophotometer. Final substrate concentrations used were oxalacetic acid 2×10^{-4} M, NADH $1{\cdot}75 \times 10^{-3}$ M.

(■) response of enzyme from plants growing on 1×10^{-2} M (Ca^{++})
(×) response of enzyme from plants growing on 1×10^{-3} M (Ca^{++})
(●) response of enzyme from plants growing on $3{\cdot}3 \times 10^{-4}$ M (Ca^{++})

(b) The response of malic dehydrogenase to calcium ions from clones of *Lemna minor* collected from a range of ecological habitats.

(■) clone S_1 (●) clone S_{22}
(×) clone S_6 (▲) clone S_{29}

It is clear that these phenotypic changes offer a means of maintaining catalytic activity over a wide range of internal calcium concentrations. Results from calculations based on measurements of growth rate, level of malic dehydrogenase and protein per frond are in agreement with this suggestion (Table 4). There appears to be no appreciable net synthesis of enzyme to overcome the inhibitory effects of the increased internal concentration of calcium ions. It is noteworthy that the differences of enzyme response to

TABLE 4

The maintenance of malic dehydrogenase activity under conditions when the growth of *Lemna minor* is varied in response to changes of calcium

Concentration of calcium ions in the medium	Specific activity of malic dehydrogenase OD 340 mμ/5 mins/100 μg	Protein content of 100 fronds in μg	Total malic dehydrogenase activity of 100 fronds	Growth constant	Total activity of enzyme/ growth constant
1×10^{-4} M	4·08	320	13·06	0·125	104
1×10^{-3} M	3·86	329	12·7	0·123	103
1×10^{-2} M	3·55	340	12·0	0·083	144
5×10^{-2} M	2·73	360	9·8	0·067	148

calcium observed in this experiment closely resemble those observed in enzyme extracts derived from populations of *L. minor* growing in natural habitats (Fig. 4).

Stewart and Sims (unpublished) have found that other allosteric enzymes of *L. minor* notably glucose-6-phosphate dehydrogenase, NADP isocitric dehydrogenase and the NAD glutamate dehydrogenase, also exhibit varied responses towards calcium ions and it seems likely that enzyme modification in response to different calcium levels may be quite general amongst regulatory enzymes of this clone.

In an attempt to elucidate the means by which calcium and/or chloride ions can produce specific changes in the enzyme two distinct mechanisms can be considered. The ability of the organism to grow in solutions which contain markedly different calcium concentrations may depend upon its capacity to produce a range of genetically distinct iso-enzymes (Kaplan 1963) and thus a differential synthesis of two or more iso-enzymes of varied response towards calcium might account for the observations. An equally plausible interpretation of the observations is that only a single enzyme is produced but that depending upon internal ionic conditions under which the protein is synthesized changes in the sensitivity of the enzyme towards calcium occur (Curdel 1966).

In an attempt to decide which of these two mechanisms is the more probable in clones of *L. minor* it is pertinent to consider additional information which has some bearing on this question. Since it appears likely that changes in response to calcium ions are general among the regulatory enzymes of *L. minor* and as similar phenotypic changes in the properties of other enzyme systems in a range of organisms are well known (Curdel 1966; Kingdom and Stadtman 1967; Harel and Mayer 1968; Sims, Stewart and Folkes 1969) it seems probable that this phenomenon is widespread in living organisms. If the explanation for all these examples lay solely in the existence of genetically distinct iso-enzymes then an organism would require considerable genetic information to adjust to even small changes of ionic conditions. In this sense it is likely that enzyme interconversions involving a single enzymic species are economical in terms of genetic information. Nevertheless, in such a flexible system, the maintenance of the integrity of metabolic pathways poses a real difficulty, since unless a large number of enzymes of a pathway adjust to a comparable extent when the internal level of an ion is altered, considerable impairment of metabolic integrity could result.

There is some evidence that malic dehydrogenase may exist as distinct iso-enzymes in plants (Yue 1966). However, by using the same ion-exchange procedures employed by other workers we find no evidence that changes in the properties of the enzyme extracted from duckweed are due to the synthesis of distinct iso-enzymes. For example, the products of fractionation are always more senstitive to calcium inhibition than the original extract

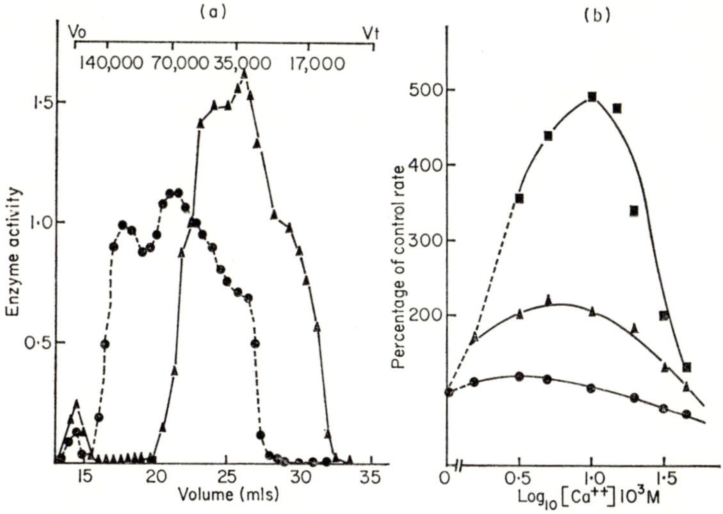

FIG. 5. The fractionation and properties of different molecular weight aggregates of malic dehydrogenase.

(a) Determination of the molecular weight of enzyme originating from cells grown on different concentrations of calcium. A column (31·5 × 1·1 cm) of SAGAVAC 10 was eluted with 0·1 M TRIS-HCl (pH 7·25) containing 1×10^{-5} M $CaCl_2$.

(●) enzyme originating from plants growing on $2{\cdot}5 \times 10^{-2}$ M (Ca^{++}).

(▲) enzyme originating from plants growing on 1×10^{-3} M (Ca^{++}).

(b) The response of the various sub-unit aggregates to calcium ions.

(■) response of the aggregate of MW 136,000 ∓ 8,000

(▲) response of the aggregate of MW 68,000 ∓ 4,000

(●) response of the aggregate of MW 34,000 ∓ 2,000

and rechromatography of a single enzyme component results in the re-appearance of several peaks of activity throughout the chromatogram. It seems likely that the original protein structure is unstable and can dissociate into a number of smaller structures during fractionation procedures. In contrast to these ion-exchange methods gel filtration avoids strong surface absorption and permits a fractionation of proteins in accordance with their molecular weight. Use of this technique (Fig. 5) shows that enzyme originating from material grown on $2{\cdot}5 \times 10^{-2}$ M (Ca^{++}) is in a higher molecular weight state than the corresponding enzyme found in plants growing at a

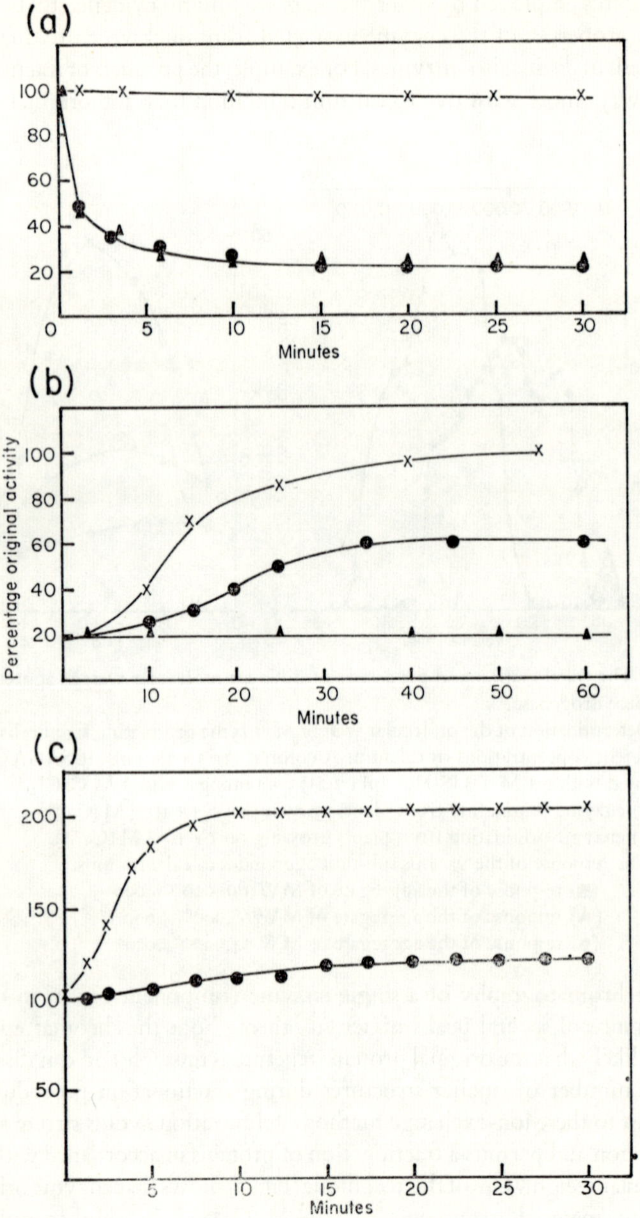
(a)
100
80
60
40
20
0
5
10
15
20
25
30
Minutes
(b)
Percentage original activity
100
80
60
40
20
10
20
30
40
50
60
Minutes
(c)
200
150
100
50
5
10
15
20
25
30
Minutes

FIG. 6

lower calcium concentration (1×10^{-3} M (Ca^{++})). It is now well established that many proteins are composed of sub-units, most dehydrogenase enzymes being based on molecular weight units between 14–20,000 (Whitehead, 1965). The most probable explanation for observed differences in the two elution profiles is that the degree of aggregation of malic dehydrogenase can vary. Stewart (1968) has shown that the lowest detectable molecular weight of malic dehydrogenase from *Lemna minor* is $17{,}000 \pm 2000$ and from the elution profile shown in Fig. 5 it is possible to calculate that the peaks of activity correspond to exact multiples of this molecular weight. The enzyme synthesized in cells growing at the higher level of (1×10^{-2} M) calcium is predominantly in the form of octamers (i.e. $8 \times 17{,}000$) tetramers and dimers, whereas the enzyme from plants growing at 1×10^{-3} M calcium never exceeds the tetrameric state. Quantities of malic dehydrogenase in its various molecular aggregate forms have been isolated and measurements made of the response of these aggregates to a range of calcium concentrations (Fig. 5). It is evident that the differences in behaviour of the octamer, tetramer and dimer forms of the enzyme closely parallel the activity of the extracts which are obtained from plants grown in different calcium solutions. It is therefore highly probable that

---

FIG. 6—*continued*

FIG. 6 *In vitro* studies with malic dehydrogenase; the participation of calcium in determining the state of aggregation of the enzyme.

(a) The effect of calcium in preventing the dissociation of the enzyme into sub-units.

The material was grown on 1×10^{-2} M [Ca^{++}] and the enzyme incubated at 25° C in pH 9·0 TRIS buffer alone or containing calcium or sodium ions. Aliquots were removed at intervals and assayed at pH 7·25 in the presence of $1{\cdot}66 \times 10^{-2}$ M [Ca^{++}].

(×) Enzyme incubated in TRIS containing 1×10^{-3} M $CaCl_2$

(●) Enzyme incubated in TRIS containing 5×10^{-2} M NaCl

(▲) Enzyme incubated in TRIS alone

(b) The re-aggregation of sub-units of malic dehydrogenase.

The enzyme, dissociated by alkali treatment, was adjusted to pH 7·25 and treated in several ways. All assays were carried out in the presence of $1{\cdot}66 \times 10^{-2}$ M [Ca^{++}]

(×) Enzyme incubated with 3×10^{-3} M NADH + 1×10^{-3} M [Ca^{++}]

(●) Enzyme incubated with 3×10^{-3} M NADH

(▲) Enzyme incubated with 3×10^{-3} M NADH + 5×10^{-2} M [Na^{+}]

(c) The involvement of calcium ions in the formation of high molecular weight aggregates of malic dehydrogenase.

Lemna was grown on 5×10^{-3} M [Ca^{++}]. Enzyme, originally giving 230% activation over control when assayed in presence of $1{\cdot}25 \times 10^{-2}$ M [Ca^{++}], was incubated at 40° C and treated in several ways. All assays were carried out in $1{\cdot}25 \times 10^{-2}$ M [Ca^{++}].

(×) incubated in the presence of 3×10^{-3} M NADH + 8×10^{-3} M [Ca^{++}]

(●) incubated in presence of 3×10^{-3} M NADH.

the changes in the response of the enzyme to calcium levels are a result of differences in the number of sub-units in the functional enzyme and that this number depends upon the concentration of calcium ions within the tissue when the enzyme was synthesized.

There is also experimental evidence to suggest that changes in the configuration of the enzyme are directly mediated by calcium ions. This evidence can be summarized as follows:

1. The presence of calcium ions in the elution buffer helps to stabilize the higher molecular aggregates of malic dehydrogenase, whereas Na^+ and K^+ are not effective.

2. When malic dehydrogenase is incubated in alkaline TRIS buffer it can be shown to lose part of its activity, as estimated by assay in the presence of a constant level of calcium, this being a result of its dissociation into smaller sub-units each having a reduced calcium activation (Yoshida 1965). Enzyme preparations incubated under similar conditions but in the presence of calcium show no decrease in activity with time when assayed at the same final concentration of calcium ions (Fig. 6); calcium ions alone are effective in protecting the enzyme from dissociation under these conditions.

3. It can be shown that enzyme preparations previously dissociated by treatment with alkali buffer can only be completely re-associated if calcium ions are present (Fig. 6).

4. It can also be shown that the presence of calcium ions is obligatory to produce an increase in the state of aggregation of malic dehydrogenase. On warming the enzyme in the presence of calcium ions it has been possible to convert a preparation which was predominantly in the form of tetrameric units into one composed largely of octameric units. No other ion tested could bring about a comparable modification of the enzyme (Fig. 6).

Finally, there is direct experimental evidence to indicate that reversible changes in sub-unit configuration can occur *in vivo* in response to changes of internal calcium concentrations. Plants were grown in a continuous culture apparatus in a medium containing 5×10^{-4} M (Ca^{++}) and allowed to reach a steady state condition. The inflow medium was switched to one containing a higher concentration of calcium ions (1×10^{-2} M (Ca^{++}) and in this way, the plants were subjected to a sudden increase of both external and internal concentration of calcium. By studying the kinetic response of the malic dehydrogenase enzyme throughout this period it has been possible to establish that modifications of the sub-unit configuration of the enzyme were occurring (Fig. 7). The immediate response of the cells to an

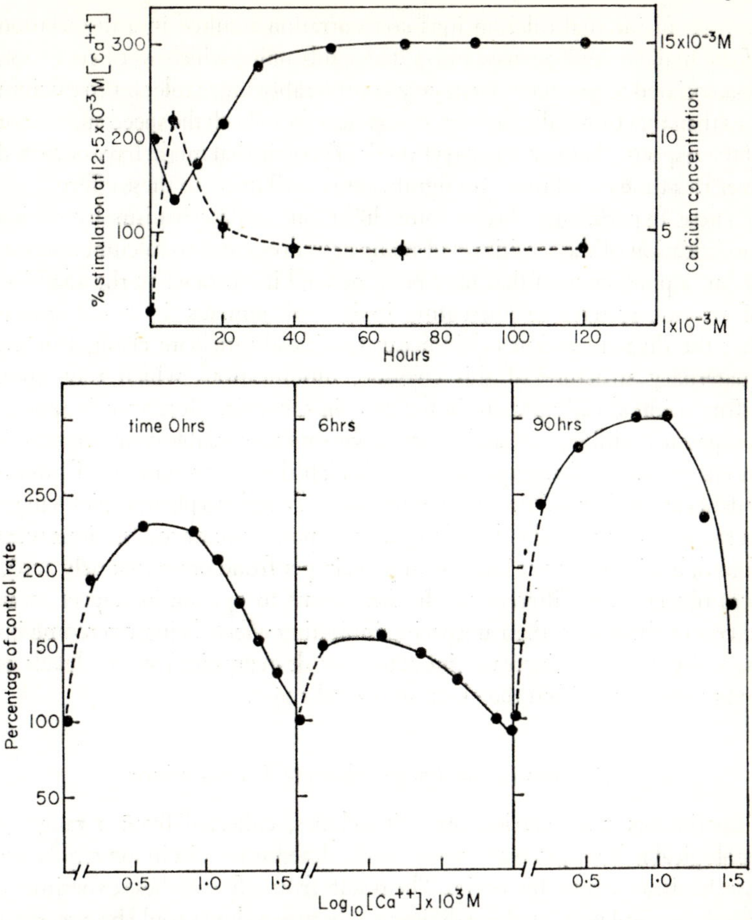

FIG 7. The involvement of aggregation and disaggregation phenomena *in vivo* as revealed by the effects of sudden changes of internal calcium concentration.

Lemna minor was grown in continuous culture in a medium containing 5×10^{-4} M (Ca^{++}). At times 0 hr the inflow medium was switched to one containing 1×10^{-2} M (Ca^{++}) and measurements were made over a period of time on the response of malic dehydrogenase to calcium and of the internal concentration of calcium ions in the duckweed.

Graph (*a*)

(⧫) Relative internal concentration of Ca^{++}

(●) Activity of malic dehydrogenase when estimated in the presence of $1{\cdot}25 \times 10^{-2}$ M Ca^{++}

Graph (*b*)

(●) Changes in the response of malic dehydrogenase to calcium ions at times after the medium has been changed.

increase in internal calcium ions concentration resulted in a dissociation of the malic dehydrogenase into smaller sub-units, which were very soon re-associated to produce an enzyme considerably more tolerant to calcium ions than that originally present. It was clear from both the speed and extent of the response that any *de novo* synthesis of protein that might have occurred over this time could make no significant contribution to these effects.

These experiments throw some light on the mechanism by which modification of the sensitivity of malic dehydrogenase to calcium ions can occur, a phenomenon that may be of general importance in the adaption of enzyme systems to fluctuating ionic environments. It would appear that the divalent cation calcium can specifically mediate changes in the quaternary structure of this enzyme, modifications which rely upon information already present in the protein sub-unit. Moreover, it appears to operate rapidly *in vivo* and to occur without appreciable *de novo* synthesis of enzymes. In other experiments in which the concentrations of monovalent cations (potassium and sodium) were altered, no phenotypic changes in the response of malic dehydrogenase to potassium or sodium ions were found, but instead a net increase of protein per frond of up to nearly 300% was observed. A different mechanism seems to operate in respect to an excess of these ions, their unspecific inhibiting effects being overcome instead by a general increase (induction or de-repression) in the levels of many enzymes within the plant (see Orgel 1964).

A comparison of different clones of Lemna minor

Experiments were carried out with clones, collected from a range of ecological habitats, to see to what extent differences exist in ion regulatory mechanisms within the species. The results from a few of these experiments are shown in Fig. 8. Striking differences between clones could be discerned, the behaviour of clone S_2 closely resembled that of S_1 both in terms of growth and enzymic response whereas S_4 and S_8 are appreciably less influenced by high levels of calcium in the culture media. It is clear that both the reduced response of the enzyme to calcium ions and the sustained growth are indicative of the ability of S_4 and S_8 clones to control the entry of calcium ions into their cells. There is circumstantial evidence to support this view (Stewart 1968). To take but one instance, whereas measurements of the rate of respiration of clones S_1 and S_2 show considerable variation in relation to changes of external calcium concentration, plants of clone S_8 remain unaffected. Moreover in clones S_1 and S_2, the concentration of calcium which gives maximal rate of respiration is variable, the maximal

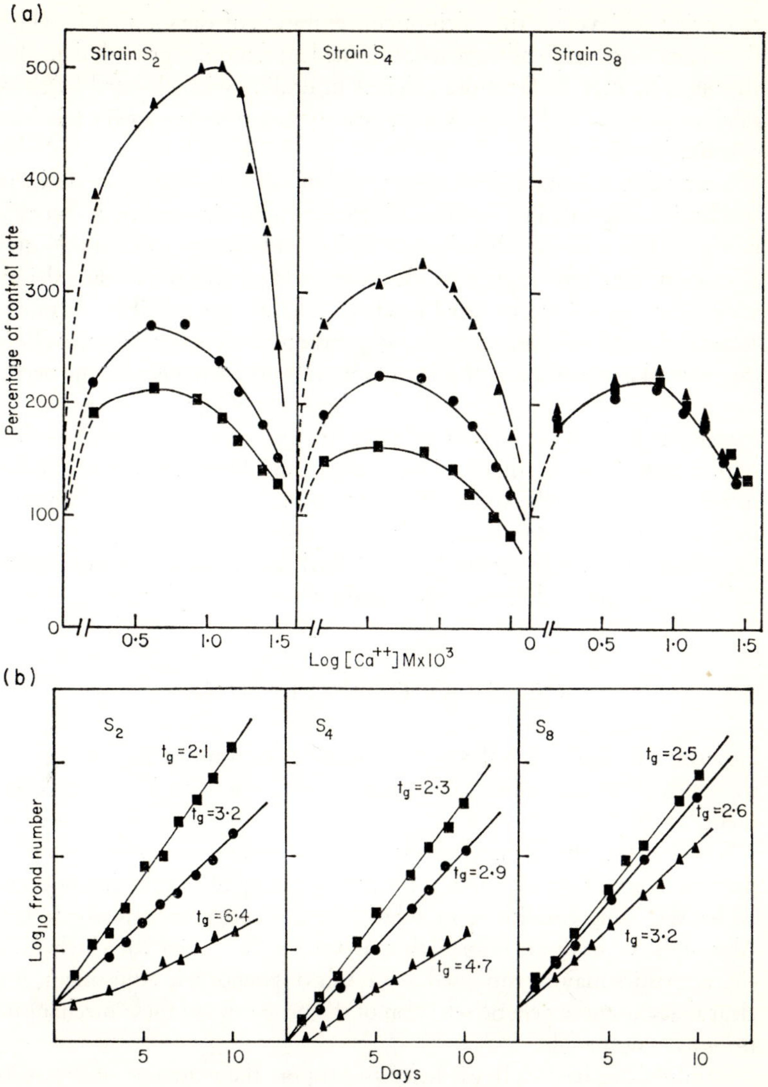

FIG. 8. Variations in the behaviour of clones of *Lemna minor* when cultured on a range of calcium concentrations.

(a) The response of malic dehydrogenase obtained from different clones.

(▲) enzyme was obtained from plants growing in medium containing 1×10^{-2} M Ca^{++}

(●) enzyme obtained from plants growing in medium containing 1×10^{-3} M Ca^{++}

(■) enzyme obtained from plants growing in medium containing 3×10^{-4} M Ca^{++}

(b) Measurements of the growth rate of various clones of *Lemna minor*.

Symbols as above except they indicate measurements of frond number. *tg* indicates the time in days for a doubling of frond number to have occurred.

rate being achieved at the calcium concentration of the solution in which the plants were originally grown. It is evident from the growth data that although S_8 may be far more efficient in dealing with elevated levels of calcium ions, S_2 and S_4 grow significantly faster on the lower levels of calcium (see Fig. 8).

It would thus seem likely that even within a single species quite different mechanisms operate in adapting cells to varying ionic conditions. It could be argued from the data that clones which exert little control over the entry of calcium ions, have a growth advantage at the lower end of the calcium scale, whereas at high external levels of calcium any possible advantage associated with economies in the expenditure of energy in controlling the internal concentration of calcium ions is more than offset by the necessity of efficiently adjusting a very large number of enzyme systems to an extreme range of ionic conditions. Clearly any interpretation of these data is speculative but nevertheless, since calcium can inhibit such a large number of enzyme systems the successful adaption of a plant to a wide range of ionic conditions, in the absence of ionic control, will necessitate extensive modification of metabolic machinery. Clearly any defect in this mechanism will impose a severe limitation on the distribution of a plant.

GENERAL CONCLUSIONS

Experiments discussed in this paper have revealed some of the mechanisms by which plants can adjust to fluctuations of ions in the external environment.

It is clear that the different properties of the potassium uptake system in the two liverwort species permit the maximum uptake of this ion to occur under the ionic conditions in which these plants are normally found. Although we have no information as yet on the extent to which these characteristics may be modified as a result of phenotypic adjustment, it is clear that extensive genetic selection of the properties of these mechanisms must have occurred.

Whilst there is already evidence to support the existence of inwardly directed potassium pumps in plants (Dainty 1962), the significance of the maintenance of a high internal level of potassium in many plant tissues has not been sufficiently stressed. It is now well known that potassium may participate in the functioning of a large number of metabolic pathways, the ion being involved with the maintenance of an active configuration of many enzymes (Evans and Sorger 1966) and hence factors which influence the assimilation of this ion are likely to be of major importance to a plant. It is clear that inwardly directed pumps do not operate in isolation of the external environment and are influenced to different degrees by the presence of other ions. In this and many other respects the properties of potassium pumps seem to resemble those of allosteric enzymes (Monod, Changeux and Jacob 1963). For example, under certain conditions the carrier appears to exhibit co-operative kinetics in that the rates of potassium uptake at very low concentrations is less than would be predicted from the apparent affinity at higher concentrations. (Jefferies, unpublished). Moreover, the very extensive activation of potassium uptake by sodium ions observed with several halophytic plants (Parham and Jefferies, unpublished) could also be readily accounted for in terms of sodium-induced configurational changes which affect the affinity of the carrier for potassium ions.

If carrier systems do resemble allosteric proteins and possess both regulatory and catalytic sites it is clear that considerable genetic and phenotypic modification of their characteristics can occur, changes that can be quite independent of the ability of the carrier to concentrate potassium ions.

In some environments ions such as calcium and sodium are present in concentrations that are toxic to many plants. However, in many instances, plants that grow in these habitats appear to possess ion extrusion pumps to regulate the internal levels of these ions. The existence of calcium extrusion pumps has been postulated in barley (Pitman and Saddler 1967) and in oats (Higinbotham *et al.* 1966). It is quite clear however that certain clones of *L. minor* show marked differences over the entry of this ion into their tissues, nevertheless considerable modifications of the properties of the enzyme machinery can occur which enable the plant to grow over a relatively wide range of calcium ion concentrations. Some plants, including *Cephalozia connivens* and *Leiocolea turbinata* may be unable to regulate adequately the internal level of calcium and sodium ions and if certain of their enzyme systems have only a limited capacity to adjust to calcium and sodium it is not surprising that such plants should be restricted to a narrow range of ionic environments. The combination of an ion extrusion pump coupled with enzyme adjustment, however, offers what can be energetically the most economical means by which plants may adapt to habitats in which an ion is present in high concentration. The two mechanisms by operating in tandem will reduce the high energy expenditure associated with maintaining the internal concentration at a low level, remove the necessity for a rigid control over the ion concentration and circumvent the need of enzyme systems to adjust to a wide range of ionic concentrations. There is now evidence that both these mechanisms are operative in some halophytic species (Dainty 1962; Sims, Stewart and Folkes 1969).

The competitive ability of a species must depend, in part, upon its making the most efficient use of protein and ions essential for metabolic functioning with a minimum outlay of energy. Laboratory experiments of the type described here can reveal how metabolic systems adjust to maintain a high efficiency in plants growing under ideal light and nutritional conditions, but it must be remembered that in natural habitats overall efficiency may be maintained by a variety of means. For example, it is clear from determinations of the electrical potentials in young and adult leaves and from measurements of the response characteristics of enzymes extracted from different tissues of a plant that a continuous adjustment can occur in the relative contributions made by different regulatory mechanisms. Such changes indicate that the effectiveness of a particular regulatory system will vary with the age of the tissue and be dependent upon internal and external factors. Although alterations in the conditions under which plants are grown should make it possible to obtain information on the relative contributions and efficiences of these regulatory mechanisms, a full appreciation of their importance as ecological determinants can come only from extensive studies of a species growing over a range of natural conditions.

ACKNOWLEDGEMENTS

The authors would like to thank Miss J. Henry and Mrs J. Stannard for their valuable technical asistance and Professor B.F. Folkes for many helpful suggestions and for a critical reading of the manuscript. Miss D. Brereton kindly typed the paper. Two of us (A.P.S. and G.R.S.) would like to thank Seravac Laboratories for gifts of SAGAVAC. Part of this work was supported by a N.A.T.O./S.R.C. Fellowship awarded to R.L.J., and G.R.S. is indebted to the S.R.C. for the award of a Research Studentship.

REFERENCES

BOLLARD E.G. and BUTLER G.W. (1966). Mineral nutrition of plants. *Ann. Rev. Pl. Physiol.* **17**, 77–112.

CURDEL A. (1966). Influence of the nature of the metallic prosthetic group on the biosynthesis and enzymatic properties of the D-lactic dehydrogenase in yeast. *Biochem. biophys. Res. Comm.* **22**, 357–363.

DAINTY J. (1962). Ion transport and electrical potentials. *Ann. Rev. Pl. Physiol.* **13**, 379–402.

EPSTEIN E. and HAGEN C.E. (1952). A kinetic study of the absorption of alkali cations by barley roots. *Pl. Physiol.* **27**, 457–474.

EPSTEIN E. and JEFFERIES R.L. (1964). The genetic basis of selective ion, transport in plants. *Ann. Rev. Pl. Physiol.* **15**, 169–184.

EVANS H.J. and SORGER G.J. (1966). Role of mineral elements with emphasis on the univalent cations. *Ann. Rev. Pl. Physiol.* **17**, 47–76.

GRIME J.P. (1963). Factors determining the occurrence of calcifuge species on shallow soils over calcareous substrata. *J. Ecol.* **51**, 375–390.

HAREL E. and MAYER A.M. (1968). Interconversion of sub-units of catechol oxidase from apple chloroplasts. *Phytochem.* **7**, 199–204.

HIGINBOTHAM N., ETHERTON B. and FOSTER R.J. (1967). Mineral ion contents and cell trans membrane electropotentials of pea and oat seedling tissue. *Pl. Physiol.* **42**, 37–46.

KAPLAN N.O. (1963). Symposium on Multiple Forms of Enzymes and Control Mechanism. I. Multiple Forms of Enzymes. *Bacteriol. Rev.* **27**, 159–169.

KINGDOM H.S. and STADTMAN E.R. (1967). Two *E.coli* glutamine synthetases with different sensitivities to feedback effectors. *Biochem. biophys. Res. Comm.* **27**, 470–473

MONOD J. and JACOB F. (1961). In: 'Cellular Regulatory Mechanisms'. *Cold Spr. Harb. Symp. quant. Biol.* **26**, 389–401.

MONOD J., CHANGEUX J.P. and JACOB F. (1963). Allosteric proteins and cellular control systems. *J. molec Biol.* **6**, 306–329.

ORGEL L.E. (1964). Adaptation to widespread disturbances of enzyme function. *J. molec Biol.* **9**, 208–212.

PITMAN M.G. (1964). The effect of divalent cations on the uptake of salt by Beetroot tissue. *J. exp. Bot.* **15**, 444–456.

PITMAN M.G. and SADDLER H.D.W. (1967). Active sodium and potassium transport in cells of barley roots. *Proc. Nat. Acad. Sci. U.S.* **57**, 44–49.

SCOTT R. (1961). A pumpless circulator and geiger counter holder. *J. Sci. Instrum.* **38**, 31.

SIMS A.P., STEWART G. R., and FOLKES B.F. (1969). The effect of inorganic cations on the activity of plant enzymes; differences in enzyme properties in relation to species and habitat. *New Phytol.* (In press.)

SINCLAIR J. (1965). The ionic relations of the leaves of *Hookeria lucens*. Ph.D. thesis. University of East Anglia.

SNAYDON R.W. and BRADSHAW A.D. (1961). Differential response to calcium within the species *Festuca ovina* L., *New Phytol.* **60**, 219–234.

STEWART G.R. (1968). The property of malic dehydrogenase in relation to ionic adaptation. Ph.D. Thesis. University of Bristol.

UMBARGER H.E. (1961). In: 'Cellular Regulatory Mechanisms'. *Cold. Spr. Harb. Symp. quant. Biol.* **26**, 301–311.

VIETS F.G. (1944). Calcium and other polyvalent cations as accelerators of ion accumulation by excised barley roots. *Pl. Physiol.* **19**, 466–480.

WHITEHEAD E.P. (1965). A theory of the quaternary structure of dehydrogenases, dehydrogenating complexes and other proteins. *J. Theor. Biol.* **8**, 276–306.

YOSHIDA A. (1965). Enzymic properties of malic dehydrogenase of *Bacillus subtilis*, *J. biol. Chem.* **240**, 1118–1124.

YUE S.B. (1966) Isoenzymes of malate dehydrogenase from barley seedlings. *Phytochem.* **5**, 1147–1152.

THE UPTAKE OF PHOSPHATE AND ITS TRANSPORT WITHIN THE PLANT

B. C. Loughman
Dept. of Agricultural Science, University of Oxford

INTRODUCTION

Before discussing the factors which influence the way in which sources of phosphorus in the soil are utilized by roots one must consider the wide range of organic forms of the element which may be present in addition to inorganic orthophosphate. These organic forms arise by decomposition of plant residues and are usually present as derivatives of sugars. Although simple sugar phosphates can be absorbed directly, the distinction between direct absorption and absorption after hydrolysis is difficult to make with existing experimental techniques.

The presence of organic forms of phosphorus in the root environment raises the question of the need for mechanisms of utilization of these forms. Surface phosphatases are readily demonstrated in roots of many species and when inorganic phosphate is in short supply the activity of surface phosphatases increases, this activity being also affected by the molybdenum level in the environment (Spencer 1954). Clearly, such changes may be of considerable significance to the economy of the plant, if hydrolysis to inorganic phosphate is a prerequisite for entry into the root. The utilization of organic phosphorus depends on micro-organisms as well as enzymes on the root surface and pH is clearly a major factor in such utilization.

The work of Jeffery (1964) with *Banksia ornata* suggests that young plants are able to absorb orthophosphate at high rates in time of plenty, store it as inorganic polyphosphate and release it within the plant when the external supply is low. Such seasonal rhythm in the storage form of phosphorus is of obvious ecological importance since it enables *B. ornata* to survive under low phosphate conditions after prior absorption under more favourable conditions. The turnover of phosphorus within the root is reduced because of incorporation into polyphosphate and at the same time transport to the shoot is lowered.

It seems that *Banksia* exhibits a special case of storage in time of plenty for use in time of shortage. This is, of course, only a variation of the more normal situation of orthophosphate storage in the vacuole when the normal cytoplasmic level is exceeded.

I shall confine myself to a consideration of the more usual situation of plants absorbing orthophosphate from the soil solution, but the ecological advantage of a particular species probably involves the development of special systems as shown by *Banksia*. One must also consider the wide range of organic compounds known to be released from roots into the environment which may act specifically on mechanisms of phosphate utilization by other plants. Little work has been done on this aspect of the problem but it is clear that the addition of organic acids, sugars, hormones and other metabolites to the root environment can have a marked effect on uptake and utilization of orthophosphate and other ions. Consequently, in mixed populations, release of metabolites from one species may markedly affect the absorption of ions by others and evidence of this kind may be very helpful to our understanding of plant interaction.

Before attempting to provide information concerning the utilization of phosphate by whole plants which might be useful to those whose primary interest lies in the ecological aspects, it is necessary to give a brief outline of current views on the relationship between the component parts of the growing plant with respect to the absorption and translocation of ions. It is clear that the major pathway of ion movement from the root to the shoot is via the xylem and as a consequence ions in the external solution must traverse the cortex in order to reach the first effective barrier to free diffusion, the endodermis. Since the thickened walls of the endodermal cells effectively prevent free diffusion of ions into the stele, the path of movement is likely to be via cytoplasmic bridges between the cells of the inner cortex and those of the stele. Although the initial absorption of the ion into a cell may occur outside, but adjacent to, the endodermal layer after free diffusion through the walls of the cortical cells, absorption can occur throughout the cortex, some fraction of the ion being effectively diverted into vacuoles from which it is relatively unavailable to other parts of the plant. It is possible that the ions reaching the xylem are accumulated into the cytoplasm or cytoplasmic organelles of cortical cells, subsequently being conveyed to inner cells via plasmadesmata. The capacity of the cytoplasm of these latter cells for retention of the accumulated ions may well be less than that of the cells in which the accumulation took place and as a result the ions may be released into the xylem prior to transfer in the transpiration stream to the shoot. Alternatively an endodermal pumping mechanism could transfer ions into the xylem. The pattern of utilization has special features in the case of rapidly metabolized ions like phosphate. If the entering phosphate is incorporated into an organic form during the accumulation process then there may be a dephosphorylative step at the point of entry into the xylem leading to the release of inorganic orthophosphate into the transport system supplying the shoot. Such a system might be modified by many factors, chemical, biochemical or physiological and I wish to present an experimental approach designed to investigate the possibility that the entry of phosphate into the xylem after active metabolic absorption by the cortical cells may be controlled in this way.

Although it is possible that phosphorus travels to the cells of the leaf as the phosphate ion without having been involved in ester formation, the evidence presented favours the interpretation that phosphorylative reactions are a prerequisite for the transport process.

The biochemical mechanisms involved in the utilization of phosphate by higher plants have been studied by a number of research groups and the effects of changes in the environment on phosphate uptake and movement have also received attention. Commonly, this work has been carried out with fragments of plants and the conclusions reached have been of relatively little value in aiding our understanding of how the whole plant functions. I would like to briefly outline an attempt to combine the biochemical and physiological approaches, in a study of phosphate absorption and transport in whole plants.

The process can be divided into the following sequence of components:

(*a*) The primary step of absorption.

(*b*) The transfer across the cortex of the root.

(*c*) The presentation of phosphate to the xylem for transfer to the shoot.

(*d*) The utilization of phosphate by the photosynthetic tissues of the leaf.

More is known about (*a*) and (*d*) than the other processes but there is no doubt that our understanding of the overall process cannot be improved without some attempt at an assessment of the intervening stages.

INITIAL ABSORPTION

Whether metabolic accumulation occurs primarily in the peripheral cells of the root or throughout the cortex is open to doubt; autoradiographs of frozen sections sometimes show an increase in concentration of ^{32}P from the epidermis to the endodermis after short periods of absorption (Crossett 1967). The elegant use of the X-ray microanalyser for examination of ionic distribution within roots by Läuchli (1967) is currently producing very important results in this field. The high resolving power of this instrument enables a distinction to be made between phosphate accumulated in the wall of a xylem vessel and that in the lumen and further refinement using a smaller beam diameter is likely to be of great value in advancing our knowledge of ion transport.

The major point that must be made is that when plants are exposed to concentrations of KH_2PO_4 similar to those occurring in the soil solution the metabolic incorporation into the phosphorylative turnover system in the cytoplasm of the root cells can be very rapid indeed. Within 10 sec of entry 30% of the phosphate absorbed by barley plants is incorporated into nucleotides and after a subsequent 50-sec period up to 80% is in the form of organic phosphates. The evidence indicates that the primary sites of phosphate absorption are those reactions of the electron transport pathway in which ATP is formed; subsequently, labelled phosphate appears in all the intermediates of the normal metabolic pathways. The rapidity with which reactions occur make it extremely difficult to establish the precise order in which compounds are labelled. It is possible to show that there can be an absorption of very small quantities of inorganic phosphate without any involvement of the esterification system in some tissues in which metabolism is kept at a low rate, e.g., fresh potato slices. However, in actively metabolising roots of whole plants it seems likely that the processes of mitochondrial oxidative phosphorylation are essential for the maintenance of phosphate entry (Loughman 1960). At this point it should be mentioned that there is evidence that only a very small proportion of the inorganic orthophosphate in the mature plant cell is in the cytoplasm and thereby actively involved in metabolism; the greater part is to be found in the vacuole and is unlikely to be involved in any system which is turning over rapidly (Loughman 1960; Bieleski and Laties 1963; Jennings 1964).

Evidence for the metabolic pathway by which phosphate is incorporated in the root has been provided by Loughman and Russell (1957), Kursanov and Vyskrebentseva (1960), Miettenen and Savioja (1958) and Jackson and Hagen (1960). Although some experimental differences have given rise to slight variations in the detailed picture there is general agreement as to the overall sequence in which intermediates are labelled.

THE SEPARATION OF ABSORPTION AND TRANSPORT

As in all biochemical pathway investigations much of the information gained has been the result of work with metabolic inhibitors. Hormones and herbicides have been used for this purpose and it is worth pointing out that in addition to obtaining information about the pathway of phosphate incorporation such experiments often give an insight into the mode of action of the growth regulator. An example of this approach is provided by experiments using the defoliant aminotriazole. This compound is capable of forming an aminoglucoside within plant tissues and it has been claimed that formation of this glucoside occurs when aminotriazole and glucose-1-phosphate are mixed in the presence of plant extracts (Frederick and Gentile 1960). Since glucose-1-phosphate is considered by Jackson and Hagen (1960) to be one of the earliest intermediates to become labelled when ^{32}P orthophosphate is fed to barley roots it was thought worthwhile to examine the effect of aminotriazole on the utilization of phosphate. (Wort and Loughman 1961). It was found that although aminotriazole has no effect on the absorption by the root, it brings about a very marked increase in the rate at which the absorbed phosphate is transferred to the shoot, amounting to six-fold increase in the first hour. This phenomenon is accompanied by a marked decrease in the rate at which the absorbed phosphate is incorporated into nucleic acids in the root and subsequent work has established that a primary effect of this particular defoliant is on the synthesis of nucleic acid precursors.

Results of this kind show that compounds provided from outside the plant can have a selective effect on the process of uptake and transport. Under conditions where there is little effect on the uptake process, the transport may nevertheless be very significantly increased or decreased.

Seth and Wareing (1967) have shown that the transport of phosphate in young bean plants can be enhanced almost 100-fold by treatment with a mixture of indoleacetic acid, gibberellic acid and kinetin and it is clear that such hormone directed transport is an important factor during the development of seeds.

A good example of selection between uptake and transport is given by results of experiments in which sugars are fed with the orthophosphate. As might be expected, glucose and fructose cause an increased uptake and no modification of the phosphorylative pattern within the root occurs. Mannose has only a small inhibitory effect on the overall uptake of phosphate by the roots of barley plants but alters the metabolic incorporation pattern, most of the ^{32}P being found in mannose-6-phosphate. However, as a result, the subsequent transport of phosphate to the shoot is almost completely blocked (Loughman 1966). In the normal plant it can be shown that although rapid esterification of incoming phosphate occurs, only orthophosphate is present in the xylem sap. The results obtained with mannose suggest that some organic phosphate compound in the root may be carried to the stele, there to be hydrolysed with the release of inorganic phosphate into the xylem. The fact that certain phosphatase inhibitors such as fluoride can prevent transport rather than uptake lends support to this idea (Loughman 1966).

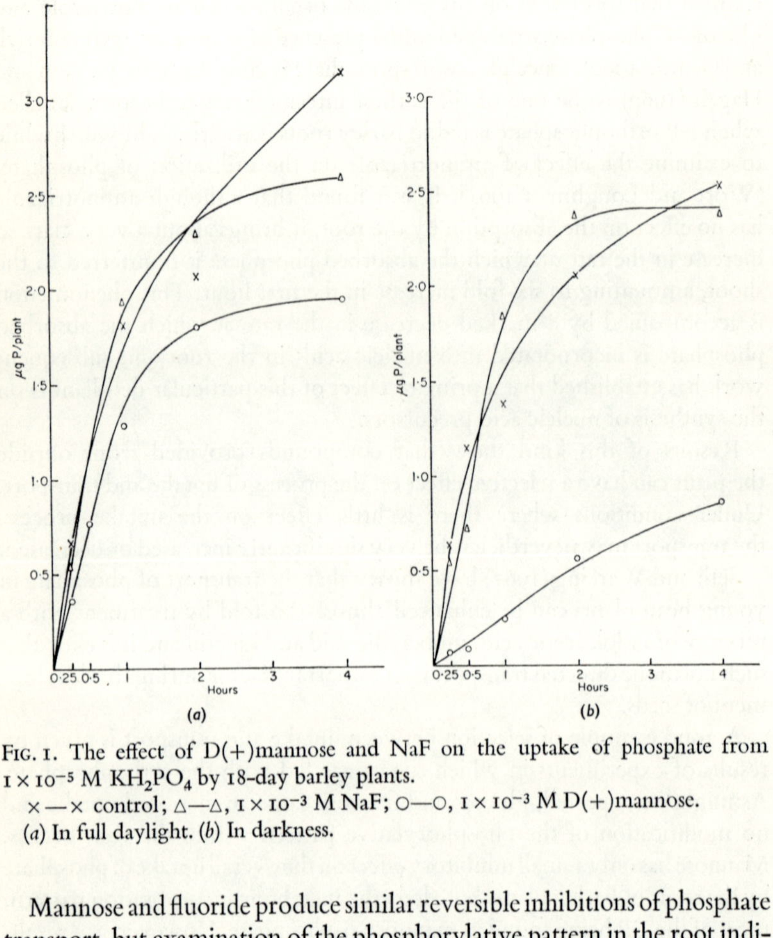

FIG. 1. The effect of D(+)mannose and NaF on the uptake of phosphate from 1×10^{-5} M KH_2PO_4 by 18-day barley plants.
×—× control; △—△, 1×10^{-3} M NaF; ○—○, 1×10^{-3} M D(+)mannose.
(*a*) In full daylight. (*b*) In darkness.

Mannose and fluoride produce similar reversible inhibitions of phosphate transport, but examination of the phosphorylative pattern in the root indicates that only in the presence of mannose is there a demonstrable biochemical change. The presence or absence of light complicates the picture still further. Control plants absorb phosphate at similar rates in the dark and light, although transport occurs more slowly in the dark. Fluoride has little effect on absorption by the root, but reduces transport significantly, this inhibition of transport being more readily demonstrated in young plants than in old. Mannose has a much greater effect on total uptake in the dark than in the light, the rate of uptake over the first hour being reduced to one fifth in darkened plants (Fig. 1).

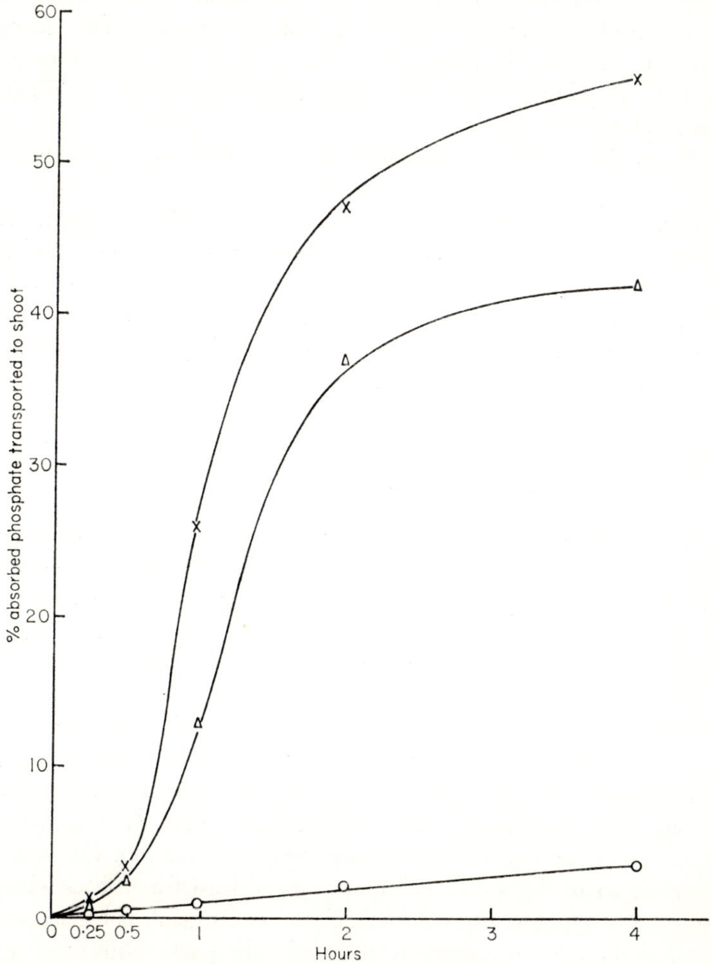

FIG 2. The effect of D(+) mannose and NaF on the percentage transport of phosphate to the shoots of 18-day barley plants in full daylight.

×—× control; △—△ 1×10^{-3} M NaF; ○—○, 1×10^{-3} M D(+)mannose.

The percentage of absorbed phosphate transferred to the shoot in the light is shown in Fig. 2 and similar results are obtained in darkness.

Such observations confirm the complex interaction between the environmental conditions of the plant as a whole and the biochemical mechanisms involved in utilization of phosphate.

Experiments carried out with plants of differing calcium status indicate that plants deprived of calcium do not lose their ability to absorb phosphate but are unable to transfer it from the root to the shoot. It is of considerable interest to find that the transport system of plants of low calcium status shows quite different responses to mannose from that already described for normal plants.

Changes in oxygen tension or temperature as well as calcium status and light intensity, can modify the selective effect of externally applied compounds. Perhaps the most important factor is the age of the plant. These factors affect the relative capacities for uptake and transport and it is possible that the biochemical changes which occur are similar to those induced by hormones or cell metabolites like mannose. If such mechanisms exist it is to be hoped that the combined physiological and biochemical approach will be of value in helping in their elucidation.

EXPERIMENTS WITH SINGLE PLANTS

Precise experimental methods for the investigation of ion absorption and transport by individual plants in a strictly controlled environment have been devised by Crossett (1966) and Larkum (1967). Both these processes can be followed over long periods during which water movement through the plant and oxygen tension in the root and shoot environment can also be continuously measured. Experiments using these methods have added much to the information gained in the experiments described earlier, and made it possible to follow alterations in absorption and transport of phosphate in a single plant brought about by controlled changes of environment.

Using these methods Crossett and Loughman (1966), showed that only a small proportion of the total orthophosphate of the root was available for transport to the shoot. The experimental basis for such conclusions rests upon the finding that the total amount of inorganic phosphate transferred to the shoot was independent of the phosphate status of the root. Data of this type confirm the conclusions of earlier experiments that the rapidity of esterification is compatible with a small cytoplasmic orthophosphate fraction with which the phosphate entering becomes equilibrated within minutes of entry. It is from this fraction that the subsequent transport material is provided, the pool itself being maintained at constant level in balance with the processes of absorption and transport.

Having established a steady rate of transport in a single plant by continuous monitoring of the leaf, an inhibitor can be introduced into the root environment and its effect followed. When mannose is used at a concentration of 1×10^{-3} M in this way, almost instantaneous cessation of transport occurs indicating that the metabolic changes leading to incorporation of the orthophosphate into sugar phosphates are confined to the cytoplasmic pool which is turning over rapidly.

THE EFFECT OF OXYGEN AVAILABILITY ON UTILIZATION OF PHOSPHATE

Since the work of Hoagland and Broyer (1936) it has generally been assumed that there is negligible absorption of salts in absence of oxygen. Many plants are, however, able to absorb ions from an environment which is virtually anaerobic and although it is often assumed that the oxygen for this active uptake of ions by the root is provided by transfer from the shoot it is also possible that genuine anaerobic energy production is involved. Larkum (1967) has shown a clear cut difference in sensitivity of the processes of uptake and transport from 1×10^{-5} M KH_2PO_4 to changes in oxygen concentration of the medium surroundings the root (Fig. 3). Whereas the total ^{32}P content of the root alters very little over the range 21–0% oxygen, there is a marked drop in the shoot content representing a fall in percentage transport from 55–25%. It is probable that the transport mechanism in the stele is dependent on oxygen supply, whether from outside the root or by transfer through the cortical regions from the shoot. Under conditions of limited oxygen supply, the absorption mechanism is more favourably placed and is therefore less affected by oxygen lack even if only aerobic metabolism is involved. This view is confirmed by experiments with 2,4 dinitrophenol and other uncouplers of oxidative phosphorylation.

Low concentrations of 2,4 dinitrophenol (1×10^{-5} M) cause the cessation of transport whereas the inhibition of absorption by the root is only temporary. The effect is reversible and after transfer of the plant from the DNP solution to normal culture solution absorption and transport are restored almost to the original rates (Fig. 4). Such concentrations of DNP can be shown to have negligible effect on water flow through the plant. These results suggest that the transport process is closely associated with oxidative metabolism although presumably, a small amount of phosphate absorption can be provided for by glycolytic metabolism. Evidence for this view has been presented by Larkum (1967) working with barley plants adapted to an environment lacking in oxygen. He showed that such plants possess a high capacity for active absorption from oxygen free solutions and concluded that this absorption is maintained primarily by glycolysis.

Evidence for the involvement of anaerobic metabolism by normal plants is given by an experiment in which a 14-day normal barley plant was

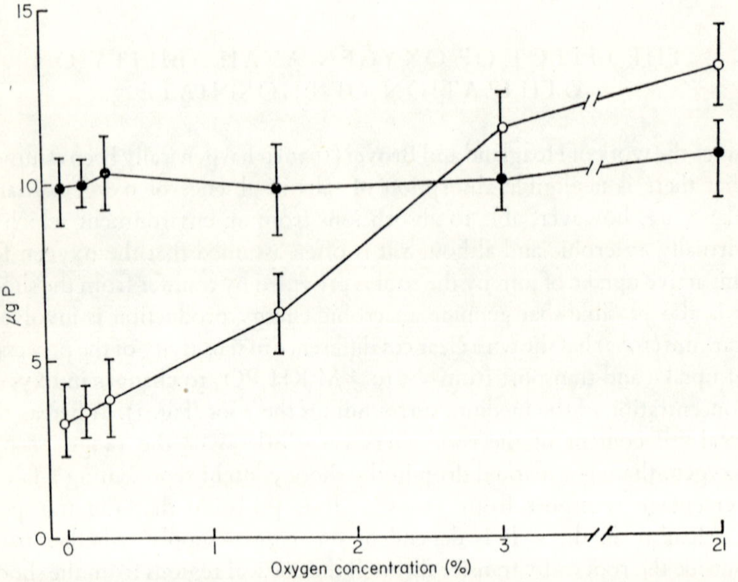

FIG. 3. The effect of oxygen concentration of the nutrient solution on the absorption and transport of phosphate by illuminated barley plants over a 6-hr period in 1×10^{-5} M KH_2PO_4. The limits shown represent ± standard deviations. (n = 12). (Larkum 1967.) ○—○, Shoot; ●—●, Root.

placed in the continuous flow equipment and uptake into the root and shoot followed for a total of 12 hr, from 1×10^{-6} M $KH_2{}^{32}PO_4$. After 3 hr in aerated solution, the root medium was exchanged for an equivalent anaerobic solution while the shoot was left in air, these conditions holding for 6 hr. For the remaining 3-hr period the shoot environment was changed to 97% N_2/3% CO_2 at 30 l per hr (Fig. 5). Since none of the treatments materially affected the rate of uptake and transport even under conditions where any photosynthetically produced oxygen would be rapidly flushed out in the high gas flow it is reasonable to conclude that both absorption and transport of phosphate from 1×10^{-6} M solution can be supported by anaerobic glycolysis.

The greater sensitivity of phosphate transport rather than uptake to low oxygen concentrations from 1×10^{-5} M KH_2PO_4 described earlier is much more strikingly shown in illuminated plants. This interaction between light and the selection mechanism is also shown in the mannose

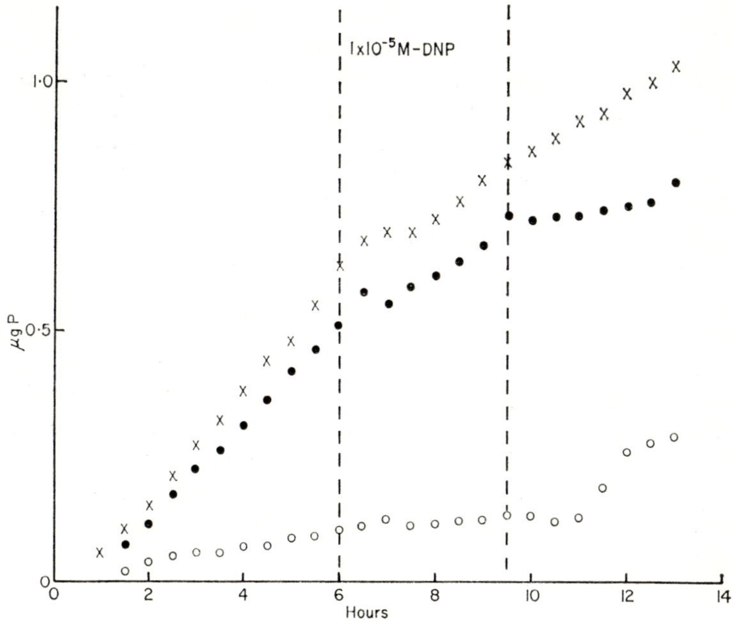

FIG. 4. The effect of addition of 2,4-dinitrophenol (1×10^{-5} M) on the absorption and transport of phosphate from 1×10^{-6} M KH_2PO_4 after establishment of steady state conditions and the recovery on return to a fresh supply of the original nutrient solution. (Larkum 1967.) ×—× Total; ●—● Root; ○—○ Shoot.

experiments with relatively old plants (18-day) of low phosphate status as shown in Fig. 1. thus emphasizing the importance of taking into account the effect of environmental changes on the whole plant when considering the role of the various biochemical pathways in phosphate utilization. The experiments of Bowling (1965) show that ringing of *Ricinus communis* plants immediately affected potassium absorption even though water movement was unimpaired and it is clear that rapid changes in the absorptive capacity of roots can be brought about by modification of their carbohydrate supply. This may account for the fact that in young illuminated barley plants the steady state transport of phosphate is rapidly reduced on darkening the plants.

SELECTION BY CHANGES IN THE OSMOTIC VALUE OF THE ROOT ENVIRONMENT

Marschner, Saxena and Michael (1965) showed that uptake and transport of phosphate by young barley plants are differentially sensitive to changes in the osmotic potential of the culture solution. Using polyethylene glycol at 0·12 M in the medium from which the phosphate was being absorbed, they showed that, whereas the total uptake of phosphate was greater than that of control plants in normal culture solution by about 10%, the amount transported to the shoot was almost halved. They interpret the results on the basis of a change in the physical dimension of the diffusion path of phosphate across the cortex to the stele. This is an excellent example of selection between uptake and transport. They were careful to show that although polyethylene glycol also reduced transpiration, under the conditions used changes in transpiration rate brought about by alteration in the humidity of the shoot environment had no significant effect on phosphate transport.

One may conclude from such results that the reduction in transport of phosphate brought about by increased osmotic potential is not due to diminished water flow but to a specific effect on the distribution mechanism within the root.

CONCLUSION

It is clear that the metabolic activity of roots of whole plants depends on the availability of carbohydrate from the leaves and it is not surprising that the presence of the growing shoot has a marked influence on this metabolism. Again, those environmental factors which affect the shoot directly can be shown to have a rapid effect on root activity, indicating the importance of attempting to combine the physiological and biochemical approach when examining the effect of any change in the environment whether momentary or of long duration. The ecological implications of such studies lie in the fact that capacities for absorption and transport are biochemical phenomena which are subject to normal genetic control. The distribution and survival of species under differing conditions of nutrient supply, and the tolerance of unfavourable environments by particular species depend to some extent on capacities for absorption and transport of ions. By making a detailed study of the basic mechanisms involved in these processes one can more readily obtain information concerning such mechanisms in the plant growing under field conditions.

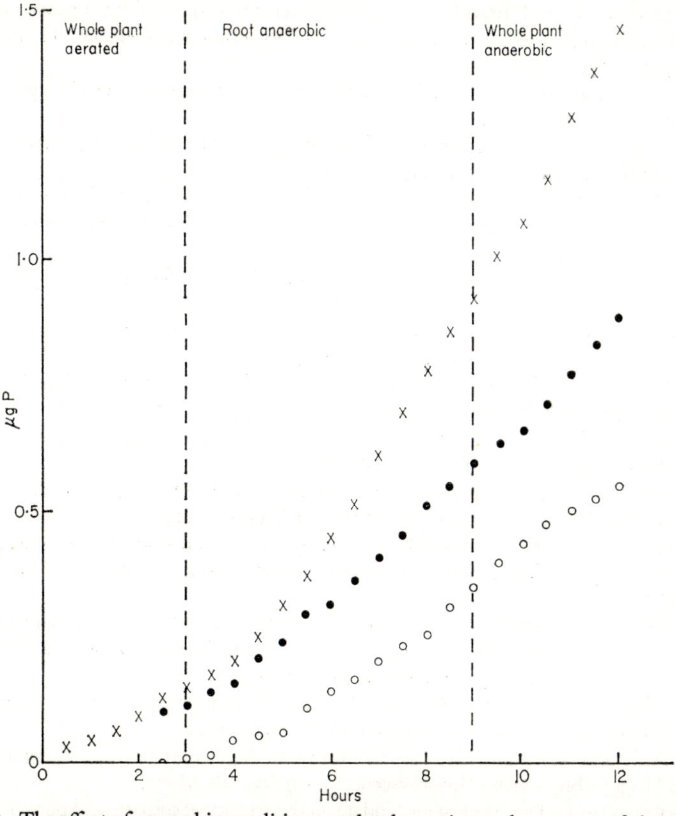

FIG. 5. The effect of anaerobic conditions on the absorption and transport of phosphate from 1×10^{-6} M KH_2PO_4 by a 14 day barley plant. (Larkum 1967.)
×—× Total; ●—● Root; ○—○ Shoot.

REFERENCES

BIELESKI R.L. and LATIES G.G. (1963). Turnover rates of phosphate esters in fresh and aged slices of potato tissues. *Pl. Physiol.* **38,** 586–594.

BOWLING D.J.F. (1965). Effect of ringing on potassium uptake by *Ricinus communis* plants. *Nature* **206,** 317–318.

CROSSETT R.N. (1966). The movement of rubidium across the root of *Hordeum vulgare* (L). *New Phytol.* **65**, 443–458.

CROSSETT R.N. (1967). Autoradiography of ^{32}P in maize roots. *Nature* **213**, 312–313.

CROSSETT R.N. and LOUGHMAN B.C. (1966). The absorption and translocation of phosphorus by seedlings of *Hordeum vulgare* (L). *New Phytol.* **65**, 459–468.

FREDERICK J.F. and GENTILE A.C. (1960). The formation of the glucose derivatives of 3-amino-1,2,4-triazole under physiological conditions. *Physiologia Pl.* **13**, 761–765.

HOAGLAND D.R. and BROYER T.C. (1963). General nature of the process of salt accumulation by roots with description of experimental methods. *Pl. Physiol.* **11**, 471–507.

JACKSON P.C. and HAGEN C.E. (1960). Products of orthophosphate absorption by barley roots. *Pl. Physiol.* **35**, 326–332.

JEFFREY D. W. (1964). The formation of polyphosphate in *Banksia ornata*, an Australian heath plant. *Aust. J. biol. Sci.* **17**, 845–854.

JENNINGS D.H. (1964). Changes in the size of orthophosphate pools in mycorrhizal roots of beech with reference to the absorption of the ion from the external medium. *New Phytol.* **63**, 181–193.

KURSANOV A. and VYSKREBENTSEVA E.I. (1960). Primary inclusion of phosphates in root metabolism. *Fiziol. Rast.* **7**, 276–282.

LARKUM A.W.D. (1967). Factors affecting the uptake and transport of ions in higher plants. D.Phil. Thesis. University of Oxford.

LAUCHLI A. (1967). Investigations on the distribution and transport of ions in plant tissue with the X-ray microanalyser. *Planta* (*Berl.*) **75**, 185–206.

LOUGHMAN B.C. and RUSSELL R.S. (1957). The absorption and utilization of phosphate by young barley plants. *J. exp. Bot.* **8**, 280–293.

LOUGHMAN B.C. (1960). Uptake and utilization of phosphate associated with respiratory changes in potato tuber slices. *Pl. Physiol.* **35**, 418–424.

LOUGHMAN B.C. (1966). The mechanism of absorption and utilization of phosphate by barley plants in relation to subsequent transport to the shoot. *New Phytol.* **65**, 388–397.

MARSCHNER H., SAXENA M.C. and MICHAEL G. (1965). Aufnuhme von phosphat durch Gerstenkeimpflanzen in Abhängigkeit vom osmotischen Druck der Nährlösung. *Z. Pflanzenernähr. Düng., Bodenkunde* **111**, 82–94.

MIETTENEN J.K. and SAVIOJA T. (1958). Uptake of orthophosphate by normal and P-deficient pea plants. *Suomen Kemistilehti.* B. **31**, 84–90.

SETH A.K. and WAREING P.F. (1967). Hormone-directed transport of metabolites and its possible role in plant senescence. *J. exp. Bot.* **18**, 65–77.

SPENCER D. (1954). The effect of molybdate on the activity of tomato acid phosphatase. *Aust. J. biol. Sci.* **7**, 151–160.

WORT D.J. and LOUGHMAN B.C. (1961). The effect of 3-amino-1,2,4-triazole on the uptake, retention, distribution and utilization of labelled phosphorus by young barley plants. *Can. J. Bot.* **39**, 339–351.

ION MOVEMENT WITHIN THE PLANT AND ITS INTEGRATION WITH OTHER PHYSIOLOGICAL PROCESSES

P. E. WEATHERLEY
Department of Botany, University of Aberdeen

INTRODUCTION

Our understanding of the fundamentals of ion transfer across cell membranes has largely and understandably come from a study of the simplest systems: single cells or storage tissues. It is not easy, however, to predict from such knowledge the behaviour of the whole plant with its complex structure and physiological integration. Even less easy is it to picture from studies at a cell or lower level of organization the functioning of the plant in an environment highly heterogeneous with respect to space and time. The other side of the coin is that the study of the salt relations of whole plants in a natural environment presents data which are the outcome of such complex interactions that they may defy analysis. Thus some degree of simplification is expedient, either of the environment by controlling conditions in chambers, etc. and using water culture instead of soil, or of the plant by studying one physiological aspect in isolation or simplifying the structure as in the use of the decapitated root system.

From such studies it is now evident that ion uptake and translocation are not only highly complex and little understood processes in themselves, but are inter-related with other physiological processes in ways even less understood. In this article no attempt will be made to discuss the nature of the processes involved in salt fluxes through the plant, but attention will be drawn to some aspects of the way in which the movement of ions is influenced by the other two main fluxes in the plant: the transpiration stream and translocation in the phloem. Even the data accumulated so far reveal physiological patterns which differ from ion to ion and from species to species and each pattern is no doubt influenced in its own way by the environment. The impact of all this on ecological thought is perhaps not obvious, but undoubtedly its ecological significance will increase as the picture becomes more defined.

THE FLUX OF IONS THROUGH THE ROOT SYSTEMS OF TRANSPIRING PLANTS

The earliest view was that the soil solution was simply drawn into the plant by the transpiration stream. However, permeability studies led to the inference of the existence of cell membranes and the independent diffusion of water and each ionic species. The implication of this was that although water and salts traverse the same pathway from root surface to xylem, they do so independently. It was not surprising therefore that transpiration was often found to have little influence on salt uptake. However some workers have found that varying the transpiration rate has a considerable influence on salt uptake, and the undoubted reliability of the work on both sides has established that both dependence and independence occur according to the plant and its previous history. A mass of often conflicting data makes generalization almost impossible, but it seems that on the whole plants with a high salt status manifest a transpiration dependent salt uptake, whereas salt starved plants do not.

A deeper insight into ion movement through the root is obtained from a consideration of the decapitated exuding root system. Root exudation which can exert a pressure and do work, clearly indicates that an active process is involved somewhere. The early work of Sabinin (1925) and the more recent outstanding contribution of Arisz, Helder and van Nie (1951) have made it clear that the root system behaves like an osmometer, i.e., the osmotic pressure of the xylem sap is higher than that of the external medium and water movement is merely passive and occurs in response to this difference. Now a steady rate of exudation must mean, therefore, the maintenance of a constant osmotic pressure in the stele and this in turn implies a continuous flux of ions into the stele across the osmotic barrier lying somewhere between the stele and the external medium. It is this centripetal ion transport which at some stage needs metabolic energy since it proceeds against an electro-chemical potential gradient and leads to osmotic work being done.

The question which arises at once is where is the osmotic barrier? The cells of the cortex are each bounded by differentially permeable membranes and would collectively act like a single osmotic barrier. However there is evidence that many solutes can penetrate the cortex rapidly as far as the endodermis (Scott and Priestly 1928; De Lavison 1910), and these older conclusions are supported by more recent concepts of an outer space in the cells which rapidly comes to diffusional equilibrium with the outside medium (Bernstein and Nieman 1960). Furthermore there is evidence that water can move more rapidly through cell walls than from vacuole to vacuole (Weatherley 1963). The suggestion is, therefore, that there is a mass flow and diffusion of water and solutes from the surface of the root to the endodermis via the outer regions of the cortical cells, probably through the cell walls. At the endodermis, however, this pathway is blocked by the Casparian strips and further centripetal progress must be through the cytoplasm of the endodermis. Inside the stele it might be supposed that movement would revert to the cell wall system. On this view the osmotic barrier of the root osmometer is the endodermis which is in virtual contact with the outside medium and has on its inside a free space volume continuous with the lumina of the xylem vessels. Thus the transfer of ions across the endodermis would be the essential step in maintaining exudation.

An additional feature of this picture of flux through the root is the property of the vacuoles of the cells of the cortex and stele to accumulate and release solutes according to conditions. Thus as the soil solution is drawn through the cortical cell walls so may ions be abstracted from it or released into it. A similar state of affairs will exist inside the stele, and according to the balance between vacuolar uptake and release, so will the salt content of the root fluctuate. The fact that this fluctuation is distinct from the transport of ions through the root has long been recognized. However this does not mean that vacuolar accumulation and release are not important. Leakage of previously accumulated ions into the transpiration stream would tend to stabilize supply to the shoot if external sources were cut off. That this occurs in the laboratory with exuding plants is demonstrated if they are transferred to a salt free medium, when exudation may continue for many hours (the so called 'tissue exudation' of van Andel 1953). Such mobilization of root stores has been demonstrated in transpiring plants by Bange and van Vliet (1961). In a natural environment presumably this might occur after leaching of the soil by heavy rain.

A further important aspect of both root cell accumulation and transfer to the shoot is their selectivity. Thus sodium may be accumulated in the root whilst potassium is passed on to the shoot (Bange and van Vliet 1961). Certainly large stores of sodium have been found in the roots of tomato very little of which ever found its way to the shoot (Jackson and Weatherley 1962). Thus the root cells may have a 'protective' function as well as a stabilizing one.

In the exuding plant water moves into the stele in response to a difference in osmotic potential between medium and xylem. The transpiring plant differs in that there is the additional factor of the transpiration pull causing a hydrostatic tension in the xylem vessels. This will produce an extra flux of water through the root. The nature of this movement should be the same as in the exuding root system. Mass flow up to the endodermis, osmotic diffusion through the endodermis and mass flow again within the stele. The endodermis is the one layer through which mass flow cannot occur and thus the soil solution should not be sucked into the stele of transpiring plants any more than it can be by exuding plants. There are, however, three sites where 'all-through' mass flow could occur. First in the region behind the root tip where the Casparian strips are not differentiated. This is a restricted area more associated with accumulation by the cells than transfer to the shoot. Second through passage cells in the endodermis and third where endogenous laterals have broken through the endodermis. Mees and Weatherley (1957) concluded from experiments using positive pressure gradients that as much as 25% of the total water flux could be mass flow. However they used large tomato root systems where the third of the above factors might be important. There appears to be less evidence for mass flow with younger root systems. Certainly the major part of the water flux seems to be 'osmotic'.

This being so the question arises as to how a change in water flux *could* alter the salt transfer. An increase of transpirational water flow through the root would be expected to reduce the concentration of ions in the xylem. This reduction would favour flux of ions into the xylem either by increasing the centripetal 'down' gradient for leakage of previously accumulated ions or reducing the steepness of the 'up' gradient for a direct transfer process. Such a dilution undoubtedly occurs. A second suggestion, made by Bernstein and Nieman (1960), is that when the solution, drawn through the cortex, arrives at the surface of the endodermis, it may become concentrated there if the flux of water through the endodermis exceeds the flux of a particular ion. The greater the rate of transpiration the greater the concentration of the ion, and so its transfer across the endodermis is favoured. Such a concentration of an ion certainly does seem possible. If the ion is not, or only slowly transferred across the endodermis its concentration will build up and it will diffuse outwards across the cortex against the stream and a steady state will be reached when outward diffusion equals inward arrival by mass flow. Precise calculation of the magnitude of such a build-up is not possible since measurements of flow rates per unit surface of root do not seem to have been made. Brouwer (1953a) however, measured the flow rate into a bean root and expressed the data in mm^3 per unit length of root. If it is assumed that the root was 1 mm thick, his data give an influx of water per unit area which would lead to the concentration of the ion being three or four times that of the external medium (see Appendix). The extent of this build up is related exponentially to the length of the pathway involved. In water culture this is the width of the cortex but in a soil-rooted plant this extends out into the soil and the build up could increase enormously. In effect the root cortex might be considered an inward extension of the soil situation as examined by Nye and Tinker elsewhere in this volume. Of course for many ions such a build up would not occur because of rapid transfer across the endodermis or into the vacuoles of the cortical cells. However concentration of ions at the endodermis does seem a possible factor in transpirational enhancement of ion flux through roots.

A third factor which may link transpiration and salt uptake is the effect of transpiration on root permeability. Over the years it has become increasingly clear that a transpirational flux of water through the root increases its permeability to water (Tinklin and Weatherley 1966). The mechanism of this phenomenon is not yet known, though it may be associated with an increase in the absorptive area of the root (Brewig 1935–36; Brouwer 1953b). It appears, in addition, to be accompanied by an increase in influx of ions (Brouwer 1953c). This is illustrated by an experiment (Jackson and Weatherley 1962) in which the pressure gradient across the decapitated root system of a tomato plant was raised but the water potential difference maintained constant by the addition of mannitol to the medium. As will be seen in Fig. 1 no change in water flux through the root system resulted, but there was a rise in potassium flux accompanied by a rise in the concentration of the xylem sap. Clearly this result is inexplicable in terms of a dilution of the xylem sap nor can it be explained in terms of a concentration at the endodermis since similar results were obtained with a potassium free medium. This picture has subsequently been confirmed by experiments on transpiring *Ricinus* plants (Bowling and Weatherley 1965). A linear relationship was found between the rate of transpiration and potassium uptake but this was clearly not due to a dilution of the xylem sap or a mass flow of medium directly to the xylem since the xylem concentration remained virtually constant at widely different water fluxes and was about twenty times more concentrated than the external medium. It appears that transpiration has a profound effect on the physiology of the root, not only causing a decline in its resistance to water flow, but facilitating ion transfer as well.

The relative importance of these ways in which transpiration may facilitate salt uptake in any particular case is not known. Even less is it known whether the transpiration controlled increase in salt uptake has any ecological significance.

However important transpiration may be in facilitating movement of ions within the plant, or ion pools in regulating supply to the shoot, in the end it is the arrival of ions at the root surface which may set a limit to the mineral content of the plant. To what extent is this itself dependent on transpiration? When plants are rooted in water culture or in soils saturated

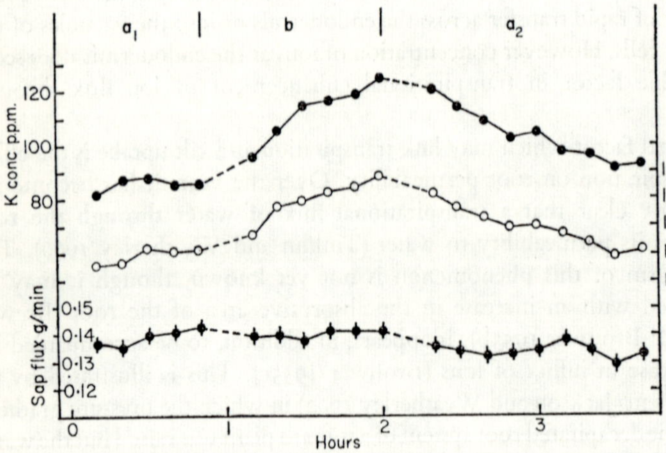

FIG. 1. The effect of raising the hydrostatic pressure difference across the root cortex and balancing this with a rise in O.P. of the medium. Although the difference in water potential across the cortex and the flux of water remained unchanged, the flux of potassium and hence the concentration of the sap increased. ⸸——⸸, sap flux; ○——○, potassium flux; ●——●, concentration of potassium in the sap; periods a_1 and a_2 pressure 1·5 atm, O.P. 0·25 atm, period b pressure 2 atm, O.P. 0·75 atm. Medium contained KNO_3 (200 ppm K) throughout. (From Jackson and Weatherley 1962).

with water, the roots are virtually bathed by a bulk solution of salts and the cross sectional area of the diffusion pathway up to the root surface is equal to or not so much less than the surface of the root itself. In this situation the diffusional arrival of ions and transpiration might be independent of one another. However with unsaturated soils where water is restricted to surface films, the diffusional pathway is greatly restricted and arrival of salts at the root surface might possibly be dependent on mass flow of soil solution round the soil particles. If so, transpiration would play an important part in bringing ions to the root. Whether this could have any ecological significance it is difficult to say. A humid habitat associated with low transpiration rates would usually be combined with near saturated soil in which low transpiration would not restrict ion movement to the root. In contrast, conditions of water stress leading to reduced or even a cessation of transpiration would imply little arrival of ions at the root surface and therefore a reduction of uptake. Of course under these conditions growth and other 'requirements' are likewise reduced.

DEPENDENCE OF ION UPTAKE BY ROOTS ON A SUPPLY OF CARBOHYDRATES FROM THE LEAVES

A root system cannot continue to function without a supply of respirable substrates. Thus the decapitated root system may continue to exude for some time, but this is a transient phenomenon and gradually there is a decline. That such a decline is a result of a cutting off of supplies from the leaves has been demonstrated by Bange (1965). He found that the transfer of potassium from the roots to shoots of maize seedlings was considerably reduced by killing the phloem of the stem by steam. That this was due to cutting off of supplies of carbohydrate to the roots rather than some specific agent was suggested by the fact that a 1% glucose solution round the roots of decapitated plants largely restored their capacity to transfer potassium to the xylem.

The effect of cutting off sugar supplies from the shoot can have a surprisingly immediate effect on salt uptake. This is well illustrated by some recent work of Bowling (1968) using sunflower plants. Downward translocation was reduced or perhaps even stopped by a cooling coil round the stem beneath the lowest leaves. As shown in Fig. 2 this had a rapid effect on potassium uptake by the roots. On removing the cold coil, uptake recovered to its original value. Again this has been shown to be due to a cutting off of sugar supply in the absence of which root respiration and salt uptake both decline. These experiments were carried out with vigorously growing plants in water culture the roots of which probably had little by way of carbohydrate reserves. However, plants growing under natural conditions might well be in a similar state and it is interesting to speculate how fluctuations in sugar from the shoot might regulate salt uptake by the root. Presumably the relative sizes of the shoot and root become adjusted so that their exchange of materials—salts to the shoot, photosynthate to the root—are balanced. It is not difficult to see how the sugar sent down from the leaves is some measure of their activity and the growth potential of the shoot. No doubt this directly induces extra salt transfer to the shoot and indirectly causes increased potential supply via extra root growth. If the balance is upset, say by the shoot being grazed, supplies of photosynthate to the roots may fall catastrophically and salt uptake will decline and transfer to the shoot be curtailed in proportion to its reduced need. The diurnal sequence of light and dark causes diurnal fluctuations in carbohydrate contents of leaves and in turn leads to fluctuations in translocation. This

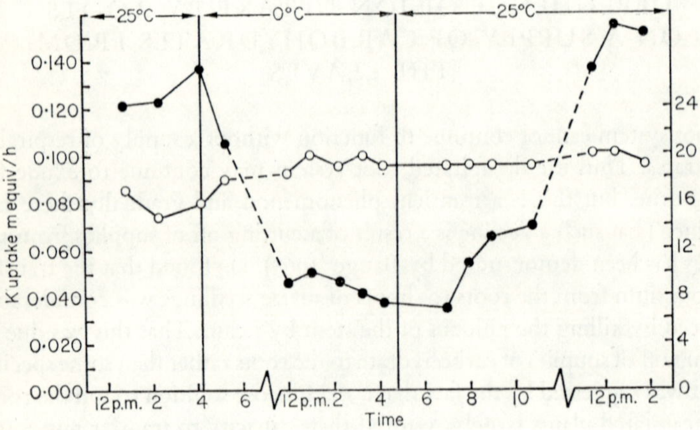

FIG. 2. The effect of cooling the stem of an *Helianthus annuus* plant to 0° C on the uptake of water and potassium. Solid circles, potassium uptake: open circles, water uptake.

might be expected to have its effects on salt transfer which would be minimal at night when incidentally transpirationally induced arrival of ions at the root surface would be minimal also.

MOVEMENT OF IONS IN THE TRANSPIRATION STREAM

Analyses of the tracheal sap of woody plants (Anderssen 1929; Bollard 1953) shows that it is a solution of salts the dilution of which varies according to the season and conditions. As might be expected it is most concentrated in the winter and early spring and most dilute during the summer when transpiration is at its height. It is evident therefore that at least some salts are transported from the root to the shoot via the xylem vessels. That transport is almost exclusively in the wood was demonstrated by Stout and Hoagland (1939) using radioactive isotopes. When radioactive phosphate was supplied to the roots of woody plants, phosphate was found in the wood and bark of the stem above. However, when waxed paper was inserted between the bark and wood, almost all the isotope was found in the wood. It was concluded that transport occurred in the wood with lateral transfer to the bark.

We can envisage the ions then, moving up the xylem in the transpiration stream but with ionic exchange occurring with surrounding tissues on the way. In woody stems this means that ray cells and xylem parenchyma may act as ionic stores as they do for carbohydrates and the same could be true of the phloem parenchyma. It is significant that the newest formed tracheary elements of the xylem and youngest sieve tubes of the phloem, i.e. the most actively functional elements in each case, are in close proximity separated only by the cambium, a highly active tissue (Steward and Sutcliffe 1959). Thus we have the spectacle of the major upward and downward transport streams positioned close together with any ions transferred from one to the other being swept back in the opposite direction. This is not necessarily as wasteful as it might appear. Firstly, the ions need not simply return whence they came; those transferred from xylem to phloem in the roots or stem would be carried down with the nutrient stream of the sieve tube to the growing region of the roots which may be less active than the maturer proximal regions in absorbing nutrients from the soil. Those passing from the phloem to the xylem might be transported not to the leaves from which they were exported, but to younger growing leaves higher up the stem. Secondly, the transfer into and out of these pathways is likely to be active and selective, in other words regulatory. Clearly this exchange leads to a circulation of ions within the plant about which more will be said later.

The question arises as to the adequacy of the supply of mineral ions to a growing leaf via the transpiration stream. I have not found data pertaining to a single species which would make a precise calculation of this possible but the following is an approximation based in part on data for cotton. Cotton leaf tissue has a dry weight of about 6 mg/cm^2 of leaf area. Since dry matter of plants usually contains around 2% of potassium (Baumeister 1958) cotton leaves will contain about 120 μg/cm^2. If the concentration of potassium in the tracheal sap is taken as 60 μg/cc (data for apple trees, Bollard 1953) it follows that 2 cc of transpiration per unit leaf area are required during the growth of the leaf. In the field in Uganda midday transpiration rates as high as 40 mg/cm^2/hr were found and if half this value is taken as an average over a 10-hr day, this would mean 0·2 cc/cm^2/

day. Thus the equivalent of the mineral content of the mature leaf would arrive in the transpiration stream in 10 days. In fact it would take longer than this since the leaf is expanding and only attains its full transpiring area at the end of the period under consideration. Assuming a linear expansion in time it can be shown that our hypothetical leaf would take about 14 days to receive its full mineral content. This rate of supply might appear to be barely adequate, however it could well vary widely in either direction. On the one hand it could be more rapid since transpiration is usually faster from leaves during expansion than from mature leaves. (In this department we have found the transpiration per unit fresh weight of young expanding cotton leaves to be three times that of mature leaves.) Also the concentration of the tracheal sap may be higher; for water culture *Ricinus communis* plants the concentration of potassium was around 600 μg/cc. On the other hand transpiration under less hot and dry contitions would be considerably less than that found for cotton in Uganda, and the potassium concentration in tracheal sap in the summer was found by Bollard to be half the figure used. Thus it could well be that under different climatic conditions the period of 14 days might be extended by a factor of 2 or 3. This would be much less than the time taken for the leaf to expand and it suggests that an additional source of ionic supply is necessary. Such a suplementary supply no doubt comes from the mature leaves below the expanding leaves. As will be discussed below, mature leaves export ions along with sugars in the sieve tubes and this stream may pass upwards into the developing leaves. Thus the transpirational catchment area for ions for the expanding leaves may be not only their own surface, but that of the crown of mature leaves below them.

THE EXPORT OF IONS FROM LEAVES AND IONIC CIRCULATION

The ions which are transported to the leaves are used in growth, i.e. either they are incorporated into the materials in the cells or are accumulated in the vacuoles and there function osmotically. When, however, the leaves have attained their full size, the need for continued supplies of ions is presumably less. Yet the arrival of ions in the transpiration stream continues and clearly they must be exported from the leaf or continue to accumulate in its tissues. Which of these occurs depends on the mobility of the ion in question.

There is evidence (Weatherley 1963) that water moves from the xylem to the transpiring surface via the cell walls. At the transpiring surface the concentration of the solution will tend to rise and the ions will diffuse back against the oncoming mass flow. Slatyer (1966) has calculated that a rise of concentration of less than 15% would be expected at the evaporating surface. However diffusion would not be rapid enough to cause an ionic flux against the mass flow stream in the xylem elements, so that a gradual build up of salts would take place in the cell wall system of the leaves were these salts not disposed of.

Now just as in the vascular system of the stem and veins there are two conducing pathways, the xylem and phloem, usually transporting in opposite directions, so at an intracellular level in the mesophyll the cell wall system conducts water and salts 'outwards' and the symplasm via the plasmodesmata conducts synthesized products from the photosynthesizing cells 'inwards' to the sieve tubes. These two pathways are in very intimate contact though separated by the plasmalemma, and transfer from one to the other is probably selective and under metabolic control. Within the symplasm ions may be transferred to the vacuoles and there be immobilized, or they may diffuse or be carried by streaming (or however one imagines movement to occur in the symplasm) to the sieve tubes. Once in the sieve tubes the ions would be carried along passively in the stream of sugar solution (or by whatever mechanism one imagines movement to occur in the sieve tubes). Thus a circulation of ions arises: transpiration carrying ions up the shoot and translocation carrying any excess back into the stem and up to the growing region of the shoot or down to the root where once again they may enter the transpiration stream.

Mason and Maskell (1931) first suggested the possibility of such a circulation of ions, but its demonstration and characterization has been largely due to Biddulph and his co-workers. Using autoradiography they have shown that the pattern of circulation depends on the particular ion. Salts containing radioactive phosphorus, sulphur and calcium were applied in separate experiments to the roots of bean plants and the movement of the tracer followed by autoradiography of sample plants selected at intervals through a 96-hr period. They found (Biddulph *et al.* 1958) that phosphorus was rapidly carried to the leaves where it accumulated. But as younger leaves expanded they imported phosphorus from the older leaves which therefore lost their radioactivity. Thus radioactive phosphorus gradually moved upwards as the plant grew, always being found in the currently young leaves. Sulphur was also mobile but became incorporated in non-mobile compounds in expanding leaves where it remained fixed. Calcium was not mobile in the phloem so it was carried in the transpiration stream to the leaves and there accumulated. The immobility of calcium in the phloem has been demonstrated by a number of workers and may be related to the insolubility or calcium compounds at the high pH found in sieve tube sap. This has interesting implications for developing regions such as

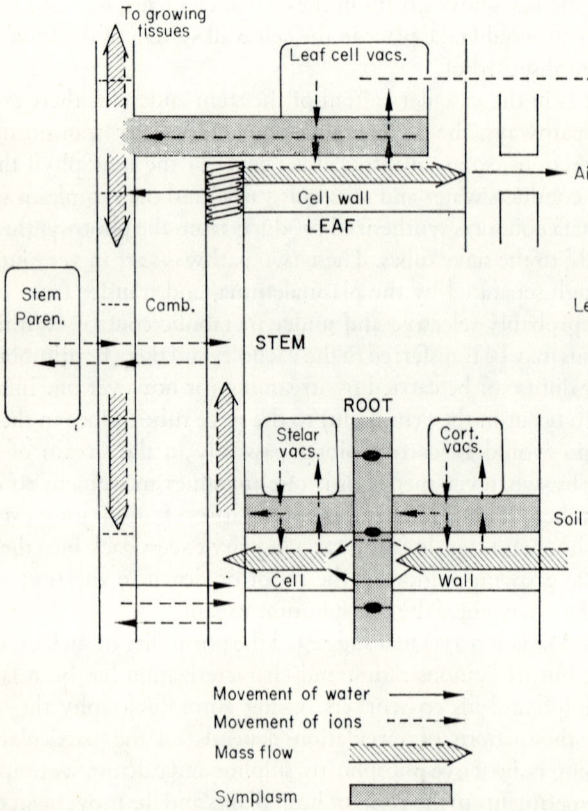

Fig. 3. Diagram showing the channels of uptake and circulation of ions and sites of vacuolar pools.

fruits and tubers the transpiration from which is small and which therefore depend on the phloem for their mineral supply. As expected their calcium content is low relative to other organs (Stenlid 1958; Wittwer and Teubner 1959). However in some situations calcium is mobile, for example during germination calcium can move out of storage organs to the new root and shoot (see Bollard and Butler 1966).

Another form of circulation is that resulting from the leaching out of ions from leaves by rain or dew (Stenlid 1958; Wittwer and Teubner 1959). In this way salts are returned to the soil and may be reabsorbed by the roots. Experimentally leaching has been mostly studied by dipping leaves or shoots into water which is analysed after a given interval. In this way Tukey and Mecklenburg (1964) have shown that 80–90% of the potassium content and 50–60% of the calcium content is leached from squash leaves in 24 hr. It is, however, difficult to assess the ecological significance of such experiments and as pointed out by Stenlid, the only way of discovering the extent of leaching under natural conditions is to collect rain water beneath plants and compare this with the mineral content of rain collected in the open. Such data as have been obtained in this way indicate that significant quantities of salts, especially potassium, may be lost by natural leaching. Loss appears to be greatest from the upper surface and from older leaves. This is in contrast to foliar absorption which occurs most in young leaves.

Leaching then constitutes an additional if rather uncontrolled circulation which involves not only those ions mobile in the phloem but also calcium and magnesium. It may have considerable ecological significance in returning ions to the soil to be reabsorbed by roots. This may result in a lifting by deep rooted plants of ions from the lower soil layers and their deposition in the surface layers where they would become available to superficially rooting species. Furthermore transfer from one species to another can occur via the leaves, the leachate from the leaves of tall plants dripping onto the leaves of those beneath to be absorbed directly (foliar absorption).

SUMMARY

The general picture of ion uptake and circulation is summarized in Fig. 3. Water is drawn through the plant by transpiration and probably moves through the soil and the plant to the evaporating surface of the mesophyll cells by mass flow except at the root endodermis across which most of it passes by osmosis. Thus it is a movement of solution all through, except at the endodermis where presumably ions and water can move independently. Here ions are actively transferred and may move faster than the water if there is no or little transpiration. In these circumstances water moves in response to ion flux (exudation or guttation). Under conditions of more rapid transpirational water flux the permeability of the root to water and ions is increased and there may be facilitation of ion transport by transpiration.

The solution flowing through the plant from soil to leaf exchanges with surrounding cells all the way along; with the root cortex, the stele, the phloem and xylem parenchyma. With these the stream is in dynamic equilibrium controlled no doubt by active and passive forces. Thus pools which are off the stream even out supply to the leaves and growing points both by additions to and withdrawals from the upward moving stream. Ions arriving in the leaves in excess of requirements accumulate or are exported if they are mobile in the phloem in which case they are sent either up to young expanding leaves and growing points or down the stem to the pools already mentioned and the growing roots. This leads to a circulation of ions in the plant, as does exchange between the xylem and phloem pathways. Circulation is enhanced by leaching from leaves by rain and dew and this leads to transfer from plant to plant and a wider circulation within the soil profile.

It is interesting that, at first sight, the rapidity of the main countercurrents of the circulation within the plant, the xylem and phloem streams, appear to be influenced by rather unrelated processes: transpiration and photosynthesis. However, broadly speaking it is true to say that high rates of transpiration occur under environmental conditions favouring high rates of photosynthesis, and as was mentioned earlier, transfer through the root relies to some extent on supplies of carbohydrate down the phloem. Thus it would appear that high plant activity is accompanied by rapid circulation and constant adjustment of ionic content between shoot: root, young leaves: old leaves and so forth.

The ecological significance of the physiological processes and the interactions between processes which have been discussed in this article are not easy to assess. To a large extent the physiology, although expressed in quantitative terms, is understood only in a qualitative way. For example the maintenance of salt uptake by roots needs a supply of photosynthate from the shoot, but how far this is important in the day to day life of the plant is not known. Transpiration may increase salt flux into and through the plant, but how far do the naturally occurring variations in transpiration affect the plant significantly in this respect? Leaching may demonstrably occur from leaves, but how important is this loss to a plant, and is it a significant gain to another? If the answer to these physiological questions is still little known, so much the less known is the answer to the ecological question of how these factors affect the relationship of a given species to its environment and its neighbours.

REFERENCES

ANDEL O.M. VAN (1953). The influence of salts on the exudation of tomato plants. *Acta bot. neerl.* **2**, 445.

ANDERSSEN F.G. (1929). Some seasonal changes in the tracheal sap of pear and apricot trees. *Pl. Physiol.* **4**, 459.

ARISZ W.H., HELDER R.J. and NIE R. VAN (1951). Analysis of the exudation process in tomato plants. *J. exp. Bot.* **2**, 257–297.

BANGE G.G.J. and VLIET E. VAN (1961). Translocation of potassium and sodium in intact maize seedlings. *Plant and Soil* **15**, 312–328.

BANGE G.G.J. (1965). Upward transport of potassium in maize seedlings. *Plant and Soil* **22**, 280–306.

BAUMEISTER W. (1958). Die Aschenstoffe. *Encyclopedia of Plant Physiology*. (Ed. W. Rhuland), **IV**, 5–34. Springer-Verlag.

BERNSTEIN L. and NIEMAN R.H. (1960). Apparent free space of plant roots. *Pl. Physiol.* **35**, 589.

BIDDULPH O., BIDDULPH S.F., CORY R. and KOONTZ H. (1958). Circulation patterns for P^{32}, S^{35} and Ca^{45} in the bean plant. *Pl. Physiol.* **33**, 293–300.

BOLLARD E.G. (1953). The use of tracheal sap in the study of apple-tree nutrition. *J. exp. Bot.* **4**, 363–368.

BOLLARD E.G. and BUTLER G.W. (1966). Mineral nutrition of plants. *Ann. Rev. Pl. Physiol.* **17**, 77–112.

BOWLING D.J.F. (1968). Translocation at 0°C in *Helianthus annuus J. exp. Bot.* (In press.)

BOWLING D.J.F. and WEATHERLEY P.E. (1965). The relationship between transpiration and potassium uptake in *Ricinus communis. J. exp. Bot.* **16**, 732–741.

BREWIG A. (1935–36). Die regulations erscheinungen bei der wasseraufnahme und die wasserleitgeschwindigkeit in vicia fabawurzeln. *Jahrb. f. Wiss. Botanik* **82**, 803–828.

BROUWER R. (1953a). Water absorption by the roots of *Vicia faba* at various transpira-strengths. I. Analysis of the uptake and the factors determining it. *Proc. Kon. Ak. Wet. C* **56**, 105–115.

BROUWER R. (1953b). Water absorption by the roots of *Vicia faba* at various transpiration strengths. II. Causal relations between suction tension, resistance and uptake. *Proc. Kon. Ak. Wet. C* **56**, 129–136.

BROUWER R. (1953c). Transpiration and ion uptake. *Proc. Kon. Ak. Wet. C* **56**, 639–649.

DE LAVISON J. DE RUFZ (1910). Du mode de pénétration de quelques sels dans la plante vivante. *Revue gén. Bot.* **22**, 225.

JACKSON J.E. and WEATHERLEY P.E. (1962). The effect of hydrostatic pressure gradients on the movement of sodium and calcium across the root cortex. *J. exp. Bot.* **13**, 404–413.

MASON T.G. and MASKELL E.J. (1931). Further studies on transport in the cotton plant. I. Preliminary observations on the transport of phosphorus, potassium and calcium. *Ann. Bot.* **45**, 125–173.

MEES G.C. and WEATHERLEY P.E. (1957). The mechanism of water absorption by roots. II. The role of hydrostatic pressure gradients across the cortex. *Proc. Roy. Soc. B* **147**, 381.

SABININ D.A. (1925). On the root system as an osmotic apparatus. *Bull. Inst. Recherche biol. Univ. Perm.* **4**, Suppl. 2, 129–136.

SCOTT, L.I. and PRIESTLEY H.J. (1928). The root as an absorbing organ. I. A reconsideration of the entry of water and salts in the absorbing region. *New Phytol.* **27**, 125–140.

SLATYER R.O. (1966). Some physical aspects of internal control of leaf transpiration. *Agric. Meteorology* **3**, 281–292.

STENLID G. (1958). Salt losses and redistribution of salts in higher plants. *Encyclopedia of Plant Psysiology.* (Ed. Rhuland), **IV**, 615–634. Springer-Verlag.

STEWARD F.C. and SUTCLIFFE J.F. (1959). Plants in relation to inorganic salts. *Plant Physiology.* (Ed. F.C. Steward), **II**, 253–478. Academic Press.

STOUT P.R. and HOAGLAND D.R. (1939). Upward and lateral movement of salt in certain plants as indicated by radioactive isotopes of potassium, sodium and phosphorus absorbed by roots. *Am. J. Bot.* **26**, 320.

TINKLIN R. and WEATHERLEY P.E. (1966). On the relationship between transpiration rate and leaf water potential. *New Phytol.* **65**, 509–517.

TUKEY H.B. (JR.) and MECKLENBURG R.A. (1964). Leaching of metabolites from foliage and subsequent reabsorption and redistribution of the leachate in plants. *Am. J. Bot.* **51**, 737–742.

WEATHERLEY P.E. (1963). The pathway of water movement across the root cortex and leaf mesophyll of transpiring plants. *The water relations of plants.* (Eds. A.J. Rutter and F.H. Whitehead.) pp. 85–100. Blackwell. Oxford.

WITTWER S.H. and TEUBNER F.G. (1959). Foliar absorption of mineral elements. *Ann. Rev. Pl. Physiol.* **10**, 13–32.

APPENDIX

ACCUMULATION OF A SOLUTE AT THE ENDODERMIS OF THE ROOT OF A TRANSPIRING PLANT

If a solute is carried by mass flow through the cortex of the root, its rate of arrival at the endodermal surface will be the product of the rate of transpirational water movement and the concentration of the solution. Thus:

$$Q = F . C_s$$

Where Q = the rate of arrival of the solute in moles/sec, F = the flux of water in cc/sec, and C_s = the concentration of the solute in moles/cc. If it is assumed that the solute cannot penetrate the endodermis, its concentration will rise and back diffusion of the solute will occur against the mass flow stream. This diffusion is given by Fick's Law:

$$J = Da\frac{dc}{dl}$$

Where J = the back diffusion of moles/sec, D = the diffusion coefficient of the solute, a = the cross sectional area of the pathway in cm², dl = the length of the pathway between the epidermis and the endodermis in cm, and dc= the difference in concentration between the external solution (C_s) and that at the endodermis (C_m) in moles/cc.

At a steady state the back diffusion will equal the rate of arrival, i.e., $J = Q$. Thus:

$$F.C = Da.\frac{dc}{dl} \text{ or } \frac{dc}{C} = \frac{Fdl}{Da}$$

Thus:

$$\int_{C_s}^{C_m} \frac{dc}{C} = \int_{e}^{0} \frac{Fdl}{Da}$$

Hence:

$$\ln\frac{C_m}{C_s} = \frac{Fl}{Da} \quad \text{or} \quad \frac{C_m}{C_s} = \text{Exp.}\left\{\frac{Fl}{Da}\right\} \qquad \ldots \text{(i)}$$

This equation is of the same form as that given by Slatyer (1966) for the accumulation of solutes at the evaporating surface of leaf cells. It gives the ratio of the concentration at the endodermis relative to that in the external solution in terms which can be measured or estimated.

F (flux of solution across the cortex)

Brouwer (1953a) found a rate of absorption by the roots of transpiring bean plants of 10 mm^3/hr/cm length of root. This gives a value of $F = 2{\cdot}8 \times 10^{-6}$ cc/sec/cm.

l (length of the cortical pathway)

As a simplification it is assumed that the cortex is made up of regular cubical cells along the walls of which the solution flows. Clearly the minimum length of the pathway is the radial distance from the root surface to the endodermis. But if there is alternate packing of the cells the flow must pass along tangential as well as radial walls. In the simplest pattern this would increase the length of the pathway to $1{\cdot}5 \times$ the distance from surface to stele. If Brouwer's root was 1 mm in diameter and the diameter of the stele one-third that of the root, the length of the pathway was 5×10^{-2} cm.

a (cross sectional area of the pathway)

If we assume that the flow is through the cell wall system and that this represents the apparent free space of the cortical tissue, the cross sectional area of the cell wall pathway is a function of the *a.f.s.* If the tangential and radial walls are equally part of the pathway, as they would be with alternate packing, the *a.f.s.* gives approximately the fraction of the total area of a tangential surface occupied by cell walls. The question now arises as to what surface to take. The external surface has three times the area of the endodermis (if the diameter of the stele is one-third that of the root as suggested above) so that there is presumably centripetal acceleration of the flow. As an approximation it would seem reasonable to take an imaginary cylindrical surface midway between outer surface and stele. Using the above dimensions this will have a diameter of 0·67 mm, and its surface per cm length of root will be 0·21 cm^2. If the *a.f.s.* is 10%, the mean area of the pathway will be $2{\cdot}1 \times 10^{-2}$ cm^2.

D (diffusion coefficient of the solute)
A value of 5×10^{-6} cm^2/sec is assumed.
Substituting the above values in equation (i)

$$\frac{C_m}{C_s} = \text{Exp}.\ 1{\cdot}3 = 3{\cdot}7$$

DISCUSSION (A) ON MECHANISMS OF MINERAL NUTRITION

Recorded by DR D. H. JENNINGS and MR J. F. HANDLEY

There were a number of comments about the model presented by Dr Jennings. Mr Nye queried the choice of rates of increase in cell area as being equal to the enzyme kinetic equation for the boundary conditions. He felt that increase in dry weight or volume would be better for defining these conditions. Dr Jennings in reply said that, at the time, increase in area seemed the simplest mathematically but he agreed that there were a large variety of equations which could be used to define the boundary conditions. Professor Harley thought that there were virtues in emphasizing area when considering the uptake of solutes, although, as Dr Jennings pointed out there could be problems in measuring surface area. The model was criticized by Dr Woolhouse and Dr Scott Russell in that it appears to bear little relation to a growing root, where there is not only an increase in cell size but also cell number, and furthermore cell differentiation is taking place. Dr Jennings however showed that, with a model like this, cell division could be readily considered. When the cell is fully grown and division takes place, there is a transformation because the internal concentration remains the same but the surface area is halved. Since this is so there should be a dramatic change in fluxes. The data of Jung and Rothstein for ion movements in synchronously growing animal tissue-culture cells shows that there can be marked change in fluxes when a cell divides and it is likely these changes could be explained in this way.

With regard to the examination of the electrochemical potential gradient across the membranes of plant cells Dr Woolhouse felt that some of the conclusions drawn from the data presented by Dr Jennings concerning the presence of ion pumps in higher plant cells seemed suspect, since the direction of the pumps could result in these cells being totally depleted of cations. Dr Jennings agreed with this and said that was why he had been careful to emphasize that, in the instance under consideration, the whole of the electrical driving force may not have been measured. Professor Epstein thought that the experiments concerned might not be very reliable, as they had been carried out with media in which calcium was absent. Membranes under such conditions tended to be leaky. Dr Jennings said that there is certainly evidence from *Nitella* which showed that the electrical characteristics of the cells differed according to whether calcium was present or absent in the external medium.

There was also discussion about the Viets effect. Professor Mengel presented evidence to show that calcium could reduce the efflux of potassium and phosphate from roots. He thought that one could explain the Viets effect in terms of an active influx which was hardly affected and a passive permeability which is very sensitive to calcium ions. Dr Jennings felt that not all the Viets effect could be explained in this way and he drew attention to the findings of Pitman who showed that calcium could increase the influx of chloride into beet tissue. In this instance it would seem that calcium was not penetrating, in consequence of which there was, among other things, an effect due to an increase in the chloride concentration in the external medium. Dr Jefferies reported some of his own recent work using low concentrations of calcium in which activation of potassium uptake was not at all anion dependent. He felt that calcium may well be acting in an allosteric manner on a potassium pump.

Dr Grubb criticized the use of essentially non-growing cells in the studies by Dr Jefferies. However, it was pointed out by Dr Jefferies that one could not interpret the electrochemical data using the Nernst equation unless equilibrium or near-equilibrium conditions prevailed.

Professor Epstein thought that the effectiveness or lack of effectiveness of fluoride on phosphate uptake and translocation might well be due to the fact that the halide absorption system does not have a high affinity for fluoride and therefore this ion does not penetrate very readily. Dr Loughman agreed, saying that with this inhibitor it was not always possible to get measurable effects and it may well be that the age of the plant or some other factor was affecting fluoride uptake. In response to a question by Professor Harley, Dr Loughman said that phosphate uptake under anaerobic conditions could be kept going for a number of days, though of course the rate of uptake was decreasing all through that period.

With regard to higher plant tissues, Mr Westlake wanted to know if there was anything known about the mechanisms of uptake of nutrients by leaves. Professor Epstein spoke about work with Professor Rains using leaf slices, in which the mesophyll cells are bathed in the external solutions which enter the cut ends. The results obtained are closely comparable with data for root material. Dr Glentworth wondered if there was any value in using variegated leaves. Dr Jennings felt that there was, because it was now well established that for large algal cells active chloride uptake is related to photosynthesis, and directly related by what is thought to be electron flow. Since this was so, it would be valuable to compare chloride uptake by similar green and non-green higher plant tissues.

Dr Sims was able to present more information to show that the dehydrogenase preparations from *Lemna* did represent the situation *in vivo*. Essentially, if the preparations were maintained at low temperatures, they were quite stable in configuration. Clues as to whether association or dissociation had occurred come from the most part from the analysis of zero-order kinetics of the enzyme. Dr Sims also talked about studies on the various strains of *Lemna*. It was quite clear from such studies that the enzymes differed markedly in their response to calcium, some having an optimum at 10^{-5} M and others working extremely well at 10^{-2} M. Some thirty strains had been examined, of which at least a dozen showed quite different distribution characteristics in relation to growth. The most subtle situation which had been examined was that where calcium can compete with magnesium for isocitric dehydrogenase. The problem is how cells cope with calcium where magnesium is the co-enzyme. For populations growing on limestone, the enzyme opted for manganese not magnesium and calcium no longer had any effect.

Concerning these enzyme studies, Dr Jennings felt that for malic dehydrogenase it should be remembered that mitochondria can be heavy accumulators of calcium. Lehninger's work has shown that, when mitochondria have been in media containing phosphate and calcium, you can see deposits of calcium phosphate within them. Dr Jennings wondered whether the adaptability of malic dehydrogenase was not in fact a symptom of the ability of mitochondria to remove calcium ions from the cytoplasm. This would mean that there ought also to be studies on the effect of calcium ions on glycolytic enzymes. Dr Jennings pointed out that ideas of this sort had already been suggested by Whittam with regard to the sensitivity of K^+—Na^+—ATPase to calcium. Mitochondria removing this ion which would otherwise inhibit the ATPase.

Following the query by Dr Kenworthy, there was considerable discussion on the movement of ions in submerged angiosperms. The general feeling as evidenced by statements by Dr Evans and Professor Weatherley, was that in these plants there is essentially a root exudation situation. There is ion transport into the xylem and the creation of a hydrostatic pressure. Because of this, there is an upward movement under quite inverse conditions from the transpiring plant, in that there is a positive pressure rather than a tension. Professor Weatherley said that it would be interesting to know whether there is not only a circulation of ions but also a circulation of water, which the Mass Flow theory of sieve tube movement predicts must happen.

Dr Martin discussed the evidence in favour of such a viewpoint. Most of the evidence came from his work on *Zostera* and other aquatic angiosperms. Most of these plants have reduced xylem but do have roots and they are strongly influenced by the substrate. In the laboratory, it is possible to show that the influence of the substrate is due to ions being taken up by the roots and being translocated through the plant in some way. There are a lot of reports of submerged transpiration—in other words an upward flow of water—and in many cases this appears to be due to root pressure. However, Dr Martin thought that most of the classical work is unsatisfactory for establishing this with any great certainty. So far all that Dr Martin and his colleagues had been able to do was show, by separating the root and leaf environment, that there is a mass flow of water in very small pipes—presumably the xylem.

MINERAL METABOLISM OF HALOPHYTES

EMANUEL EPSTEIN
Department of Soils and Plant Nutrition, University of California,
Davis, California 95616, U.S.A.

INTRODUCTION

The oceans and their shores are salty, and so is the water of many rivers and lakes and the soil solution of many soils of the arid and semi-arid regions of the earth. Yet most species of terrestrial higher plants are sensitive to even moderate levels of salinity. Almost all crop plants belong in this category. Nevertheless, the ocean and saline terrestrial environments are the habitat of a varied flora of salt-tolerant bacteria, algae, and higher plants (halophytes).

Three considerations shall be basic to our discussion of the mineral metabolism of halophytes.

1. At least sixteen chemical elements are essential for the growth of higher plants. With the exception of carbon, they are all derived in the main from the mineral substrate on which the plants grow.

2. The metabolic machinery of the cell consists of an organized, highly integrated system in which numerous chemical reactions must proceed in a manner which is coordinated in both time and space. The system could not function in the promiscuous, continuous presence of all kinds of mineral ions, in all manner of concentrations and ratios, at the sites of all the individual reactions.

3. Cells and their components can only function in a matrix of water.

As a metabolic machine, a plant must therefore deal with its chemical environment in three ways. It must selectively acquire essential nutrient elements from it, it must cope with elements present in excess, and it must acquire water.

The acquisition of nutrients

Apart from carbon and the elements of water all higher green plants require the following elements: potassium, calcium, magnesium, nitrogen, phosphorus, sulfur, iron, manganese, zinc, copper, chlorine, boron, and molybdenum (Epstein 1965). Perhaps cobalt will have to be added to this list (Wilson and Nicholas 1967).

The relative amounts in which different elements are acquired by plants differ enormously. Such differences are by no means a direct reflection of the respective concentrations of these elements in the soil. For example, in calcareous soils, the ratio of calcium to potassium available for absorption by plants may be on the order of 40/1. In plants growing on such a soil this ratio may be 1/2. Faced with a very large preponderance of calcium over potassium the plant nevertheless absorbs more potassium than calcium. Many such examples could be given, all documenting the remarkable ability of plants to absorb nutrient elements selectively.

Although the capacity of plants for selective or discriminating absorption is remarkable, it is not absolute. Sodium is a case in point. It is not known to be generally required by plants (Epstein 1965; Gerloff 1963). Yet all plants contain measurable amounts of sodium which is a ubiquitous element in the biosphere. Both ecological observation and laboratory experimentation show nevertheless that discriminating mechanisms are at work in sodium absorption by plants. More about that later. Sodium is of interest in another connection. While most land plants do not require it or have a requirement for it too small to be demonstrated, halophilic bacteria, marine algae, and at least some halophytic land plants require sodium as an essential element, as of course do animals.

Coping with excess

Excessive concentrations of salts in the nutrient medium, that is, concentrations much greater than those that would result in deficiency, pose two distinct problems. An element present in excess may cause metabolic derangements by competing for entry with other elements present at lesser concentrations, and once absorbed, by poisoning enzymes, by displacing other essential elements from their normal, functional sites, by precipitating other essential elements, by disrupting the structure of water, and by other mechanisms. Actual concentrations producing some of these effects need not be very high.

The other effect through which high external concentrations of salt may be detrimental to the growth of plants is indirect and depends on the depression of the water potential in the external medium. This matter is therefore discussed under a separate heading.

The acquisition of water

Unlike the toxic effects mentioned before, this osmotic effect is only provoked by high external concentrations of salt, in the order of 50 mM and above, corresponding to osmotic pressures in the vicinity of 2 atmospheres and higher in salt-sensitive plants, and at much higher concentrations in salt-tolerant plants and halophytes. Plant cells bathed by a medium of high osmotic pressure must maintain internal osmotic pressures higher than the one prevailing in the external solution, else they will lose water. It is this requirement for maintenance of high internal osmotic pressures that has led to the elaboration of mechanisms of transport capable of building up and maintaining within the cells high concentrations of selected ions and other solutes.

SELECTIVE ION TRANSPORT

Dual carrier mechanisms

The foregoing brief discussion serves to emphasize that in the salt economy of all plants—not only of halophytes—selective ion transport is of central importance. It is responsible for the import, from the environment, of essential nutrients in amounts conducive to growth and function; it regulates the flux of ions present in the external medium at high concentrations; and it contributes to the maintenance of internal osmotic pressures higher than those of the surroundings, and thereby, to the maintenance of tissues in a hydrated state.

It will therefore serve our purpose to discuss the present ideas concerning selective ion transport in plants and some of the evidence upon which they are based. In the present context, examples will be drawn from work with salt-tolerant or halophytic species.

Briefly, the current ideas on membrane transport of mineral ions, and of many organic molecules also, involve the operation of entities variously called carriers, pumps, translocators, permeases, or transport enzymes. The term, carriers, is common in work on absorption of mineral ions by higher green plants, and will be used here. In broad outline carriers are supposed to function as follows (Epstein 1965). At the outer surface of the cell membrane, the ion combines with the carrier to form a carrier-ion complex. This complex is transitory and after reorientation within the membrane dissociates, releasing the ion into the compartment at the far side of the membrane. Return of the ion is impeded by the relative impermeability of the membrane and the configuration of the carrier at the far side of the membrane which favors dissociation rather than binding of the ion. Calcium ions must be present in the external solution if the membrane and its carrier apparatus are to function normally (Epstein 1965). In halophytes as in glycophytes, unequal absorption of mineral cations and anions is compensated by adjustment of the intracellular concentrations of organic acids (Williams 1960).

Figure 1 shows the results of an experiment of the kind which, more than any other, has made the carrier hypothesis attractive to many investigators. This is an experiment on absorption of chloride by excised roots of tall wheatgrass, *Agropyron elongatum* (Host) Beauv. (Elzam 1966). For such an experiment, seedlings are grown in a dilute solution of calcium sulfate

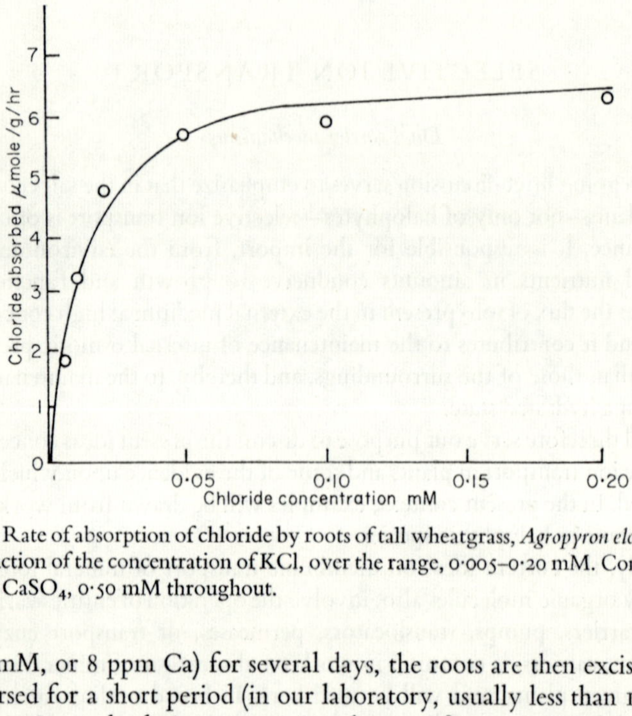

FIG. 1. Rate of absorption of chloride by roots of tall wheatgrass, *Agropyron elongatum*, as a function of the concentration of KCl, over the range, 0·005–0·20 mM. Concentration of $CaSO_4$, 0·50 mM throughout.

(0·20 mM, or 8 ppm Ca) for several days, the roots are then excised and immersed for a short period (in our laboratory, usually less than 1 hr) in the experimental solutions containing the ion of interest in radioactively labelled form. Such experiments yield rates of absorption of the ion. For details of the technique, reference is made to Epstein *et al.* (1963).

In this experiment with roots of tall wheatgrass, the ion was chloride labelled with ^{36}Cl. The rate of its absorption is plotted as a function of the external concentration of chloride. At the lowest concentrations, the rate of uptake rises steeply as the external concentration is raised, but at higher concentrations, the rate of absorption becomes independent of the concentration of chloride.

Many reactions involving the binding of molecules or ions by a receptor or ligand show this same type of concentration dependence, including the rate of transformation of a substrate by an enzyme as a function of the concentration of the substrate. The interpretation is that a limited number of binding sites is available. At progressively higher concentrations a point is eventually approached where all the active sites available for binding are occupied, and an upper limit is therefore reached for the rate of the process which depends on this transient binding by an active agent, be it an enzyme or a carrier.

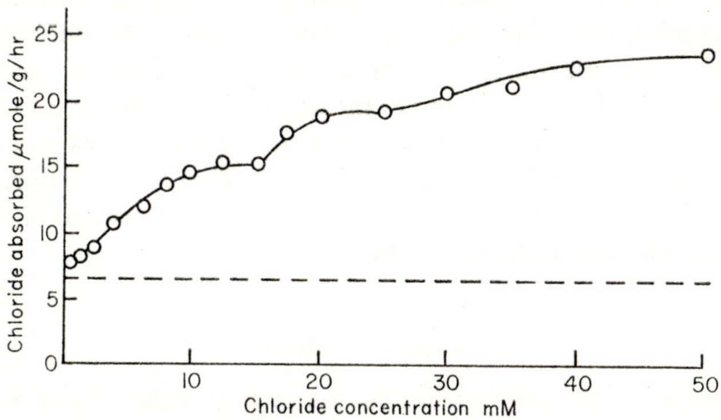

FIG. 2. Rate of absorption of chloride by roots of tall wheatgrass, as a function of the concentration of KCl, over the range 0·5 to 50 mM. Concentration of $CaSO_4$, 0·50 mM throughout. The dashed line represents the maximal rate of absorption via mechanism 1 (cf. Fig. 1).

In the present experiment, this upper limit was closely approached at a chloride concentration of 0·20 mM, or 7 ppm. In fact, there is a wide plateau between 0·05 and 0·20 mM, and the rate of absorption is barely higher even at 0·50 mM (Fig. 2). These are not high concentrations. Tall wheatgrass is quite a salt-tolerant grass which is often found on soils in which the chloride concentration of the soil solution is far in excess of these low values. It is therefore proper to ask what happens, in terms of the absorption of chloride, when its concentration is raised much above the top concentration in this experiment.

Figure 2 supplies the answer. This is an experiment like the previous one, except that the range of concentrations explored is from 0·50–50 mM chloride (18–1800 ppm). The plateau reached in the previous experiment at a chloride concentration of 0·20 mM lay at about 7 μmole chloride absorbed per gram tissue (fresh weight) per hour. Over the present, high range of concentrations, the rate of absorption of chloride becomes much higher, and at 50 mM reaches a value over three times as much. If the previous experiment did indeed indicate that at 0·20 mM chloride, a carrier mechanism was operating at full capacity, then it follows that at the high concentrations of the present experiment a second mechanism participates in the absorption of chloride.

This is in fact the interpretation which has been put on this and many similar experiments. At high concentrations, a given ion, chloride in this case, is absorbed not via a single carrier mechanism but via two. Of the two, only the first (mechanism 1) operates at low concentrations, mechanism 2 not coming into play except at concentrations where mechanism 1 operates at near-maximal rate. Such a dual pattern of ion absorption has been shown for ions of so many elements, and in so many and so diverse plant materials, as to lead to the surmise that it may be virtually universal in mature tissues of higher plants (Epstein 1966). The two mechanisms differ not only in their responses to different external concentrations of the ions but in several other features as well.

Adaptive significance

There are two features of a saline environment that the cellular transport mechanisms must be adapted to cope with if essential nutrients are to be acquired in physiologically adequate quantities. First, even in saline substrates the concentrations of some mineral nutrients required in large amounts may be low. The mechanism of absorption must therefore be able to import these ions from a region of low concentration and build up internal concentrations much higher than those in the medium. Second, there are present in such an environment other ions at high concentration which are potential competitors of the ions of the essential nutrient element. The transport mechanism must therefore possess a sufficiently high degree of specificity for the essential ion to absorb it selectively; otherwise it might be excluded through competition by the chemically related ion present in the environment in excess.

We shall focus on the two alkali metals, potassium and sodium, to show how this twin requirement is in fact met. Of all the mineral nutrient cations potassium is the one needed by plants in largest amount (Epstein 1965). Yet in the soil solutions of many soils its concentration is very low. In a survey of nutrient ion concentrations in many soils it was found that 44·5% of the soils had soil solutions with potassium concentrations of 1 mM (40 ppm) or less (Reisenauer 1966). It is from such dilute solutions that the roots of plants growing on these soils absorb the potassium essential for growth.

These observations draw attention to the type 1 mechanism of potassium absorption, for only it has a sufficiently high affinity for potassium to absorb this element at appreciable rates from solutions of such low concentration. The roots of many species absorb potassium via this mechanism at half the maximal rate when the external concentration of potassium is about 0·02 mM, or 0·8 ppm (Epstein 1966; and cf. the minus sodium run in Fig. 3). At 1 mM potassium (40 ppm), absorption via mechanism 1 is maximal, while mechanism 2 does not yet contribute appreciably.

Evidently, mechanism 1 fulfills the first of the two requirements, that of a high affinity for potassium. Does it also possess the second, that of selectivity *vis-à-vis* a potentially competing ion? Sodium is the ion posing the greatest physiological threat, for two reasons. Chemically, it is closely related to potassium, the two elements being neighbours in the group of alkali metals of the periodic table, and ecologically, it is the mineral element which preponderates in the environment of halophytes.

Figure 3 shows the results of an experiment on potassium absorption by roots of tall wheatgrass over the low range of potassium concentrations, i.e., the range where the type 1 mechanism is the only one operating, in the presence and absence of 10 mM (230 ppm) sodium. At the lowest potassium concentration, 0·005 mM (0·2 ppm), sodium ions in the 10 mM sodium treatment outnumbered potassium ions in the solution by a factor of 2000/1. Yet this large excess of sodium ions only diminished the rate of absorption of potassium to about half that in the control (minus sodium) treatment. Evidently, the potassium absorbing mechanism 1 shows a high degree of resistance to interference by sodium. The type 1 mechanisms of absorption of other elements also are highly selective. This specificity or selectivity of the type 1 mechanisms is found in both salt-sensitive and salt-tolerant species (Epstein 1966, and references given there; Rains and Epstein 1967).

Since the distinguishing characteristics of halophytes have to do with their ability to thrive at high salt concentrations the question arises whether the type 2 mechanisms of absorption—the mechanisms which come into play at high salt concentration—play a crucial role in the salt economy of halophytes. The evidence suggests that this is so. For example, in roots of tall wheatgrass, a salt-tolerant species, the type 2 mechanisms of absorption of chloride and the alkali cations contribute much more to the total rate of absorption than is the case in the much less tolerant intermediate wheatgrass, *Agropyron intermedium* (Host) Beauv. (Elzam 1966). In leaf tissue of the mangrove, *Avicennia marina* (Forst.) Vierh., absorption of potassium at high concentrations is predominantly via the type 2 mechanism (Rains and Epstein 1967).

It might be argued that conclusions based on short-term experiments with excised tissues may not bear closely on the regulation of salt uptake by growing plants (Pitman 1967). There is, however, ample evidence from long-term experiments for the validity of the dual pattern of transport outlined above and its significance in the regulation of the salt economy of growing plants, including halophytes.

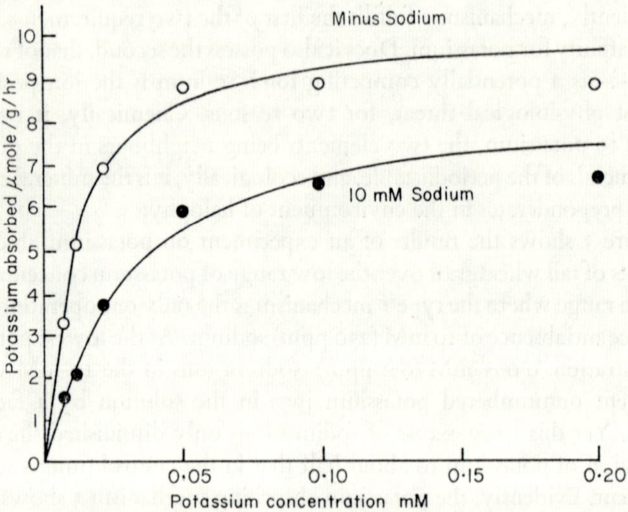

FIG. 3. Rate of absorption of potassium by roots of tall wheatgrass, as a function of the concentration of KCl, over the range 0·005–0·20 mM, and the effect of sodium at 10 mM. Concentration of $CaSO_4$, 0·50 mM throughout.

Black (1960) grew plants of the salt-tolerant Australian xerophyte, *Atriplex vesicaria* Heward, in solutions of varying sodium concentrations, up to 1 M NaCl (twice the salt concentration of sea water). The conclusion emerged that there are two mechanisms of absorption of alkali cations, one highly selective for potassium, the second one mainly functioning at high concentrations of either sodium or potassium.

Osmond (1966) did long-term experiments in which plants of *Atriplex* species were grown in nutrient solution. His findings also led to the conclusion that there is a dual system of absorption mechanisms, one specific, the other operating at high external concentrations and referred to as a mechanism of 'luxury absorption'. The term is probably a misnomer, since absorption of large amounts of salt from saline media is an adaptation by which the plants cope with the problem posed by high external osmotic pressures (see below).

Long-term solution culture experiments by Elzam (1966) with *Agropyron* produced results concordant with those of the short-term kinetic experiments using excised roots discussed above.

There is other evidence to the effect that the system of dual absorption mechanisms discovered in short-term experiments with excised tissues operates in the cells of normally growing plants.

It develops from the above discussion that both the acquisition of essential nutrients and the regulation of the flux of ions present in the medium at high concentrations are governed by dual systems of carrier mechanisms. We now turn to the third factor listed in the introduction: the need for water.

UPTAKE OF WATER

The water potential of a solution is progressively depressed the more concentrated the solution (Kramer *et al.* 1966; Slatyer 1967). This realization led to the classical idea that the stress imposed on plants in saline media is a water deficit; in other words, plants in saline media suffer from 'physiological drought'. This view has been much championed by agricultural scientists (Bernstein and Hayward 1958). However, Eaton (1927), Greenway (1962), and Slatyer (1961) doubted this interpretation, maintaining instead that plant cells, through solute absorption, maintain internal osmotic pressures sufficiently higher than that of the external medium to effect an osmotic adjustment which prevents osmotic loss of water to the medium, or physiological drought. It might be said that, faced with high external osmotic pressures, plants fight fire with fire and build up internal osmotic pressures higher yet. The work of Bernstein (1961) has provided firm evidence for osmotic adjustment even in roots.

What distinguishes halophytes, then, is not the ability to withstand physiological drought, but rather their capacity for selective accumulation of ions to very high internal concentrations, and their metabolic tolerance of such high concentrations. It is this realization that has led to the emphasis, in the present paper, on selective ion transport as the principal device by which halophytes cope with the salinity of their substrates. Even their metabolic tolerance of salt may be in part the result of selective transport or distribution of ions among the compartments and particulates within the cells.

OTHER MECHANISMS

Selective accumulation and toleration of salt are universals in the mineral metabolism of halophytes. There are other adaptations possessed by some but not all of these plants. Particularly noteworthy are the salt glands by which some halophytes excrete salt (Helder 1956; Atkinson *et al.* 1967). Another adaptation found in some plants is the development of succulence (Biebl and Kinzel 1965). Succulence increases the volume of the leaf per unit area of surface, thus resulting in a dilution of the intracellular salt.

ACKNOWLEDGEMENTS

I thank D.W. Rains for reading the manuscript and Beecher Crompton for tracking down the authorities for the Latin binomials. This research was supported by grants from the Office of Saline Water, United States Department of the Interior.

REFERENCES

Atkinson M.R., Findlay G.P., Hope A.B., Pitman M.G., Saddler, H.D.W. and West K.R. (1967). Salt regulation in the mangroves *Rhizophora mucronata* Lam. and *Aegialitis annulata* R. *br. Aust. J. biol. Sci.* **20**, 589–599.

Bernstein L. (1961). Osmotic adjustment of plants to saline media. I. Steady state. *Am. J. Bot.* **48**, 909–918.

Bernstein L. and Hayward H.E. (1958). Physiology of salt tolerance. *Ann. Rev. Pl. Physiol.* **9**, 25–46.

Biebl R. and Kinzel H. (1965). Blattbau und Salzhaushalt von *Laguncularia racemosa* (L.) Gaertn. f. und anderer Mangrovebäume auf Puerto Rico. *Öster. Bot. Z.* **112**, 56–93.

Black R.F. (1960). Effects of NaCl on the ion uptake and growth of *Atriplex vesicaria* Heward. *Aust. J. biol. Sci.* **13**, 249–266.

Eaton F.M. (1927). The water requirement and cell-sap concentration of Australian saltbush and wheat as related to the salinity of the soil. *Am. J. Bot.* **14**, 212–226.

Elzam O.E. (1966). Absorption of sodium, potassium, and chloride by two species of *Agropyron* differing in salt tolerance. Doctoral thesis, University of California, Davis.

See Elzam O.E. and Epstein E. (1969). *Agrochimica*. Pisa. (In press).

Epstein E. (1965). Mineral metabolism. *Plant Biochemistry* (Ed. J. Bonner and J.E. Varner), pp. 438–466. New York

Epstein E. (1966). Dual pattern of ion absorption by plant cells and by plants. *Nature*, **212**, 1324–1327.

EPSTEIN E., SCHMID W.E. and RAINS D.W. (1963). Significance and technique of short-term experiments on solute absorption by plant tissue. *Plant and Cell Physiol.* **4**, 79–84.

GERLOFF G.C. (1963). Comparative mineral nutrition of plants. *Ann. Rev. Pl. Physiol.* **14**, 107–124.

GREENWAY H. (1962). Plant response to saline substrates. I. Growth and ion uptake of several varieties of *Hordeum* during and after sodium chloride treatment. *Aust. J. biol Sci.* **15**, 16–38.

HELDER R.J. (1956). The loss of substances by cells and tissues (salt glands). *Encyclopedia of Plant Physiology.* (Ed. W. Ruhland), **2**, 468–488.

KRAMER P.J., KNIPLING E.B. and MILLER L.N. (1966). Terminology of cell-water relations. *Science* **153**, 889–890.

OSMOND C.B. (1966). Divalent cation absorption and interaction in *Atriplex*. *Aust. J. biol. Sci.* **19**, 37–48.

PITMAN M.G. (1967). Conflicting measurements of sodium and potassium uptake by barley roots. *Nature*, **216**, 1343–1344.

RAINS D.W. and EPSTEIN E. (1967). Preferential absorption of potassium by leaf tissue of the mangrove, *Avicennia marina*: an aspect of halophytic competence in coping with salt. *Aust. J. biol. Sci.* **20**, 847–857.

REISENAUER H.M. (1966). Mineral nutrients in soil solution. *Environmental Biology* (Ed. P.L. Altman and D.S. Dittmer), p. 507. *Fed. Am. Soc. Exp. Biol.*, Bethesda, Maryland.

SLATYER R.O. (1961). Effects of several osmotic substrates on the water relations of tomato. *Aust. J. biol. Sci.* **14**, 519–540.

SLATYER R.O. (1967). *Plant-Water Relationships.* New York.

WILLIAMS M.C. (1960). Effect of sodium and potassium salts on growth and oxalate content of *Halogeton*. *Pl. Physiol.* **35**, 500–505.

WILSON S.B. and NICHOLAS D.J.D. (1967). A cobalt requirement for non-nodulated legumes and for wheat. *Phytochem.* **6**, 1057–1066.

DIFFERENCES IN THE PROPERTIES OF THE ACID PHOSPHATASES OF PLANT ROOTS AND THEIR SIGNIFICANCE IN THE EVOLUTION OF EDAPHIC ECOTYPES

H. W. WOOLHOUSE
Department of Botany, University of Sheffield

INTRODUCTION

The adaptation of a plant to its environment involves selective responses to many qualitatively and quantitatively different physical and chemical factors. For this reason the adaptive behaviour of the plant may take many forms ranging from timing devices in relation to seasonal fluctuations in climate, to morphological and physiological changes in the shoots and roots in relation to the prevailing above and below-ground conditions. The point which I wish to emphasize in the present paper is that the chemical environment in the soil may frequently be a dominant component in this selective complex acting on a plant. As examples of the pre-eminence of the chemical composition of the soil as a decisive selection factor in the distribution of the flora of the British Isles, one may cite the distinctive plant communities of podsols, brown earths, sand dunes, chalk and limestone soils and suggest that in these situations other factors such as temperature, water availability or grazing regime are frequently of secondary importance.

The unique significance of the chemical composition of the soil in the selective complex arises from the fact that there is at the root surface, or at the membrane surfaces bounding the apparent free spaces of the root, a direct confrontation of structural and catalytic proteins with the chemical constituents of the soil solution. Thus when a seedling germinates in a particular soil there will be an immediate contact between the surface proteins of its developing root system and the solutes in the soil solution. For this reason a direct and often severe selective 'sieve' is at once established and in consequence there will be a relatively rapid selection of closely adapted individuals. This situation is clearly demonstrated by the work of the Bangor school showing the rapid evolution of heavy metal-tolerant races of *Agrostis tenuis* and *Festuca ovina* (Bradshaw 1952, 1965). A further cause of the immediacy of the response of plant roots to contact with constituents of the soil solution is that the exposed proteins may have important enzymic properties such as those of carriers in ion transport systems. If these components are adversely affected they will be much more likely to evoke rapid responses in the overall metabolism of the plant than will a climatic effect on the shoot for example, since this will impinge on a large number of interlinked enzyme systems having built-in propensities for mutual compensation.

There are numerous examples of genetically controlled intra-specific variations in the nutrient responses of both wild and cultivated plants (Epstein and Jefferies 1964). The present study deals with acid phosphatases and with findings concerning these surface proteins which may have physiological and ecological implications with respect to edaphic adaptation. A relatively small proportion of the total phosphorus content of most soils is in the form of inorganic phosphate readily available for uptake by the plant from the soil solution. Much of the phosphorus may be present in the soil as insoluble phosphates of calcium, iron and aluminium or as organo-phosphorus compounds particularly inositol hexaphosphates and related compounds (Pederson 1953; Cosgrove 1962, 1964). The hydrolysis of insoluble inorganic phosphates in soils probably results mainly from bacterial action (Sperber 1958; Swaby and Sperber 1958; Muromtsev 1959; Louw and Webley 1959). Organo phosphorus compounds may be hydrolysed by both bacteria (Greaves, Anderson and Webley 1967) and fungi (Dox and Golden 1911; Jackman and Black 1951 and 1952; Casida 1959) many of which may be closely associated with the rhizosphere of higher plants (Katznelson, Peterson and Rouatt 1962; Das 1963; Subba-Rao and Bajpai 1965; Chonkar and Subba-Rao 1967). Studies using higher plants in sterile culture have shown that even in the absence of micro-organisms they also are able to utilize phytates and other organo-phosphorus compounds as sources of phosphorus for growth, probably by virtue of phosphatase enzymes associated with the root surfaces (Weissflog and Mengdehl 1933; Rogers *et al.* 1940; Saxena 1964; Wild and Oke 1966). The contribution of these root enzymes to the mobilization of insoluble phosphorus compounds in the soil under field conditions is not known and may indeed vary according to species and the prevailing soil conditions. As will be seen later however this hydrolytic activity may be only one aspect of the normal functions of these particular root enzymes.

Agrostis tenuis is a polymorphic species which occurs widely on calcareous and acid soils and also, as noted earlier, is capable of evolving races adapted to growth on soils containing heavy metals, particularly lead (Bradshaw 1952). The adaptations to these three types of soil probably involve a variety of physiological mechanisms but there are also indications from several studies that phosphorus metabolism is probably involved in part of the adaptive mechanism to each of these three types of soil, calcareous (Olsen 1935), acidic (Wright 1943) and lead spoil (Jowett 1959). With this information in mind it was decided to investigate further the comparative aspects of the phosphate metabolism of races of *A. tenuis* from these three types of soil by examining initially the response of the root surface phosphatases of each race to a range of concentrations of calcium, aluminium and lead, the dominant ions of the calcareous, acid and lead spoil soils respectively.

MATERIALS AND METHODS

Edaphic ecotypes of *A. tenuis* were collected from the following sites:

Calcareous soil race:	Rock-ledge protorendzina, pH 7·5— Coombsdale, Derbyshire
Lead spoil heap race:	Lead spoil heaps, pH 7·2— Coombsdale, Derbyshire
Acid soil race:	Podsolized millstone grit soil. pH 3·8— Houndkirk Moor, Derbyshire

For some of the studies plant roots were taken directly from the soil, 1 cm root tips were cut and shaken with five successive changes of distilled water until, even under the microscope, no adhering foreign matter could be detected. In further experiments tillers were taken from the plants at the time of field collection and grown for 6 weeks in solution culture under the conditions described in Woolhouse (1966). Phytate hydrolysis may be measured directly using intact roots but the rates are somewhat slow necessitating the use of relatively large quantities of root material; further difficulties arise from the fact that some of the phosphate released is then taken up by the roots. Fortunately the enzyme shows a high affinity for p-nitrophenyl phosphate, catalysing the reaction:

$$\text{p-nitrophenyl phosphate} \rightarrow \text{p-nitrophenol} + \text{phosphate}$$

Under alkaline conditions the p-nitrophenol released forms the yellow-coloured phenolate ion with an absorption maximum at 410 mμ. A particular advantage of this method is that the p-nitrophenol which one is measuring is only slowly taken up by the roots; this is readily demonstrated by placing root tips in a solution of p-nitrophenol for an equivalent incubation time and then measuring the residual phenolate ion following the addition of sodium hydroxide, Fig. 1. The relative rates of hydrolysis of β-glycerophosphate, inositol hexaphosphate and p-nitrophenyl phosphate by the root surface phosphatase of *A. tenuis* are shown in Fig. 2. It is perhaps worth noting that the capacity for hydrolysis of p-nitrophenyl phosphate

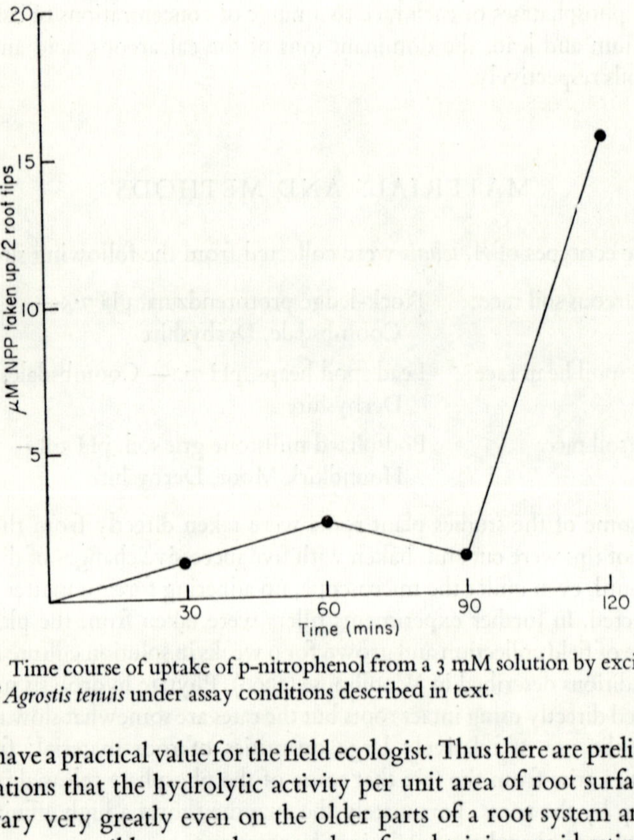

FIG. 1. Time course of uptake of p-nitrophenol from a 3 mM solution by excised root tips of *Agrostis tenuis* under assay conditions described in text.

may have a practical value for the field ecologist. Thus there are preliminary indications that the hydrolytic activity per unit area of root surface may not vary very greatly even on the older parts of a root system and so it may prove possible to use the procedure for obtaining rapid estimates of the surface area of a root system.

Assay procedure: Two 1 cm root tips were placed in a vial containing 1·0 ml 1 mM sodium citrate buffer pH 4·5 and 0·5 ml 3 mM p-nitrophenyl phosphate and incubated in a shaker both at 20° C. for 30 min. The root tips were then removed from the solution and the mean diameter of each was measured under the microscope; 4·5 ml of 0·1 N NaOH was then added to the incubation medium and the absorption at 410 mμ was measured against a blank containing the same proportions of buffer, substrate and NaOH which had been incubated without roots present. The amount of substrate hydrolysed was then read off from a calibration curve prepared from alkaline solutions of p-nitrophenyl phosphate of known concentrations. The data are expressed as μM substrate hydrolysed per mm² of root surface per hour.

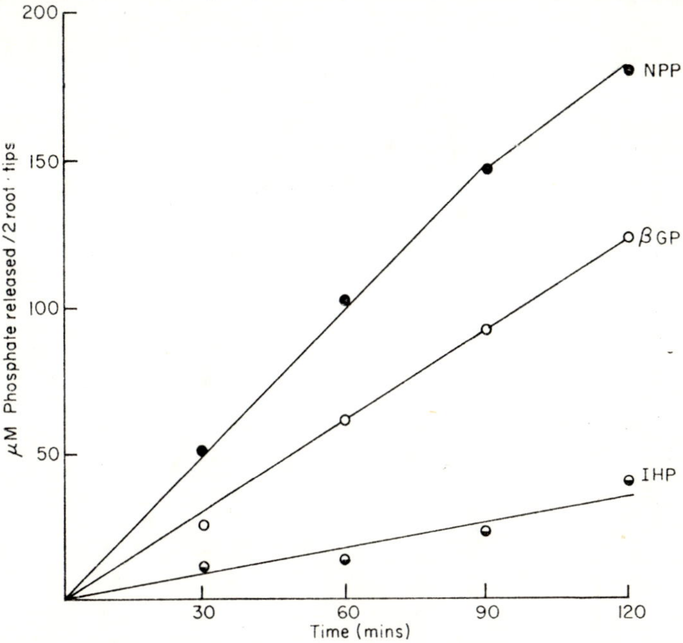

FIG. 2. Relative rates of hydrolysis of β-glycerophosphate (β-GP), inositol hexaphosphate (IHP) and p-nitrophenyl phosphate (NPP) by excised root tips of *Agrostis tenuis* under assay conditions described in text. Substrate concentration 10 mM for each substrate, pH 4·4.

RESULTS AND DISCUSSION

Phosphatases of intact roots

Figure 3 shows the amount of hydrolytic activity as a function of the pH of the incubation medium. At first sight this graph presents an unusual appearance; thus above pH 4·6 the enzyme activity is relatively low, between pH 3·8 and 4·6 there is a plateau of uniform activity whilst below pH 3·8 there is a sharp rise in the activity of the enzyme. As will be seen later this apparent increase in enzyme activity below pH 3·8 is an artefact resulting from the breakdown of cell permeability so that the substrate gains access to the internal phosphatases of the root cells.

Table 1 gives values for the phosphatase activity of roots of *A. tenuis* grown in the soil and in water culture. For purposes of comparison data for other grasses and for beech mycorrhizas are also included. It is seen that

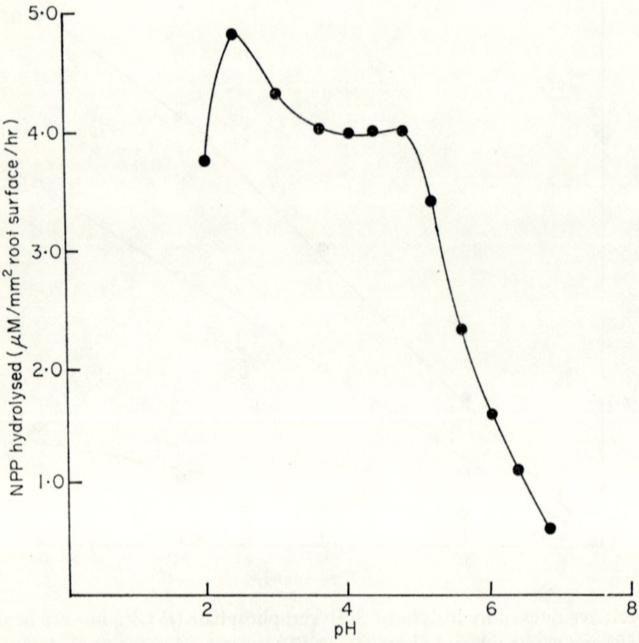

FIG. 3. Effect of pH on the rate of hydrolysis of p-nitrophenyl phosphate by excised root tips of *Agrostis tenuis*. Assay conditions as described in text.

there are no significant differences between soil and water-grown plants although the variability is greater in the former case, probably due in part to bacterial contamination and perhaps also to a certain amount of tissue damage during washing. Table 1 also shows that the phosphatase activity varies considerably between species, the highest activity so far recorded being on the surfaces of beech mycorrhizas. In most of the samples of *Agrostis* collected the phosphatase activity on the roots of the calcareous soil and spoil-heap races tends to be approximately 20% higher than that of the acid soil race. Too much significance should not however be attached to such comparisons since the figures depend upon what really constitutes the surface in terms of which the activity is being expressed. This surface measurement may be subject to variations due to the presence of root hairs, differences in the microtopography of the root surfaces, differences in the extent of the intra-wall spaces to which the substrate may have ready access and to variations in the number and relative permeability of the root cap cells.

TABLE I

Surface phosphatase activity of 1-cm root tips of *Agrostis* ecotypes and other species. Assay conditions as described in text. pH 4·4.

Species	Source	NPP hydrolysed μM/mm²/hr
Agrostis tenuis		
AR	Field	4·1 ± 1·1
CR	Field	5·5 ± 1·2
LR	Field	5·3 ± 1·0
AR	Water culture	4·3 ± 0·7
CR	Water culture	5·2 ± 0·7
LR	Water culture	5·5 ± 0·8
Arrhenatherum elatius	Water culture	6·1 ± 1·3
Holcus mollis	Water culture	2·0 ± 0·1
Fagus sylvatica (mycorrhizas)	Field	8·8 ± 0·8

AR—acid soil ecotype; LR—lead soil ecotype; CR—calcareous soil ecotype.

For the investigation of the effects of metal ions the phosphatase activity was measured on samples of the root tips of each *Agrostis* race in the presence of the chlorides of calcium and aluminium or of lead nitrate in the incubation medium at a series of concentrations in the range 0–10^{-2} M. The data summarizing the action of each metal on each ecotype are given in Figs. 4, 5 and 6. Figure 4 shows the effect of Al^{+++} on the surface phosphatases of the three ecotypes; it is seen that at the lower concentrations of aluminium the activity of the enzyme decreases in all three races but the inhibition is greater at lower concentrations of aluminium in the case of the calcareous and lead soil races. Similarly it is seen that lead (Fig. 5) and calcium (Fig. 6), although somewhat inhibitory for all three races, have the greatest inhibitory effects on the enzymes of the races which normally grow on soils not

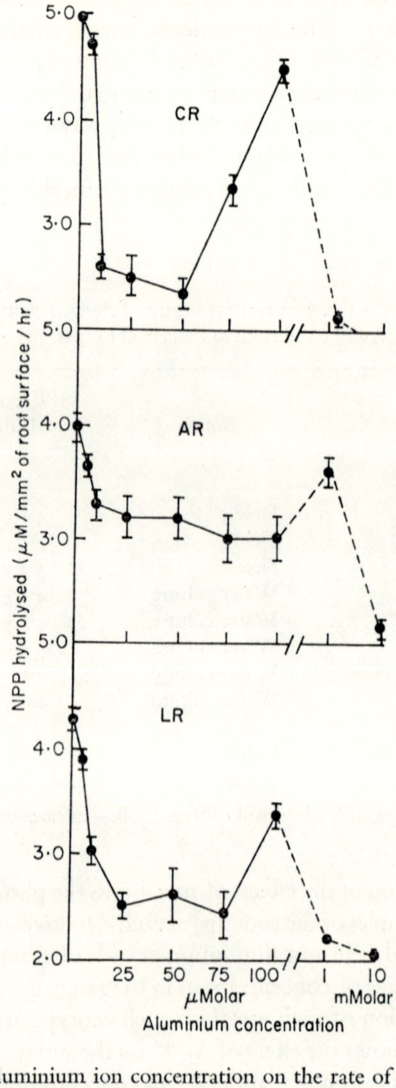

FIG. 4. Effect of aluminium ion concentration on the rate of hydrolysis of p-nitrophenyl phosphate by root tips of edaphic ecotypes of *A. tenuis*.

containing such high concentrations of these ions. Thus there appears to be differences between the enzymes of each race affecting their interactions with particular metal ions.

A peculiar feature of Figs. 4, 5 and 6 is that the initial decrease in enzyme activity with increasing metal ion concentration is followed by a secondary rise in activity, after which the activity again declines at yet higher concentrations of the particular metal. This apparent stimulation of the enzyme is

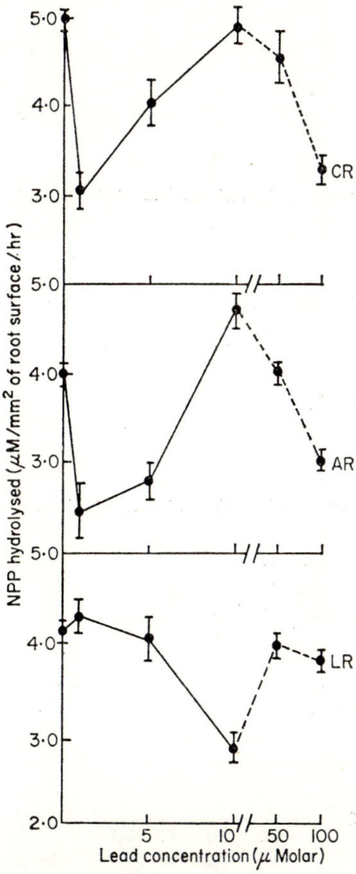

FIG. 5. Effect of lead ion concentration on the rate of hydrolysis of p-nitrophenyl phosphate by root tips of edaphic ecotypes of *A. tenuis*.

most probably due to the fact that at the particular metal concentrations at which this occurs the semipermeability of the root cell membranes is broken with the result that substrate molecules can now penetrate freely into the cells and so come into contact with internal phosphatases associated with the mitochondria and other cell components. In support

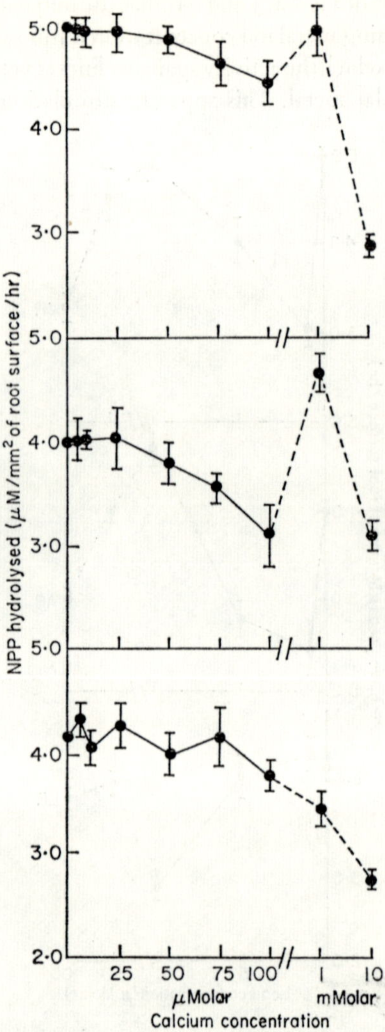

FIG. 6. Effect of calcium ion concentration on the rate of hydroysis of p-nitrophenyl phosphate by root tips of edaphic ecotypes of *A. tenuis.*

of this view one may note that there is an upward shift in the pH optimum of this increased phosphatase activity to a value of 5·0–5·5 which is nearer to that normally found for the phosphatase activity of cell-free extracts of the roots. Further, as seen in Fig. 7, flame photometric measurements of the amounts of potassium released into the incubation medium of aluminium-treated root tips, show that the increase in enzyme activity is associated with a sharp increase in the amount of potassium again pointing to an increased 'leakiness' of the root cells.

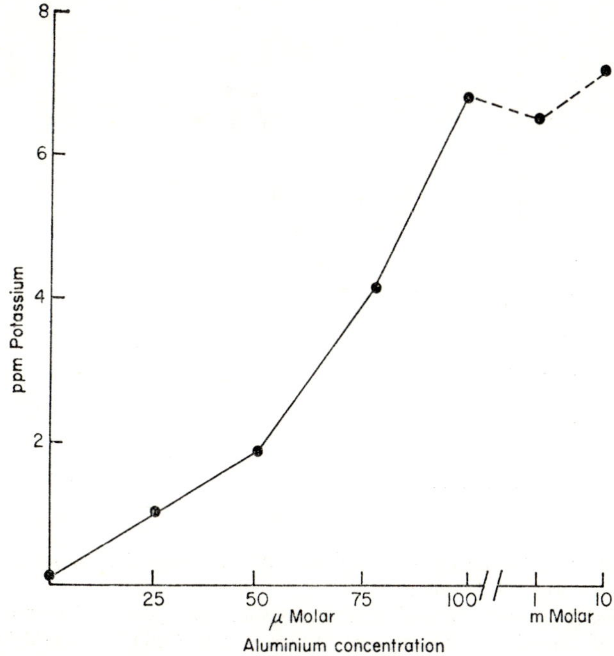

FIG. 7. Effect of aluminium on the leakage of potassium from excised root tips of the calcareous-soil ecotype of *A. tenuis*. The data are expressed as ppm K^+ in the substrate solution used for phosphatase assay after 30 min incubation at 20° C.

Phosphatases of root extracts

In an attempt to overcome some of these problems associated with the study of excised root tips, cell-free fractions were prepared from the roots of each of the *Agrostis* races grown in water culture. Some of the phosphatase activity of these preparations was associated with the cell wall fraction, whilst some remained in the soluble phase. There was a somewhat higher pH optimum (5·0–5·4) in these preparations than was found with the intact root enzyme, Fig. 8. Both the cell wall enzyme and the soluble component were particularly active in hydrolysing ATP, were stimulated by Mg^{++} ions and were further activated by Na^+ and K^+ ions, Table 2.

From Fig. 9 it is seen that, as in the case of the intact-root enzyme, there is ecotypic differentiation in the susceptibility of the ATP-ase of the cell wall fraction to the presence of Al^{+++} ions; in the case of the soluble ATP-

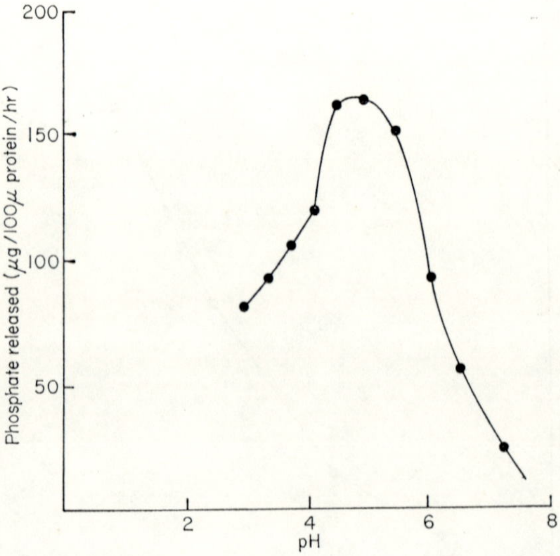

FIG. 8. Effect of pH on the rate of hydrolysis of ATP by cell walls extracted from roots of acid-soil ecotype of *A. tenuis*.

Extraction: Known weights of roots were ground in a pestle and mortar with 2 volumes of a solution containing 0·25 M sucrose, 3 mM EDTA, 10 mM Tris-HCl buffer, pH 7·4. The extract was strained through cheese cloth and then centrifuged at 1,400 ×*g* for 10 min. The supernatant was discarded and the cell wall pellet washed and recentrifuged with 2 further aliquots of extraction medium before finally suspending in 10 mM Tris-HCl buffer, pH 7·4.

Assay conditions: The reaction was started by adding 0·5 ml of enzyme extract containing 0·2 mg protein to a solution containing 3 mM ATP, 50 mM citrate buffer of appropriate pH and 1·5 mM $MgCl_2$, in a final volume of 3·5 ml. The preparation was incubated for 60 min at 30° C and the reaction then stopped by addition of 2·0 ml of 20% trichloroacetic acid. Phosphate released was estimated by the method of Fiske and Subbarow (1924) and protein by the method of Lowry *et al.* (1951).

ase activity on the other hand no firm conclusions can be drawn on this point since the response of successive preparations from the same plants has so far proved extremely inconsistent. Evidence of similar ATP-ase activity in plant cells has been obtained from cell wall preparations of roots of *Avena* seedlings (Fisher and Holmes 1966), soluble preparations from the

TABLE 2

The effect of Mg^{++}, Na^{+} and K^{+} ions on the activity of the ATP-ase from roots of the acid soil ecotype of *Agrostis tenuis*. For assay conditions see legend to Fig. 8. Each value is the mean of three replicate assays

Cell fraction	Ions added	Ion concentration in molar	Phosphate released µg/100 µg protein/hr
†Crude homogenate	—	—	207
†Soluble fraction	—	—	194
Cell walls	—	—	57
Cell walls	Mg^{++}	1·5	144
Cell walls	Na^{+}	2·0	84
Cell walls	K^{+}	2·0	65
Cell walls	$Mg^{++}+K^{+}$	1·5+2·0	156
Cell walls	$Mg^{++}+Na^{+}$	1·5+2·0	143
Cell walls	$Mg^{++}+Na^{+}+K^{+}$	1·5+2·0+2·0	137

† Values corrected for phosphate released from endogenous organo-phosphorus compounds.

roots of barley (Brown, Chattopadhyay and Patel 1967) and beans (Neumann and Gruener 1967) and in preparations from carrot, beet and *Chara australis* (Atkinson and Polya 1967); whilst McClurkin and McClurkin (1967) have provided cytochemical evidence for a similar enzyme in seedling roots of loblolly pine.

The finding of this ATP-ase activity in plant roots raises the intriguing speculation that we may in fact be dealing with enzymes involved in a 'pump' functioning in active cation transport in and out of the plant cells and probably comprising the cation 'carriers' inferred from the ion uptake studies of many workers, e.g., Epstein and Raines (1965). Although such carrier systems are now becoming known in some detail from studies using animal tissues it must be stressed that there is as yet no definitive evidence coupling these plant ATP-ases with ion transport, moreover they appear to differ, at least in certain respects, from the animal systems. Thus for example most of the studies to date using plant extracts show stimulation by Na^{+} or K^{+} ions but no further stimulation in the presence of both ions as is the case with the animal enzymes. It would be of interest to know whether this difference was related to the finding that in some plant cells at least only the Na^{+} ion is 'pumped' whilst the K^{+} ion is moved along an electrochemical gradient (Dainty and McRobbie 1963). Despite these differences it would seem pertinent in view of the great potential importance of these ATP-ases

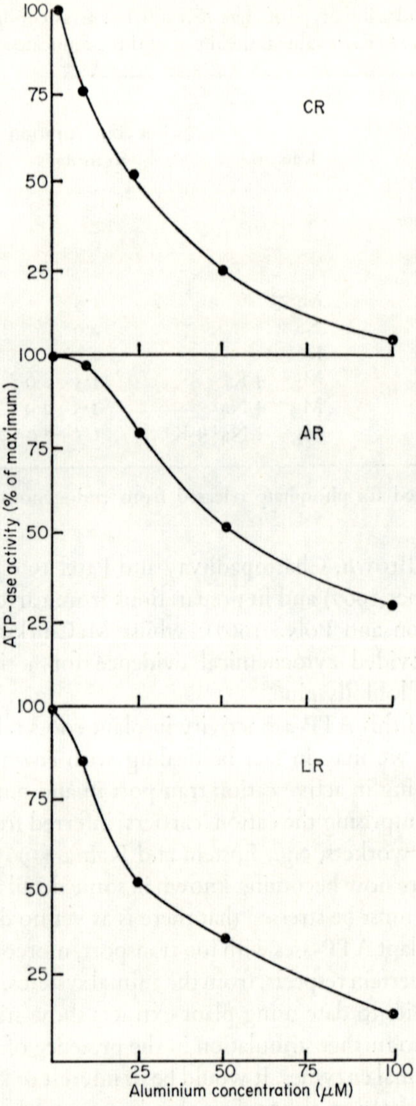

FIG. 9. Effect of aluminium concentration on the rate of hydrolysis of ATP by cell wall preparations from roots of edaphic ecotypes of *A. tenuis*. Extraction and assay conditions as for Fig. 8, at pH 4·8.

in plant nutrition studies in the future, to examine briefly the present state of knowledge concerning the animal enzymes and to consider their possible relevance for studies with plant tissues.

These (Mg^{++}–Na^{+}–K^{+})-activated ATP-ases have been intensively studied in recent years in preparations from a range of animal tissues including nerve cells of the crab *Carcinus maenus* (Skou 1957, 1960, 1961) reticulocyte membranes (Post and Jolly 1957; Post and Albright 1961) and brain and kidney tissues (Skou 1962; Post, Sen and Rosenthal 1965). On the basis of tracer exchange experiments, reciprocal competitive inhibition studies between Na^{+} and K^{+} and the use of selective inhibitors such as ouabain, azide and strophanthidin, a number of models have been developed which attempt to describe the functioning of the pump enzyme system. The scheme outlined below which is a modification of that of Skou (1967) is typical of the general pattern of models which have been propounded.

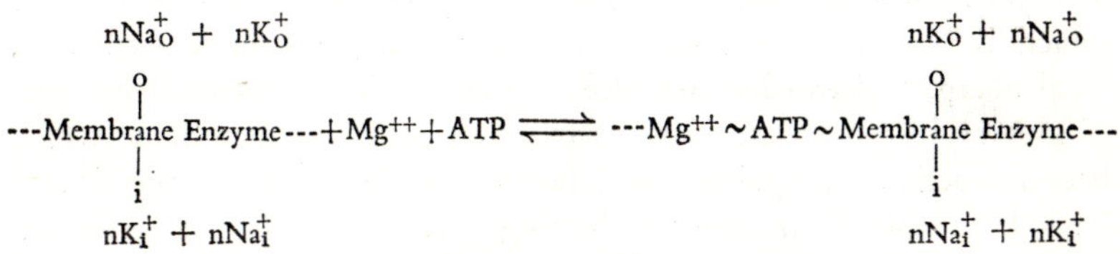

PARTIAL REACTION 1

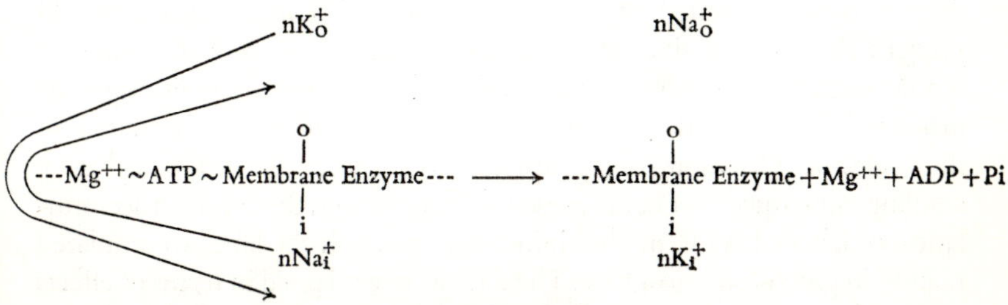

PARTIAL REACTION 2

The left-hand side of partial reaction 1 depicts a terminal state at which the pump has just completed a cycle, the membrane enzyme is now inactive having a sodium ion, which has just been pumped out, at the site o which is located on its outer surface and a potassium ion which has just been moved in, is on the site i which is located on the inner surface of the membrane enzyme. As the next cycle commences ATP reacts with the membrane enzyme causing it to increase the affinity of its external site o for K^{+} so that the Na^+ ion at o exchanges with a K^+ ion. At the same time the affinity of its internal site, i, for Na^+ also increases so that the K^+ at i exchanges with an Na^+ inside the cell (right-hand side of partial reaction 1). The presence of K^+ at o and Na^+ at i now activates the hydrolysis of the bound ATP and as this takes place the cations are transported (partial reaction 2). It is not known how the movement of ions takes place in this phase although the finding that ATP can induce allosteric changes in the configuration of these ATP-ases suggests that it may be brought about by a change in shape perhaps even involving a partial rotation of the membrane enzyme. Probably the most controversial aspect of this scheme at the present time, albeit one which has considerable potential interest for the future when the investigation of the plant enzymes achieves this level of detail, concerns the membrane enzyme$\sim$ATP complex (right-hand side of partial reaction 1). Thus work from a number of laboratories using ^{32}P–ATP has led to the isolation of a phosphorylated intermediate in the ATP-ase reaction (Post, Sen and Rosenthal 1965; Bader, Post and Jean 1967), the formation of which is Na^+-activated and the subsequent degradation K^+-activated. Kahlenberg, Galsworthy and Hokin (1967) claim to have shown that N-glutamyl γ phosphate is the specific residue in the membrane which becomes activated. Against this Schoner, von Iberg and Kramer (1967) have been able to fractionate the Mg^{++} from the K^+–Na^+-activated systems as separate fractions and have argued that a phosphorylated intermediate such as the glutamyl phosphate is not involved in the reaction since it is not inhibited by hydroxylamine which it is claimed would prevent the reaction since it would break the acyl phosphate bonds. It would seem possible however that the acyl phosphate residues might be protected from such degradation by features of the tertiary structure of the molecule. Clarkson (1966) supplied $^{32}PO_4$ to excised roots of barley seedlings and noted that in the presence of Al^{+++} ions there was an accumulation of labelled ATP, in aluminium-free controls the label accumulated mainly in glucose-6-phosphate. The data were explained in terms of effects on hexokinase activity. A further possibility is that the labelled ATP was accumulating due to a jamming of an ATP-ase system. In the light of the work on animal systems it becomes of interest to explore the possibility that the ecotypic differentiation in cation uptake capacity (Epstein and Jeffries 1964) and the differential Al^{+++} sensitivity of the *Agrostis* ecotype ATP-ases, might be related to the numbers of such intermediate phosphorylation sites per cell and to the degree of 'protection' of these sites consequent on allosteric changes in the tertiary structure of the ATP-ase molecules.

Before considering possible ecological implications of this work two further findings concerning these enzyme systems in *Agrostis* appear to be important. Thus it is subject to at least two types of control; firstly by end-product inhibition from phosphate and secondly by phosphate repression of the biosynthesis of the enzyme. Fig. 10 shows the effect of phosphate concentration on the activity of an ATP-ase extract from the roots of the

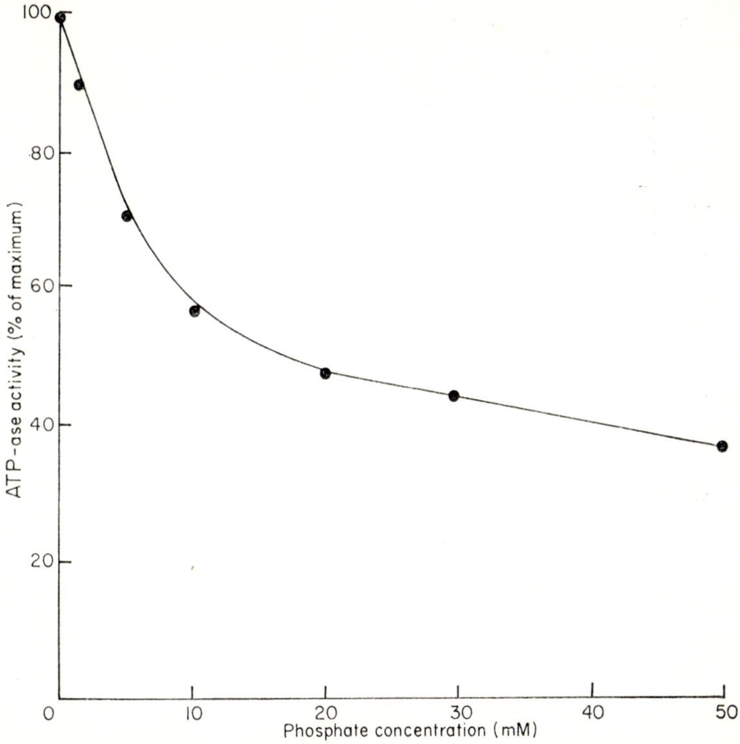

FIG. 10. Effect of phosphate concentration in the incubation medium on the rate of hydrolysis of ATP by cell wall preparations from the acid soil ecotype of *A. tenuis*. Assay conditions as in Fig. 8.

calcareous soil ecotype of *Agrostis*, the effect of phosphate is seen to become particularly pronounced at concentrations above 5 mM. Whilst phosphate concentrations of this magnitude probably rarely obtain in the soil solution they may be quite frequent in the relevant intracellular compartments. Fig. 11 shows the activity of the cell wall ATP-ase in preparations from the roots of the same *Agrostis* ecotype grown in water culture at different levels of phosphate; it is seen that above 10^{-5} M phosphate the activity of the enzyme begins to decline so that in plants grown at 10^{-3} M phosphate the specific activity of the phosphatase in the preparations is only 50% of that of plants grown at 10^{-6} M phosphate. A number of examples of repressible phosphatases are known from other groups of organisms. In bacteria (*E. coli*) the formation of the alkaline phosphatases is dependent on the level

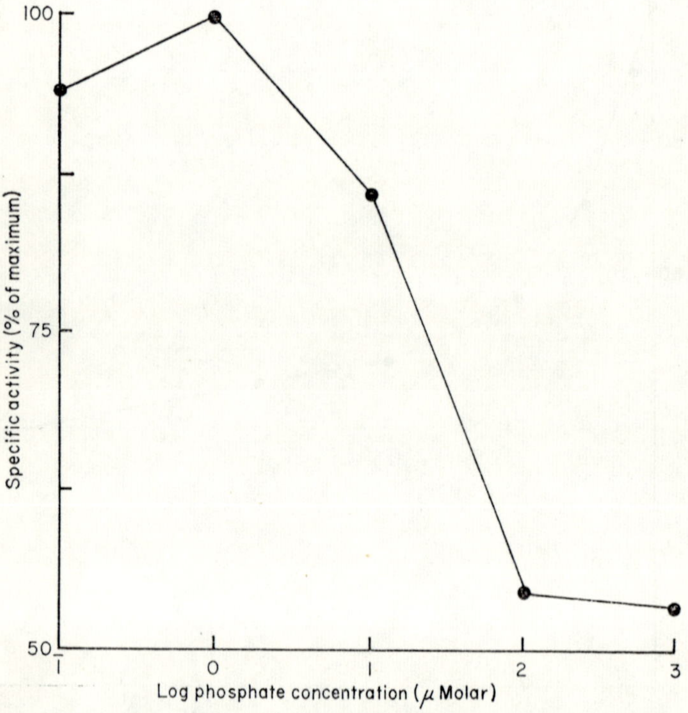

FIG. 11. Effect of phosphate concentration in solution culture medium in which plants were grown, on rate of ATP hydrolysis by cell wall preparations from the acid soil ecotype of *A. tenuis*. Assay conditions as described in Fig. 8.

of phosphate in the medium although the acid phosphatases appear to be constitutive (Horiuchi, Horiuchi and Mizuno 1959; Torriani 1960; Hofsten 1961). Suomalainen, Linko and Oura (1960) reported a six-fold increase in the acid phosphatase activity of yeast following transfer to a low phosphate medium. Subsequent work has shown that it is the external phosphatases associated with the yeast cell wall which are subject to repression–depression control, rather than the internal enzymes (Weimberg and Orton 1963, 1966; Heredia, Yen and Sols 1963; Gunther, Kattner and Merker 1967).

A phosphate-repressible acid phosphatase has also been reported from *Neurospora crassa* (Nye 1967). For higher plants the recent elegant experiments of Bianchetti and Sartirana (1967) have shown that phosphate probably acts at the transcriptional level in repressing phytase formation in germinating embryos of wheat. Higher plant species differ in the extent of the phosphate repression of phosphatase activity and under conditions of low phosphate supply the phosphatases of the shoots as well as the roots may be affected (Woolhouse, unpublished).

Ecological considerations

The conclusive demonstration of the significance of these phosphatase enzymes in the ecological relationships of higher plants clearly requires much more detailed study, several points of potential interest may however be made.

It is well known that phosphate availability varies greatly from one soil to another and there is evidence that this may be a major influence in moulding the floristics of whole regions (Beadle 1966) and in the determination of plant distribution on a smaller scale (Piggot and Taylor 1963). At the intra-specific level also there is evidence of variations in the ability of plants to take up phosphate and in their capacity for growth in the presence of different concentrations of phosphate, both in cultivated plants (Foote and Howell 1964) and in natural populations (Jowett 1959; Snaydon and Bradshaw 1962). The data presented here concerning the edaphic ecotypes of *Agrostis tenuis* suggests that differences of this kind may result from active evolution of the structure of the phosphatase enzymes in relation to the phosphate availability and the ionic composition of the soil solution. It would seem worthwhile to consider briefly the potential importance of evolutionary changes in this type of system in specific relation to the calcicole-calcifuge problem. It is clear that adaptation in plants for growth in acidic or calcareous soils is a complex process involving many different aspects of the metabolic and physiological responses of the plant. Thus in relation to plant growth on calcareous soils there is evidence that the organic acid metabolism of the roots in response to bicarbonate ions in the soil solution may be of particular importance (Brown 1961; Wallihan 1963; Miller 1963; Woolhouse 1966; Lee and Woolhouse 1969). It is also evident that the phosphorus metabolism of the plant may be important in determining chlorosis-susceptibility on calcareous soils. Thus Olsen (1935) showed that increased phosphate supply could cause chlorosis on calcareous soils and provided histochemical evidence of iron immobilization resulting from ferric phosphate precipitation around the veins in chlorosis-susceptible species. Certain chlorosis-susceptible mutants of maize (Bell *et al.* 1962) and of soybean (Brown, Holmes and Tiffin 1958) also develop chlorosis more rapidly in the presence of increased levels of phosphate supply. It would be of particular interest to make comparative studies of the control of phosphatase activity and its influence on the intra-cellular balance between inorganic and organic phosphate in relation to the tendency for iron immobilization in chlorosis-susceptible and resistant genotypes of these species.

It has frequently been observed that the incidence of chlorosis is related to the oxygen tension in the soil atmosphere. Although there are some conflicting reports in this matter the general consensus of opinion seems to be that chlorosis is most frequent in conditions of high oxygen tension and is not developed when there is a raised water table or otherwise reduced oxygen tension (Wallihan and Embleton 1960). This situation has been explained in terms of oxygen tension influencing the redox state of iron in the soil, Fe^{++}, which predominates at low oxygen tensions being more readily available to the plant than Fe^{+++} which is the major ionic form in a well-aerated soil. If, however, ATP-ases are involved in the release of inorganic phosphate at critical sites within the cell, then the oxygen tension influencing the aerobic metabolism of the roots and hence the proportions of organic to inorganic phosphate, may also be acting to imobilize the iron through these internal systems as well as by means of the redox state of the iron.

In conclusion it is of interest to consider briefly some of the wider aspects of enzyme modification in relation to environmental conditions. We are at the present time attempting to isolate the cell wall phosphatases with a view to their further purification and characterization. The work of Weiss (1943) on soybean, Bell *et al.* (1962) on maize, and of the Bangor school with *Agrostis* clearly shows that aspects of the phosphate nutrition of the plant may be under genetic control. Moreover the demonstration by Schwartz (1967) of genetic control of esterase isozymes from maize encourages the hope that we may be able to isolate and purify modified forms of the phosphatase enzymes from the roots of edaphic ecotypes. Should this prove successful it will become possible to define the physical and chemical properties of the proteins which render them functional in particular extremes of ionic environment. One would thus be in a position to define in molecular structural terms the requirements, and hence the genetic potential which a species must possess, in order to adapt to a particular set of soil conditions. In the long term this could also prove of value to the plant breeder, enabling him to carry out purposeful selections and hybridizations for particular properties of the root enzymes in relation to the prevailing soil conditions of a particular area.

ACKNOWLEDGEMENTS

The author is indebted to Mrs R.E. Spencer and Miss C. Smith for skilled technical assistance in the course of this work.

REFERENCES

ATKINSON M.R. and POLYA G.M. (1967). Salt-stimulated ATP-ases from carrot, beet and *Chara australis*. *Aust. J. biol. Sci.* **20**, 1069–1080.

BADER H., POST R.L. and JEAN D.L. (1967). Further characterization of a phosphorylated intermediate in (Na^+– K^+)-dependant ATP-ase. *Biochim. biophys. Acta* **143**, 229–238.

BEADLE N.G.W. (1966). Soil phosphate and its role in molding segments of the Australian flora with special reference to xeromorphy and sclerophily. *Ecology* **47**, 992–1007.

BELL N.D., BOGORAD L., MCIRATH W.J. (1962). Yellow-stripe phenotype in maize. Effects of yS_1, locus on uptake and utilization of iron. *Bot. Gaz.* **124**, 1–8.

BIANCHETTI R. and SARTIRANA M.L. (1967). The mechanism of the repression by inorganic phosphate of phytase synthesis in the germinating wheat embryo. *Biochim. biophys. Acta* **145**, 485–490.

BRADSHAW A.D. (1952). Populations of *Agrostis tenuis* resistant to lead and zinc poisoning. *Nature* **169**, 1098.

BRADSHAW A.D. (1965). Evolutionary significance of phenotypic plasticity in plants. *Adv. Genet.* **13**, 115–155.

BROWN H.D., CHATTOPADHYAY S.K. and PATEL A. (1967). Characteristics of an ATP-ase in membrane particles, solubilized and linked to a cellulose matrix. *Enzymologia*, **32**, 205–212.

BROWN J.C., HOLMES R.S. and TIFFIN L.O. (1958). Iron chlorosis in soybeans as related to the genotype of the rootstalk. *Soil Sci.* **86**, 75–82.

CASIDA L.E. (1959). Phosphatase activity of some common soil fungi. *Soil Sci.* **87**, 305–310.

CHONKAR P.K. and SUBBA-RAO N.S. (1967). Phosphate solubilization by fungi associated with legume root nodules. *Can. J. Microbiol.* **13**, 749–753.

CLARKSON D.T. (1966). Effect of aluminium on the uptake and metabolism of phosphorus by barley seedlings. *Pl. Physiol.* **41**, 165–172.

COSGROVE D.J. (1962). Forms of inositol hexaphosphate in soils. *Nature* **194**, 1265–1266.

COSGROVE D.J. (1964). An examination of some possible sources of soil inositol phosphates. *Plant and Soil* **21**, 137–141.

DAS A.C. (1963). The utilization of insoluble soil phosphates by soil fungi. *J. Indian Soc. Soil Sci.* **11**, 203–207.

Dox A.W. and Golden R. (1911). Phytase in lower fungi. *J. Biol. Chem.* **10**, 183–186.

Epstein E. and Jeffries R.L. (1964). The genetic basis of selective ion transport in plants. *Ann. Rev. Pl. Physiol.* **15**, 169–184.

Epstein E. and Raines D.W. (1965). Carrier-mediated cation transport in barley roots: Kinetic evidence for a spectrum of active sites. *Proc. Nat. Acad. Sci.* **53**, 1320–1324.

Fisher J. and Hodges T.K. (1966). Characterization of an ion-sensitive ATP-ase from oat roots. *Pl. Physiol.* (*Supplement*), **41**, Li.

Fiske C.H. and Subbarow Y. (1925). The colorimetric determination of phosphorus. *J. biol. Chem.* **66**, 375–400.

Foote B.D. and Howell R.W. (1964). Phosphorus tolerance and sensitivity of soybeans as related to uptake and translocation. *Plant Physiol.* **39**, 610–613.

Greaves M.P., Anderson G. and Webley D.M. (1967). The hydrolysis of inositol phosphates by *Aerobacter aerogenes. Biochim. biophys. Acta* **132**, 412–418.

Gunther Th., Kattner W. and Merker J.H. (1967). Über das verhalten und die lokalisation der sauren phosphatase von hefezellen bei repression und derepression. *Exp. Cell Research* **45**, 133–147.

Heredia C.F., Yen F. and Sols A. (1963). Role and formation of the acid phosphatase in yeast. *Biochem. biophys. Res. Comm.* **10**, 14–18.

Hofsten B. (1961). Acid phosphatase and the growth of *Escherichia coli. Biochim. biophys. Acta* **48**, 171–181.

Horiuchi T., Horiuchi S. and Mizuno D. (1959). A possible negative feedback phenomenon controlling formation of alkaline phosphomonoesterase in *Escherichia coli. Nature* **183**, 1529–1530.

Jackman R.H. and Black C.A. (1951). The hydrolysis of Al, Ca and Mg inositol phosphates at different pH values. *Soil Sci.* **72**, 261–266.

Jackman R.H. and Black C.A. (1952). The hydrolysis of phytate phosphorus in soils. *Soil Sci.* **73**, 167–171.

Jowett D. (1959). Adaptation of a lead-tolerant population of *Agrostis tenuis* to low soil fertility. *Nature* **184**, 43.

Kahlenberg A., Galsworthy P.R. and Hokin L.E. (1967). Sodium-potassium adenosine triphosphatase acyl phosphate 'Intermediate' shown to be L-glutamyl-γ-phosphate. *Science* **157**, 434–436.

Katznelson H., Peterson E.A. and Rouatt J.W. (1962). Phosphate-dissolving organisms in the root zone of plants. *Can. J. Bot.* **40**, 1181–1186.

Lee J.A. and Woolhouse H.W. (1969). A comparative study of bicarbonate inhibition of root growth in calcicole and calcifuge grasses. *New Phytol.* **68**, 1–14.

Louw H.A. and Webley D.M. (1959). A study of soil bacteria dissolving certain mineral fertilizers and related compounds. *J. appl. Bact.* **22**, 227–233.

Lowry O.H. Rosebrough N.J., Farr A.L. and Randall R.J. (1951). Protein measurement with the Folin reagent. *J. biol. Chem.* **193**, 265–275.

McClurkin I.T. and McClurkin D.C. (1967). Cytochemical demonstration of a sodium-activated and a potassium-activated adenosine triphosphatase in Loblolly pine seedling root tips. *Pl. Physiol.* **42**, 1103–1110.

Miller G.W. and Hsu W. (1965). Effects of carbon dioxide-bicarbonate on oxidative phosphorylation by cauliflower mitochondria. *Biochem. J.* **97**, 615–619.

Muromtsev G.S. (1958). The dissolving action of some root and soil microorganisms on calcium phosphates insoluble in water. *Soils and Fert.* **22**, 44.

NEUMANN J. and GRUENER N. (1967). A soluble ATP-ase from bean roots stimulated by monovalent cations. *Israel J. Chem.* **5**, 107–116.

NYE J.F. (1967). A repressible acid phosphatase from *Neurospora crassa. Biochem. biophys. Res. Comm.* **27**, 183–188.

OLSEN B. (1935). Iron absorption and chlorosis in green plants. *C.R. Trav. Lab. Carlsberg, Chem.* **21**, 15–51.

PEDERSON E.J.N. (1952). On phytin phosphorus in the soil. *Plant and Soil* **4**, 252–266.

PIGOTT C.D. and TAYLOR K. (1964). The distribution of some woodland herbs in relation to the supply of nitrogen and phosphorus in the soil. *J. Ecol.* **52** (Supplement) 175–185.

POST R.L. and ALBRIGHT C.D. (1961). Membrane adenosine triphosphatase system as part of a system for active sodium and potassium transport. In: *Membrane Transport and Metabolism.* (Eds. A. Kleinzeller and A. Kotyk), pp. 219–227. New York.

POST R.L. and JOLLY P.C. (1957). The linkage of sodium, potassium and ammonium active transport across the human erythrocyte membrane. *Biochim. biophys. Acta* **25**, 118–128.

POST R.L., SEN A.K. and ROSENTHAL A.S. (1965). A phosphorylated intermediate in adenosine triphosphate dependent sodium and potassium transport across kidney membranes. *J. biol. Chem.* **240**, 1437–1445.

ROGERS H.T., PEARSON R.W., PIERRE N.H. (1940). Absorption of organophosphates by corn and tomato plants and the mineralizing action of exoenzyme systems of growing roots. *Soil Sci. Soc. Amer. Proc.* **5**, 285–291.

SAXENA S.N. (1964). Phytase activity of plant roots. *J. exp. Bot.* **15**, 654–658.

SCHONER W., VON ILBERG C. and KRAMER R. (1967). On the mechanism of the Na^+–K^+ stimulated hydrolysis of adenosine triphosphate. 1. Purification and properties of a Na^+–K^+ activated ATP-ase from ox brain. *European J. Biochem.* **1**, 334–343.

SCHWARTZ D. (1967). Esterase isozymes of maize. On the nature of the gene-controlled variation. *Proc. Nat. Acad. Sci.* **58**, 568–573.

SKOU J.C. (1957). The influence of some cations on an adenosine triphosphatase from peripheral nerves. *Biochim. biophys. Acta.* **23**, 394–401.

SKOU J.C. (1960). Further investigations on a (Mg^{++}–Na^+)-activated adenosine triphosphatase, possibly related to the active linked transport of Na^+ and K^+ across the nerve membrane. *Biochim. biophys. Acta* **42**, 6–23.

SKOU J.C. (1961). The relationship of a (Mg^{++} – Na^+)-activated, K^+-stimulated enzyme system to the active, linked transport of Na^+ and K^+ across the cell membrane. *Membrane Transport and Metabolism.* (Eds. A. Kleinzeller and A. Kotyk), pp. 228–236. New York.

SKOU J.C. (1962). Preparation from mammalian brain and kidney of the enzyme system involved in active transport of Na^+ and K^+. *Biochim. biophys. Acta* **58**, 314–325.

SKOU J.C. (1967). The enzymatic basis for the active transport of sodium and potassium. *Protoplasma*, **63**, 303–308.

SNAYDON R.W. and BRADSHAW A.D. (1962). Differences between natural populations of *Trifolium repens* L. in response to mineral nutrients. *J. exp. Bot.* **13**, 422–434.

SPERBER J.I. (1958). The evidence of apatite-solubilizing organisms in the rhizosphere and soil. *Aust. J. agric. Res.* **9**, 778–781.

SUBBA-RAO N.S. and BAJPAI P.D. (1965). Fungi on the surface of legume root nodules and phosphate solubilization. *Experientia* **21**, 386–387.

SUOMALAINAN H., LINKO M. and OURA E. (1960). Changes in the phosphatase activity of bakers yeast during the growth phase and location of the phosphatases in the yeast cell. *Biochim. biophys. Acta* **37**, 482–490.

SWABY R.J. and SPERBER J.I. (1958). Phosphate-dissolving micro-organisms in the rhizosphere of legumes. *The Nutrition of Legumes.* (Ed. Hallsworth), pp. 289–294. London.

TORRIANI A. (1960). Influence of inorganic phosphate in the formation of phosphatases by *Escherichia coli. Biochim. biophys. Acta* **38**, 460–479.

WADLEIGH C.H. and BROWN J.W. (1952). The chemical status of bean plants afflicted with bicarbonate-induced chlorosis. *Bot. Gaz.* **113**, 373–392.

WALLIHAN E.F. (1961). Effect of sodium bicarbonate on iron absorption by orange seedlings. *Pl. Physiol.* **36**, 52–53.

WALLIHAN E.F. and EMBLETON T.W. (1960). Iron chlorosis. *Californian Citrus Grower* **46**, 67.

WEIMBERG R. and ORTON W.L. (1963). Repressible acid phosphomonoesterase and constitutive pyrophosphatase from *Saccharomyces mellis. J. Bact.* **86**, 805–813.

WEIMBERG R. and ORTON W.L. (1966). Evidence for an exocellular site for the acid phosphatase of *Saccharomyces mellis. J. Bact.* **88**, 1743–1754.

WEISS M.G. (1943). Inheritance and physiology of efficiency in iron utilization in soybeans. *Genetics* **28**, 253–268.

WEISSFLOG J. and MENGDEHL H. (1933). Studien zum phosphorstoffwechsel in den höheren Pflanzen, III. Aufnahme und Verwertbarkeit organischer Phospho-säureverbindugen durch die Pflanze. *Planta* **19**, 182–241.

WILD A. and OKE O.L. (1966). Organic phosphate compounds in calcium chloride extracts of soil: identification and availability to plants. *J. Soil Sci.* **17**, 356–371.

WOOLHOUSE H.W. (1966). Comparative physiological studies on *Deschampsia flexuosa, Holcus mollis, Arrhenatherum elatius* and *Koeleria gracilis* in relation to growth on calcareous soils. *New Phytol.* **65**, 22–31.

WRIGHT K.E. (1943). The internal precipitation of phosphorus in relation to aluminium toxicity. *Pl. Physiol.* **18**, 708–712.

METABOLIC ASPECTS OF ALUMINIUM TOXICITY AND SOME POSSIBLE MECHANISMS FOR RESISTANCE

DAVID T. CLARKSON
Agricultural Research Council Radiobiological Laboratory,
Letcombe Regis, Wantage, Berkshire

The restriction of some plant species to extremely acid soils is open to a variety of interpretations. It may be argued that the pressure of competition in such environments is less than in those where there are fewer nutritional limitations on plant growth. While this is certainly one explanation of calcifuge behaviour it begs a number of important physiological questions. Since the seedlings of most plants are unable to establish themselves on podsolic soil from heathland (Rorison 1960a; Clymo 1962; Hackett 1964 and Clarkson 1966b), it is clear that calcifuges, which will grow in this soil, must have undergone genetic and physiological adaptation to combat its unfavourable influences. The limitation on plant growth may be due to combinations of deficiencies of major plant nutrients, particularly of phosphorus, and high concentrations of ions which are demonstrably toxic to most plants. Rorison (1960b) showed that aluminium ions exert an inhibitory effect on root growth of non-calcifuge plants in acid soils—a fact well appreciated by agronomic researchers in the U.S.A. much earlier (Ruprecht 1915; Hartwell and Pember 1918), but apparently little considered by ecologists. While the toxic effect of aluminium is not the only adverse influence on plant growth in acid soils, it is an important one which is likely to be of wide occurrence (Hutchinson 1945).

There have been frequent reports in the literature which show that aluminium has inhibitory effects on plant growth, particularly on the growth of roots, but attempts to relate these observations with specific physiological processes are less common. Johnson and Jackson (1964) and Munns (1965) have shown that aluminium has an inhibitory effect on the uptake and translocation of calcium in wheat seedlings and in *Medicago sativa L.* The work of Trenel and Alten (1934), Randall and Vose (1963) and Rorison (1965) indicates that several types of interaction between aluminium and phosphate may occur in the roots of plants. While these effects of aluminium are of interest and importance the results of Rorison (1958, 1960b) and Clymo (1962) indicate that they are unlikely to provide satisfactory explanations of root failure in aluminium-treated plants.

Evidence presented in this paper indicates that cell division is a primary site of disturbance by aluminium and some suggestion is made of the stage in the mitotic cycle where aluminium acts. There are also more general metabolic changes brought about by aluminium which influence sugar phosphorylation and some consideration is given to the relationships between these effects.

RESULTS AND DISCUSSION

Effects on cell division and DNA synthesis

A convenient starting point for this discussion is to consider the abnormal morphology in a root of *Agrostis tenuis* which has been grown in a dilute solution of aluminium. Where plants are grown in a water culture without aluminium the lateral roots usually are of a smaller diameter than the main axes, and do not develop near to the root tip. The aluminium-treated root in Fig. 1 presents a different picture. Here the main axis and the laterals of different orders are of similar diameter and are developed close to the apical meristems, the whole resembling corralloid roots with mycorrhizal associations. This pattern of development may be reconstructed as follows: failure of the main axis I was followed by initiation of first order laterals IIa and IIb which in their turn ceased to develop, giving rise to secondary axes IIIa and IIIb. This process may continue to give rise to third and fourth order laterals in axes of decreasing length. The consequences of this type of development can be lethal for seedlings under natural conditions since the failure of their roots to penetrate the soil to any depth can eventually lead to their death by desiccation. This growth form strongly suggests that aluminium inhibits the development of the root apex and may have an effect on cell division.

To investigate this matter experiments were made with the adventitious roots developing from the bulbs of onion, *Allium cepa*, which provide convenient material for studying rates of root growth in relation to cell division. Prior to their use in experiments, onion roots were grown in an acid culture medium at pH 4·0, in which the following salts were supplied in mM concentrations $CaCl_2$ 2·5; KCl 0·5; NH_4NO_3 1·0; $MgSO_4$ 0·25; KH_2PO_4 0·1; ferric citrate 0·04 plus minor elements as listed in Hewitt (1952). Aluminium sulphate was added to this solution in concentrations of 10^{-3}–10^{-5} M and the pH adjusted to 4·0 with potassium hydroxide. Measurements of root elongation were made in specially designed chambers, which will be described elsewhere, and the abundance of mitotic figures in aceto-carmine squashes from root meristems was estimated using a technique described in Clarkson (1965).

The effect of aluminium at various concentrations on vigorously growing onion roots is shown in Fig. 2. Six to eight hours after the addition of 10^{-3} M aluminium sulphate the rate of root elongation was reduced to almost zero: if the aluminium solution was removed at this point, and

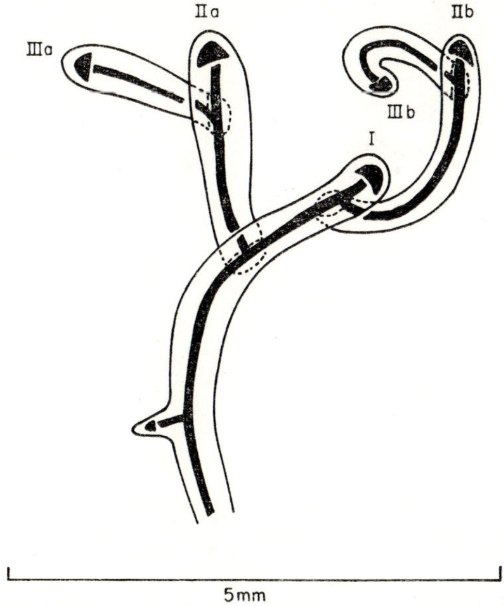

FIG. 1. Diagram of abnormal root branching in a seedling of *Agrostis tenuis* grown for several weeks in a culture solution containing 10^{-3} M aluminium sulphate.

replaced by aluminium-free culture, elongation of the root did not recommence over a period of 36–48 hr, indicating that the effect of aluminium at this concentration is irreversible. The rate of elongation in roots treated with 10^{-4} M aluminium sulphate approached zero after 8–12 hr of treatment, while 10^{-5} M aluminium sulphate caused only a slight decrease in root elongation in comparison with the untreated control. Counts of mitotic figures in root apices were made during similar experiments and are shown in Fig. 3. In the 10^{-3} and 10^{-4} aluminium sulphate treatments there is a similar time course for the inhibition of mitosis and the decline in the rate of root elongation. It is particularly important to note that the aluminium treatments do not seem to arrest mitosis itself since there was no accumulation of mitotic figures; the dividing cells pass into interphase apparently unchecked. Partial inhibition of elongation was achieved if roots were treated with 10^{-3} M aluminium sulphate for periods of time less than 6 hr and then replaced in aluminium-free cultures. If, for instance, roots were treated for 3 hr the rate of elongation decreased by approximately a half of its former value, but thereafter remained constant for at least 36 hr (Fig. 2, broken line). Fig. 4 summarizes a number of experiments of this kind and indicates a linear relationship between the duration of aluminium treatment (in the range 0–6 hr) and subsequent rate of root elongation.

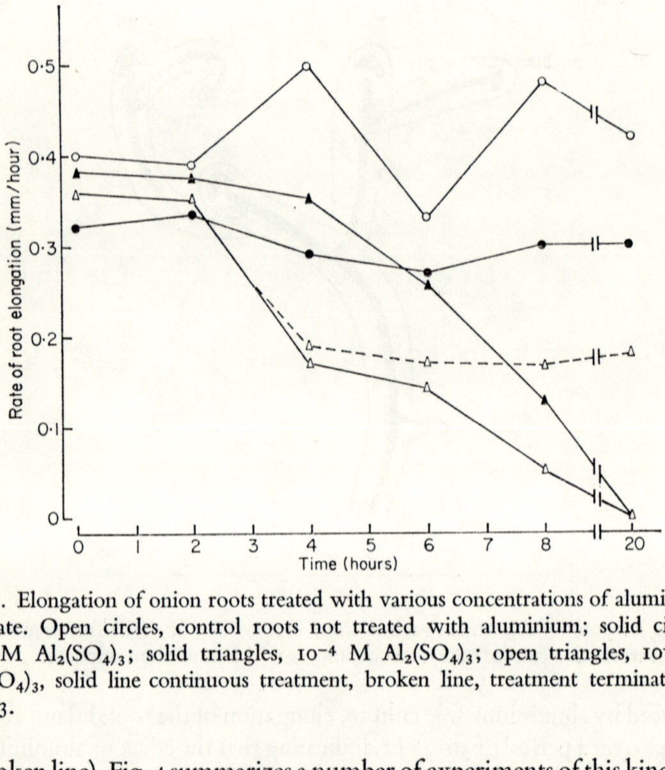

FIG. 2. Elongation of onion roots treated with various concentrations of aluminium sulphate. Open circles, control roots not treated with aluminium; solid circles, 10^{-5} M $Al_2(SO_4)_3$; solid triangles, 10^{-4} M $Al_2(SO_4)_3$; open triangles, 10^{-3} M $Al_2(SO_4)_3$, solid line continuous treatment, broken line, treatment terminated at hour 3.

This time-course of events would be expected if meristematic cells were vulnerable to the influence of aluminium at some specific stage of the mitotic cycle. The partial inhibition of root growth and cell division by short periods of aluminium treatment would depend on the fact that cells do not divide synchronously, thus they would not be at the vulnerable stage at the same time, and also that aluminium is quickly immobilized in the root by binding so that it is unable to move from one site to another.

Some suggestion of the stage in the mitotic cycle where injury occurs can be made by considering the present results jointly with the observations of Mäkinen (1963) and Van't Hoff (1965) on the length of the mitotic cycle in *A. cepa*. The speculative nature of the following discussion is stressed.

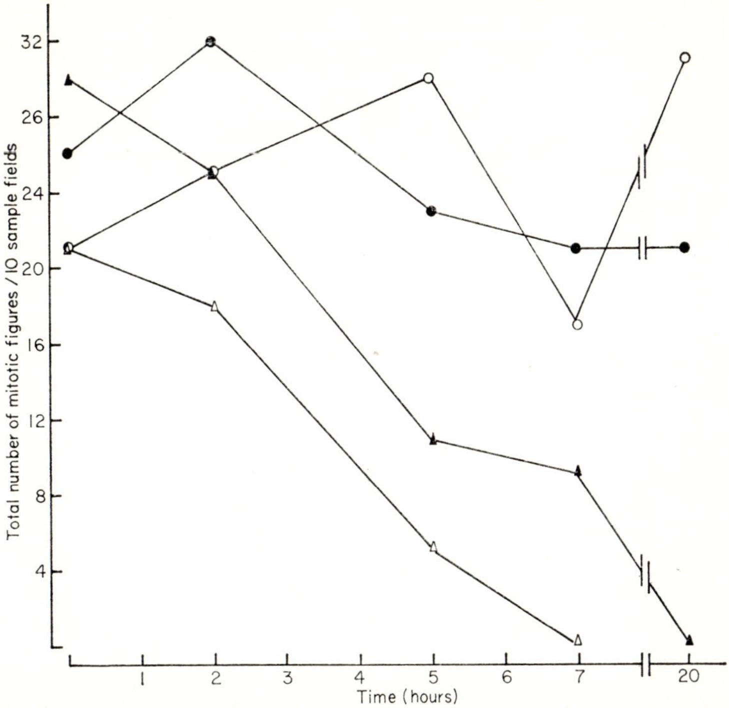

FIG. 3. Abundance of mitotic figures seen in root squashes from root tips of onion treated for various lengths of time with solutions of aluminium sulphate. Values are totals from ten selected microscope fields (× 400) and are the average of four replicate squashes. Symbols as in Fig. 2.

Figure 5 shows the approximate duration of the stages in the mitotic cycle in *A. cepa* and is based on the work of Van't Hoff and Ying (1964) and Van't Hoff (1965). Of the stages, mitosis itself is the shortest and at 25° C it lasts for about 1·5 hr. This is followed by a stage, G1, during which the various precursors involved in DNA replication are synthesized. The subsequent S period, in which DNA is synthesized, is the longest part of the cycle and takes something over 10 hr. Following DNA replication

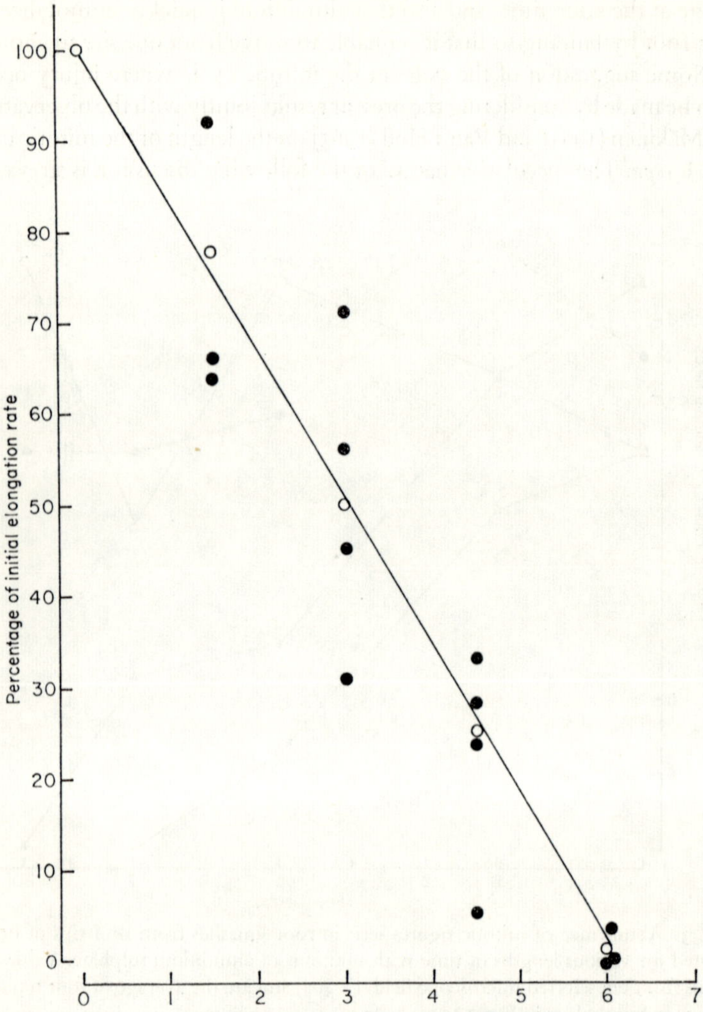

FIG. 4. Rate of root elongation after various periods of treatment with 10^{-3} M $Al_2(SO_4)_3$. Rates expressed as a percentage of that measured before treatment commenced. Open circles are average values, solid dots represent range in replicates.

there is a delay, G2, before the cells enter division. If it is assumed that the time which aluminium takes to reach its site of action, when roots are immersed in 10^{-3} M aluminium is so short that it can be ignored, it would be expected that mitotic figures would continue to be visible for at least

15 hr if aluminium caused a blockage in the G1 period. Blockage at the end of the S-period would, however, eliminate most mitotic figures within 5 hr, whereas a blockage in G2 would reduce this period to no more than 1·5–2·5 hr and would result in the formation of polyploid nuclei. The observations in the present study show that mitosis and root growth are inhibited after 6–8 hr and this timing is consistent with disturbance during

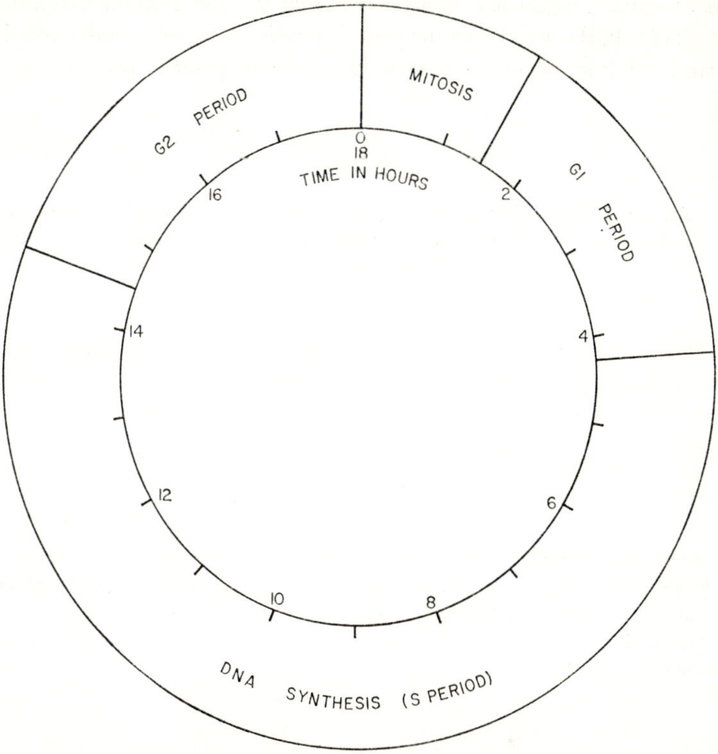

FIG. 5. Diagram to illustrate the duration of the various stages of the mitotic cycle in *Allium cepa*. Derived from Van't Hoff and Ying (1964) and Van't Hoff (1965).

the S-period of the mitotic cycle. Since DNA replication occurs in this period, aluminium should inhibit nucleic acid synthesis; evidence in this direction has been provided by Sampson, Clarkson and Davies (1965) who studied the incorporation of radioactive phosphorus into the DNA of barley roots where cell division had been stopped by treatment with aluminium sulphate. Table 1 shows that two distinct fractions of DNA can be distinguished in both aluminium treated and untreated barley roots. One of these fractions is the usual 'genetic' DNA which is stable and has a high molecular weight, the second is a DNA of lower molecular weight which is metabolically labile and is found characteristically in young actively growing tissue (Sampson and Davies 1966). In the untreated roots of the control ^{32}P was incorporated into both of these fractions to produce high specific activities during a 4-hr incubation, but in aluminium-treated roots only one fraction, the low molecular weight fraction, contained significant levels of ^{32}P, the activity in the genetic fraction not being high enough to distinguish it from contamination during the preparative procedures.

TABLE 1

Amount and specific activity of DNA-P fractions from aluminium-treated and untreated barley roots after labelling for 4 hours with carrier free ^{32}P. Compiled from Sampson, Clarkson & Davies (1965).

Treatment	DNA fraction	Amount (μg) DNA-P	Radioactivity counts/sec/μg DNA-P
Control −Al	Genetic†	12·8	6·0
+Al		18·5	1·2
Control −Al	Labile‡	3·4	23
+Al		4·8	16

† Genetic DNA with molecular weight, $4–6 \times 10^6$ and with a mole percentage of Guanine + Cytosine of 42 (Sampson *et al.* (1963)).

‡ Labile DNA with molecular weight $2–3 \times 10^5$ and with a mole percentage of guanine + cytosine of 52.

These findings are of relevance to the present discussion for they show that (a) there is a failure of genetic DNA synthesis so that the cells cannot pass through the S-period, (b) low molecular weight DNA is synthesized in similar amounts and at similar rates in treated and untreated roots, indicating that nucleic acid metabolism as a whole is not disturbed by aluminium. This latter finding is supported by evidence which shows that aluminium treatment of barley roots had only minor effects on the amount and specific activity of RNA labelled by ^{32}P (Clarkson 1966a). It thus appears that cell division is much more readily disturbed by aluminium than metabolism in general.

Effect of aluminium on phosphorus metabolism and respiration

In contrast with the results so far considered there are a number of more general effects of aluminium on metabolism. Wright (1943) and Wright and Donahue (1953) have advanced the view that the toxic effect of aluminium is an indirect one and is mediated through an internal precipitation of aluminium phosphate in the root which induces phosphorus deficiency. It is difficult to believe that this process could explain the observations on root elongation and cell division, since it was shown that the presence or absence of phosphorus in the aluminium treatment had no effect on the inhibition of mitosis, neither could this inhibition be overcome by treating the root subsequently with phosphorus in the absence of aluminium for periods of up to 1 week (Clarkson 1965). There have been reports, however, which suggest that aluminium may have more subtle effects on phosphorus metabolism (Ragland and Coleman 1962; Randall and Vose 1963; Rorison 1965).

The full significance of interactions between aluminium and phosphorus are not apparent unless the individual phosphorus fractions in the root are examined. This question was studied using barley seedlings pre-treated for 24–48 hr with 10^{-3} M aluminium sulphate and then immersed in solutions of potassium dihydrogen phosphate labelled with ^{32}P (Clarkson 1966a). The effect of aluminium was to increase the amount of phosphorus in the roots, but all of the additional phosphorus was present as inorganic phosphate (Table 3, line 6), and was found to be readily exhangeable with non-radioactive P (Table 2). This suggests that the reaction between aluminium and phosphorus is superficial and occurs either at the root surface or in the free space of the root. Further evidence that this reaction between aluminium and phosphate does not depend on cellular metabolism is shown by the fact that low temperature and toxic levels of dinitrophenol have little effect on the accumulation of extra phosphorus in aluminium treated roots (Table 2). Surface fixation of phosphate is likely to reduce the amount of phosphorus available for translocation to the shoot, and this may be particularly important in situations, like heathland, where phosphorus in the soil solution is already at a very low concentration.

There are, in barley roots, effects on phosphorus metabolism which clearly must occur *within* the cells. If we compare the amount of phosphorus which is not readily exchangeable in the treated and untreated roots (Table 2, column 5) we see that in each case it is similar. Generally this may be taken as the phosphorus that is within the cells. The total phosphorus incorporated into readily soluble organic compounds is shown in Table 3

TABLE 2

Incorporation of phosphorus by barley root segments pretreated in presence or absence of 10^{-3} M $Al_2(SO_4)_3$. Values are nano moles $P(10^{-9})$/g (fresh weight) and are means of 3 replicates ± standard deviation. From Clarkson (1966a)

Conditions	Sample (min)	Treatment	Total P	Non-exchangeable P	Exchanged P
25° C	2	Control − Al	16·0 ± 2	14·0 ± 1·8	2
		+ Al	97·0 ± 8·5	18·0 ± 1·8	79
	60	Control − Al	402·0 ± 47·0	399·0 ± 51·9	3
		+ Al	588·0 ± 76·3	365·0 ± 48·5	223
25° C	2	Control − Al	8·7 ± 1·2	0	8·7
		+ Al	28·2 ± 4·1	0	28·2
+ 10^{-3} M DNP					
	60	Control − Al	34·3 ± 3·6	2·2 ± 0·1	32·1
		+ Al	129·6 ± 14·7	7·1 ± 0·2	122·5
1° C	2	Control − Al	0·6 ± 0·1	0·2	0·4
		+ Al	32·7 ± 5·1	0·7 ± 0·2	32·0
	60	Control − Al	12·0 ± 2·4	2·9 ± 0·3	9·1
		+ Al	179·3 ± 18·2	2·5 ± 0·3	176·8

TABLE 3

Incorporation of ^{32}P into phosphorylated compounds in whole barley roots after 20 and 100 min. Values are in counts/second/root. From Clarkson (1966a)

Spot	20 min Control	20 min Aluminium	100 min Control	100 min Aluminium
UTP	7	10	10	14
ATP	19	29	37	56
ADP	4	6	11	9
Hexose-(P)	58	21	161	66
Total organic P	101	79	271	180
Inorganic P	237	804	524	1304
Total	338	883	795	1490

Plants pretreated in an acid water culture, pH 4, with or without 10^{-3} M aluminium sulphate. Root material killed and phosphorus compounds extracted in a mixture of methanol, chloroform and formic acid at −80° C and 1° C respectively. ^{32}P labelled compounds in this extract located on chromatograms by autoradiography, and counted directly.

(line 5) and it is clear that some reduction in this fraction is caused by aluminium. More interesting than this is the partition of phosphorus between a number of important metabolic intermediates. The most striking examples of this are shown in Table 3, where, at sampling periods at 20 and 100 min the amount of ^{32}P incorporated into phosphorylated hexose sugar is much reduced, while there is an apparent accumulation of nucleotide triphosphates, particularly of ATP and UTP.

What are the implications of this disturbed pattern of phosphorus distribution? Hexose phosphates, particularly glucose-6-phosphate, are the entry point of carbohydrate substrates into the respiratory process, and hence a reduction in the rate of their formation, or their amount, will have far reaching consequences on metabolism. Glucose becomes phosphorylated in a reaction with ATP in which hexokinase is the enzyme (Saltman 1953) and the products are glucose-6-phosphate and ADP. Aluminium treatment did not limit this process by reducing the rate of formation of ATP, since, in Table 3 it is clear that this actually increased, as did the total amounts of ATP in the root; in this connection it is interesting to note that Woolhouse (this volume) has shown that acid phosphatases and ATP-ase are inhibited by aluminium. Thus aluminium must act either directly or indirectly to prevent the utilization of ATP in glucose phosphorylation. This inhibition should be reflected by a decrease in respiration, and, indeed, this has been found to be so in several varieties of barley (Table 4), but only after extended aluminium treatment.

If the failure of cell division is not due to the direct action of aluminium on the process, but results, secondarily, from impaired respiratory metabolism, the decrease in oxygen uptake should either precede or at least occur at the same time as the inhibition of cell division. Sampson, Clarkson and Davies (1965) showed that inhibition of cell division in barley root apices is at an advanced stage 6–10 hr after the commencement of aluminium treatment. However, only trifling decreases in the rate of oxygen uptake by barley roots can be detected after 8–11 hr of treatment (Table 4); only after 24 hr was an appreciable reduction found. Similar disparities between the onset of mitotic inhibition and the depression of respiration have been found using onion roots (Clarkson, unpublished data). These results make it seem unlikely that impaired respiratory metabolism is the cause of the inhibition of mitosis. But reduced rates of respiration may account for the secondary effects on the uptake of calcium and of other ions, since these processes are closely linked with respiratory metabolism. In contrast with the present results with onion and barley, Norton (1967 and personal communication) has found decreases in oxygen uptake in the roots of the calcicole sainfoin (*Onobrychis sativa*). A depression of 5% was detected within 30 min of the addition of 2×10^{-3} M aluminium and this increased to around 35% in 3 hr. Inhibition of oxygen uptake and carbon

TABLE 4

Oxygen uptake by excised roots of barley seedlings pre-treated for various periods with 10^{-3} M $AlCl_3$.

Values are μl O_2 taken up by 1 g root tissue (fresh weight) per hour. Percentage inhibition due to aluminium shown in brackets.

Variety	Pre-treatment	Pre-treatment period (hr) 3	5	8	11	24
Himalaya	Control − Al	1099	1113	1086	1097	1009
	+ Al	1021(7)	1012(9)	1007(7)	941(14)	628(38)
Proctor	Control − Al		652			771
	+ Al		632(3)			431(45)
Dayton	Control − Al	635		639		600
	+ Al	710(+11)		602(6)		497(17)
Kearny	Control − Al	762		727		719
	+ Al	772(+1)		698(4)		673(6)

Seedlings pretreated in an acid culture solution, pH 4, with or without 10^{-3} M $AlCl_3$. Excised roots incubated in respirometer flask in the presence of 1% sucrose, pH 4·0, with or without 10^{-3} M $AlCl_3$. Oxygen uptake measured in a Gilson differential respirometer.

dioxide production was also detected in intact embryos of sainfoin before the onset of cell division. There were no comparable effects on the respiration of the calcifuge Lupin (*Lupinus luteus*). It would seem that further experiments are necessary to establish unequivocally, cause and effect in this work.

Resistance to aluminium toxicity

From the ecologists viewpoint the nature of resistance to aluminium toxicity may be of more interest than the mechanism of the toxicity itself. At this time it is possible only to speculate on resistance but the justification for doing so is that it may suggest lines for future work.

Difficulties in envisaging a method of resistance arise because the system must be specific for aluminium. Preliminary findings (Clarkson, unpublished) suggest that resistance to aluminium in *Agrostis setacea* and rye *Secale cereale*, does not confer on these plants resistance to other trivalent metal ions of similar chemical nature, e.g., gallium, scandium, yttrium and lanthanum, even though they may act on cell division in the same way as aluminium (Clarkson 1965). This problem of metal specificity is also discussed by Turner (this volume). Taking the most direct approach, we may propose that the resistant species prevents aluminium reaching its site of action either by preventing it entering the cell, or by waylaying the aluminium once it is within the cell.

The biochemical evidence in this paper suggests that there may be at least two sites within the cell where aluminium acts, viz. the nucleus and the mitochondria where respiratory disturbances are likely to have their origin. An intracellular resistance mechanism might prevent the aluminium reaching these sites either by providing alternative binding sites of higher affinity which preferentially accumulate aluminium without damaging metabolism, or by rendering the ion into a non-toxic form by chemical modification. Where differentiated cells possess a vacuole, it is easy to visualize this as a depository for unwanted metals, however, meristematic cells are not vacuolated in either resistant or tolerant species and it seems that aluminium is particularly disruptive in these cells. Almost any charged surface within the cell might be expected to bind aluminium, but no suggestion can be offered of one which would do so preferentially. Chemical modification of aluminium by chelation with EDTA is known to reduce its toxicity greatly and to promote its mobility within the plant (Clymo 1962). It would seem, then, if the cell could act in some way to prevent the aluminium being in a cationic form it can avoid injury; as aluminate anion it is accumulated in quantity, and apparently innocuously by calcicole plants, Jones (1961). If such a specific chelating agent exists within the cell it has thus far escaped detection.

There is evidence that a large proportion of the aluminium in barley roots is associated with the cell wall where it may be present as an amorphous precipitate of $(Al(OH)_3)_n$ (Clarkson 1967). This precipitation is brought about by the hydrolysis of aluminium ions by hydroxyl ions at the root surface. A resistant species might produce hydroxyl ions at its root surface rapidly enough to precipitate all the incoming aluminium in this way. Vose and Randall (1962) suggested that low root cation exchange capacities are associated with tolerance to aluminium and manganese in some varieties of *Lolium perenne*, and more recently Foy *et al.* (1967) have expanded this idea. Some interesting studies by Foy, Burns, Brown and Fleming (1965) have shown that an aluminium-resistant variety of wheat, Atlas 66, was able to increase the pH of the soil and water cultures in which it was grown, towards more neutral values, causing removal of aluminium from solution by precipitation. A non-resistant variety, Monon, lacked this ability or showed a tendency to make its cultures more acid.

These differential pH effects appeared to be partially dependent on the balance between nitrate and ammonium ions in the culture solution, and suggest that nitrate removal from the solution is more rapid in Atlas 66 than in Monon (Foy *et al.* 1967). Where ammonium ions are appreciably more abundant than nitrate the differential pH changes were not observable (Foy, personal communication). In relating these results to field situations it is as well to remember that in most podsolized soils, where resistance to aluminium toxicity may determine the ability of a species to establish itself, the nitrate concentration in the soil is very small in relation to the ammonium (Richardson 1938).

Magistad (1925) showed that small changes in pH in the range 4·0–4·5 will produce large, and physiologically significant, changes in the solubility of aluminium in water. The ability of plants to effect such changes in the vicinity of their roots may be the basis of varietal differences in response to soil acidity and other factors, and merits the closest attention.

It is not immediately apparent how general properties of the root such as its cation exchange capacity would permit discrimination between aluminium and, say, scandium, but specificity may be dependent on more subtle features, for instance, the spatial distribution of hydroxyl ions at the root or cell surface.

CONCLUSIONS

When aluminium reaches its site of action within cells of the root it causes disordered metabolism which results in either the total or the partial failure of cell division probably by interference with DNA synthesis in the S period of the mitotic cycle. This results in the inhibition of root growth and the development of abnormal root morphology. Disturbances in phosphorus metabolism may be detected at two levels, the first at the root surface where a reaction with aluminium apparently fixes phosphate, and the second is within the cell where the distribution of phosphorylated intermediates is altered in such a way as to depress respiratory metabolism. There is also evidence that aluminium may disturb the uptake and translocation of other ions.

From this we may attempt to define at least three properties of a hypothetical calcifuge plant growing in an acid soil:

(1) An ability to thrive in situations where available phosphorus is at a low level, and in the face of phosphorus fixation by aluminium at the root surface.

(2) An ability to grow in low concentrations of calcium and in the face of impaired calcium uptake and translocation.

(3) The possession of either specific sites within the cytoplasm where aluminium may be harmlessly accumulated, or a specific method for chelating aluminium, and/or

(4) An ability to prevent aluminium from entering the cell by precipitation at the cell surface.

ACKNOWLEDGEMENTS

It is a pleasure to acknowledge the helpful criticism of many colleagues at various stages; I would most particularly like to thank Professors T.A. Bennet-Clark, J. Dainty and D.D. Davies in whose departments at the University of East Anglia most of this experimental work was performed, Drs W.S. Hillman and J. Van't Hoff of Brookhaven National Laboratory, U.S.A., for laboratory facilities and their most instructive interest in this problem, Drs I. Rorison, and Charles Foy whose considerable contributions to the field are freely drawn upon in this paper and to Dr R.S. Russell for his patient help in preparing the manuscript.

REFERENCES

CLARKSON D.T. (1965). The effect of aluminium and other trivalent metal cations on cell division in the root apices of *Allium cepa*. *Ann. Bot. N.S.* **29**, 209–315.

CLARKSON D.T. (1966a). Effect of aluminium on uptake and metabolism of phosphorus by barley seedlings. *Pl. Physiol.* **41**, 165–172.

CLARKSON D.T. (1966b). Aluminium tolerance in species within the genus *Agrostis*. *J. Ecol.* **54**, 167–178.

CLARKSON D.T. (1967), Interactions between aluminium and phosphorus on root surfaces and cell wall material. *Plant and Soil* **27**, 347–356.

CLYMO R.S. (1962). An experimental approach to part of the calcicole problem. *J. Ecol.* **50**, 707–731.

FOY C.D., BURNS G.R., BROWN J.C. and FLEMING A.L. (1965). Differential Al tolerance of two ehat varieties associated with plant-induced pH changes around their roots. *Soil Sci. Soc. Amer. Proc.* **29**, 64–67.

FOY C.D., FLEMING A.L., BURNS G.R. and ARMINGER W.H. (1967). Characterization of differential aluminium tolerance among varieties of wheat and barley. *Soil Sci. Soc. Amer. Proc.* **31**, 513–521.

HACKETT C. (1964). Ecological aspects of the nutrition of *Deschampsia flexuosa* (L) Trin. (I) The effect of aluminium, manganese and pH on germination. *J. Ecol.* **52**, 159–168.

HARTWELL B.L. and PEMBER F.R. (1918). The presence of aluminium as a reason for the difference in the effect of so-called acid soil on barley and rye. *Soil Sci.* **6,** 259–281.

HEWITT E.J. (1952). *Sand and water culture methods used in the study of plant nutrition.* Tech. Commun. Commonw. Hort. Plantn. Crops 22.

HUTCHINSON E.G. (1945). Aluminium in soils, plants and animals. *Soil Sci.* **60**, 29–40.

JOHNSON R.E. and JACKSON W.A. (1964). Calcium uptake and transport by wheat seedlings as affected by aluminium. *Soil. Sci Soc. Amer. Proc.* **28,** 381–386.

JONES L.H. (1961). Aluminium uptake and toxicity in plants. *Plant and Soil* **13,** 297–310.

MAGISTAD O.C. (1925). The aluminium content of the soil solution and its relation to soil reaction and plant growth. *Soil. Sci* **20,** 181–226.

MÄKINEN Y. (1963). The mitotic cycle in *Allium cepa,* with special reference to the diurnal periodicity and to the seedling aberrations. *Ann. Bot. Soc. 'Vanamo'* **34,** No. 6, 1–61.

MUNNS O.N. (1965). Soil acidity and growth of a legume II Reactions of Al and H_2PO_4 in solution and effects of Al, H_2PO_4, Ca and pH on *Medicago sativa L.* and *Trifolium subterraneum* in solution culture. *Aust. J. agric. Res.* **16,** 743–755.

NORTON G. (1967). *Some aspects of aluminium toxicity on plant growth.* University of Nottingham School of Agric. Rept. 1966–67, 99–103.

RAGLAND J.L. and COLEMAN N.T. (1962). Influence of aluminium on phosphorus uptake by snap bean roots. *Soil Sci. Soc. Amer. Proc.* **26,** 88–89.

RANDALL P.J. and VOSE P.B. (1963). Effect of aluminium on the uptake and translocation of phosphorus by perennial ryegrass. *Pl. Physiol.* **38,** 403–413.

RICHARDSON H.L. (1938). The nitrogen cycle in grassland soil with special reference to the Rothamsted Park Grass experiment. *J. Agric Sci.* **28,** 73–87.

RORISON I.H. (1958). The effect of aluminium on legume nutrition. *Nutrition of legumes* (Ed. E.G. Hallsworth), pp. 43–58. London.

RORISON I.H. (1960a). Some experimental aspects of the calcicole-calcifuge problem. I. The effects of competition and mineral nutrition upon seedling growth in the field. *J. Ecol.* **48,** 585–599.

RORISON I.H. (1960b). Some experimental aspects of the calcicole-calcifuge problem. II. The effects of mineral nutrition on seedling growth in solution culture. *J. Ecol.* **48,** 679–688.

RORISON I.H. (1965). The effect of aluminium on the uptake and incorporation of phosphate by excised sanfoin roots. *New Phytol.* **64,** 23–27.

RUPRECHT R.W. (1915). The toxic effect of iron and aluminium on clover seedlings. *Mass. Agric. Expt. Sta. Bull.* **161,** 125–129.

SALTMAN P. (1953). Hexokinase in higher plants. *J. Biol. Chem.* **200,** 145–157.

SAMPSON M., KATOH A., HOTTA Y. and STERN H. (1963). Metabolically labile DNA. *Proc. Nat. Acad. Sci.* **50,** 454–463.

SAMPSON M., CLARKSON D.T. and DAVIES D.D. (1965). DNA synthesis in aluminium-treated roots of barley. *Science* **148,** 1476–1477.

SAMPSON M. and DAVIES D.D. (1966). Metabolically labile DNA in mitotic and non-mitotic cells of *Zea mays. Life Sciences* **5,** 1239–1247.

TRENEL M. and ALTEN F. (1934). Die physiologische Bedentung der Mineralischen Bodenaziditat. *Angew. Chem.* **47,** 813–820.

VAN'T HOFF J. and YING HUEN-KUEN (1964). Simultaneous marking of cells in two different segments of the mitotic cycle. *Nature (Lond.)* **202,** 981–983.

VAN'T HOFF J. (1965). Relationships between mitotic cycle duration of the S period and the average rate of DNA synthesis in the root meristem cells of several plants. *Exptl. Cell Res.* **39**, 48–58.
VOSE P.B. and RANDALL P.J. (1962). Resistance to aluminium and manganese toxicity in plants related to variety and cation exchange capacity. *Nature* (*Lond.*) **196**, 85.
WRIGHT K.E. (1943). The internal precipitation of phosphorus in relation to aluminium toxicity. *Pl. Physiol.* **18**, 708–712.
WRIGHT K.E. and DONAHUE B.A. (1953). Aluminium toxicity studies with radioactive phosphorus. *Pl. Physiol.* **28**, 674–680.

HEAVY METAL TOLERANCE IN PLANTS

R. G. TURNER
School of Plant Biology, U.C.N.W., Bangor†

INTRODUCTION

The nutritional problems of plants growing in mining regions and in acid and saline soils have been well reviewed (Bollard and Butler 1966; Baumeister, 1967). In spite of the high concentrations of various metals in these soils many plant species are tolerant and survive even though others are poisoned and therefore excluded from these habitats. The mechanisms whereby heavy metal tolerance is achieved are largely unknown and in fact at the present moment there is little information available concerning the physiology of metal tolerant plants. This paper attempts to appraise the situation. Much of the detailed evidence concerning metal tolerance in living organisms has been obtained using microbial and animal cells and this is referred to for completeness. This evidence is not included for direct extrapolation to plant metabolism but rather as an indication of the complexities of this problem.

The ensuing discussion is restricted largely to the heavy metals zinc and copper, mainly because these two metals have received the most attention. However, information about other heavy metals has been included where appropriate studies have been made.

MECHANISMS OF HEAVY METAL TOLERANCE

Metal tolerance can be demonstrated in water culture as well as soil (Gregory 1964; Turner 1967) and thus the soil effects only the form of the metal and its uptake pattern. Several studies have demonstrated that soil organic matter can alter the pattern of metal uptake. Dykeman and De-Sousa (1966) concluded that soil organic chelates reduced copper uptake by several Canadian trees and shrubs. Ernst (personal communication, 1967) has demonstrated that a similar situation exists for zinc uptake in several European plant genera. In *Agrostis tenuis*, Turner and Gregory (1967) have

† Now at Shell Research Ltd., Sittingbourne, Kent.

shown that the pattern of intra-plant metal distribution can differ depending on whether the metals were taken up as inorganic salts or as metal chelates. However these alterations in uptake and translocation patterns apply equally to both tolerant and non-tolerant plants, so offer no basis for a mechanism of heavy metal tolerance.

Soil conditions are very important in determining by natural selection both the nature and degree of metal tolerances of tolerant populations. Gregory and Bradshaw (1965) showed that the tolerances of populations of *A. tenuis* are related to the soil concentrations of the particular metals found in the mine soils. In addition Bröker (1963) showed that the tolerance of populations of *Silene vulgaris* is related to individual metals and their soil concentrations. Soil conditions thus reflect the degree of adaptation required for plants to tolerate these conditions.

Heavy metal tolerance is not usually due to a mechanism of differential ion uptake, although Earley (1943) had concluded that zinc tolerant varieties of soybeans could restrict their uptake of zinc ions. Analyses for many metals of many different plants growing in many areas of the world on mine soils have revealed that tolerant species contain large quantities of what for other species can be toxic metals (Bröker 1963 in Germany; Gregory and Bradshaw 1965 in N. Wales; Nicolls, Provan, Cole and Tooms 1965 in Australia; Dykeman and DeSousa, 1966 in Canada; Reilly 1967 in Africa; Ernst 1966 in France and Germany). These findings agree with those for other instances where plants can tolerate environments containing toxic compounds (e.g. aluminium, fluoride, sodium and selenate). It seems a general rule that plants do not exclude poisonous compounds; in other words there is true tolerance.

However, it is possible that a slow uptake rate over a long period of time could explain the high metal contents of tolerant plants. In this case the cells would never contain at any one time more than a trace amount of the toxic metals. Experiments have shown however, that the uptake of zinc (Turner and Gregory 1967) and copper (Bradshaw, McNeilly and Gregory 1965) is the same for both tolerant and non-tolerant populations of *A. tenuis*. Metal tolerance in plants is not therefore based upon restricting the uptake or in controlling the rate of uptake of large amounts of the toxic ions.

This lack of restriction of metal uptake is characteristic of the vast majority of metal tolerant cells. Copper tolerant bacteria (*Mycobacterium tuberculosis avium*) (Horio, Higashi and Okunuki 1955) various metal tolerant fungi (Ashida 1965) and cobalt tolerant animal cells (Daniel, Dingle and Lucy 1961) all contain large quantities of the heavy metals.

It would seem that heavy metal tolerance in plants and in the vast majority of other metal tolerant cells must therefore involve a specialized internal metabolism. Although only a few heavy metals serve essential functions in biological systems (Hewitt 1963) all heavy metals have a common feature in that they are toxic to living systems if present in excessive amounts. Thus the tolerant organism would appear to be adapted to withstand these metals when they are in high concentrations within the cell. Such adaptation must either prevent the metal from affecting metabolism, or from reaching metal-susceptible sites, or involve changes that allow enzymes to function normally in the presence of toxic amounts of metals.

There is evidence that such adaptations and alterations do occur and perhaps the best example concerns the role of sulphur in tolerance mechanisms in micro-organisms. Thus copper tolerant and non-tolerant yeast cells (*Saccharomyces cerevisiae*) differ in their sulphur metabolism. Copper tolerant yeasts deposit excess copper as sulphide at the cell periphery (Ashida, Higashi and Kikuchi 1964). This metal deposition as copper sulphides is a consequence of the tolerant yeasts having a higher sulphur content than non-tolerant yeasts such that the metabolism of resistant cells changed when excess copper was present (Ashida and Nakamura 1959). Further investigations (Kikuchi 1965a, b, c) revealed that the hydrogen sulphide producing activities of tolerant clones generally parallel the copper sulphide contents of their cells. The differences between metal tolerant and non-tolerant yeast cells lie in their metabolic pathway from sulphate to sulphite. These differences are due to the large capacity of the reaction sequence from sulphate to sulphite in copper tolerant yeasts and excess sulphite thus formed is liberated as sulphide rather than by way of cysteine. A reduction in the sulphur available in the growth medium of copper tolerant yeasts produces a corresponding decrease in their copper tolerance, whilst a similar reduction in the amount of sulphur in the medium of non-tolerant yeasts has no significant effects on their copper tolerance (Ashida and Nakamura 1959).

However deposition of the toxic copper as copper sulphides is only one aspect of metal tolerance in yeasts. Seno (1963) demonstrated that there is no genetic link between sulphide production and copper tolerance and that each character is under separate genetic control. Nakamura (1962) investigating other heavy metal tolerances in yeast (namely Cd, Co, Ni, Ag) concluded that these are not based upon alterations in sulphur metabolism.

There is also good evidence that sulphur is important in metal tolerance in other microbial cells. Browning, Russell, Kingsnorth and Peerless (1959) found sulphur compounds are important in the binding of mercury fungicides. Ashworth and Amin (1964) showed that mercury tolerance in *Aspergillus niger* is due to a pool of 'non-protein sulphydryl groups' which complexes excess mercury entering the fungal thallus. The internal mercury content is high but in an unavailable form.

In higher plants there are only indications that sulphur may be of importance in heavy metal tolerance. Brenchley (1938), who examined the effects of cobalt and nickel in barley, and Url (1956), who examined copper, chromium and vanadium in several plant species, found that low concentrations of these metals as sulphates are innocuous to plant growth whilst equivalent concentrations of other metal salts are toxic. On the other hand a detailed study of several aspects of the sulphur nutrition and metabolism of metal tolerant clones of *A. tenuis* gave no indications that sulphur is important in its metal tolerance mechanism (Turner 1967).

Another example of a metal complexing system, although not involving sulphur, has been demonstrated in animal cells (Porter, Johnston and Porter 1962). A protein, which was not a cytochrome, isolated from the mitochondria of immature bovine liver contained 4% copper and 3% iron. No other reported cases of heavy metal tolerance have implicated proteins as heavy metal binders although the affinity of proteins for metals is well known (Gurd and Wilcox 1956). Treatment with proteolytic enzymes of the metal rich fraction of grass roots have revealed that proteins are not involved in heavy metal tolerance mechanisms in *A. tenuis* (Turner 1967).

Turner (1967) has presented evidence from studies of *A. tenuis* for a mechanism of heavy metal tolerance in higher plants which may well be widespread. He studied the intracellular distribution of the heavy metals in tolerant and non-tolerant plants. In tolerant *A. tenuis* a greater proportion of both copper and zinc is located in the cell wall fraction of the roots and this apparent preferential localization occurs at both normal and high levels of metal nutrition. This localized intracellular distribution could indicate an exclusion of these metals from susceptible sites in intermediary metabolism. Indeed further studies revealed that deposition of zinc at the cell wall was positively correlated with the index of zinc tolerance of the populations studied. Turner also showed that zinc tolerance may be associated with alterations in the carbohydrate composition of the cell wall. Furthermore Nevins, English and Albersheim (1967) have demonstrated that the plant cell wall composition is genetically controlled. The differences in cell wall composition between zinc tolerant and non-tolerant ecotypes of *A. tenuis* could thus be genetically controlled and such differences provide the basis for the mechanism of heavy metal tolerance within this grass species. The cell wall may thus act as a selective barrier preventing the entry of excess quantities of zinc into cell metabolism.

The deposition of large amounts of heavy metals at the cell wall is known in other higher plants. Cartwright (1966) showed that 64 per cent of the total copper within subterranean clover nodules was located in the plant cell wall fraction. Diez-Altares and Boroughs (1961) and Diez-Altares and Bornemisza (1967) have shown that even at normal levels of nutrition approximately half of the total zinc found in germinating corn tissues (*Zea mays*) was associated with the cell wall.

Ernst (personal communication, 1967) has measured the amount of zinc in the expressed cell sap and vegetative tissues of metal tolerant plants of *Silene vulgaris*, *Armeria maritima* and *Thlaspi alpestre*. His measurements, which were made on plants growing on the sites of metal mines in North France and Western Germany have shown that although the zinc content of the plants increased the zinc content of the cell sap remained constant. His personal opinion was that the metal was prevented from entering cell metabolism by the cell wall.

Despite the marked differences between microbial and plant cell walls in their chemical composition (Northcote 1963), the cell wall seems to be important as a heavy metal accumulator in micro-organisms. Eagon, Simmons and Carson (1965) found 900 μg zinc per gram dry weight in the cell wall of *Pseudomonas aeruginosa* and thought that the metal was essential for cell wall integrity. Somers (1963) showed the cell walls of *Alternaria tenuis* and *Pencillium italicum* to be important in the control of copper uptake in these fungi and Ashida, Higashi and Kikuchi (1967) have presented electron micrographs of copper deposition at the cell wall in copper tolerant yeasts.

The possible mechanisms of heavy metal tolerance discussed so far are all based upon restricting the entry of the poisonous metals into susceptible sites within the metabolic system. However there is evidence that there can be alterations in the nature of what would otherwise be susceptible enzymes and there can also be alterations in metabolic sequences such that an organism can function in an apparently normal manner in the presence of large amounts of heavy metals. As an example of the latter, attention has already been drawn to changes which can occur in the pathway of sulphur metabolism in copper tolerant yeasts (see Kikuchi 1965a, b, c). Horio *et al.* (1955) found that in copper tolerant cells of *Mycobacterium tuberculosis avium* several stages in the tricarboxylic acid (TCA) cycle were not inhibited by copper. The tolerant organisms could oxidize malate, fumarate and succinate in the presence of large amounts of copper although all these stages in the TCA cycle were inhibited by copper in the non-tolerant organism. The lack of inhibition by copper in the TCA cycle in tolerant cells has also been reported by Murayama (1961a, b) for yeast cells. He concluded that in non-tolerant cells, coenzyme I (NAD) linked reactions were important in the TCA cycle, whilst in tolerant cells, coenzyme II (NADP) linked reactions were important. The coenzyme I linked reactions were more sensitive to copper than coenzyme II linked reactions and were therefore inhibited. In addition to these differences between tolerant and non-tolerant cells, Murayama proposed that the nature of the enzymes involved in these reactions were also altered.

No similar studies have as yet been reported for metal tolerant plants. There are indications that adaptations within the metabolic machinery may have occurred (Gregory 1964; Turner 1967; Baumeister and Burghardt 1956) but positive proof is lacking. Repp (1963) has reported generalized changes in the protoplasm of metal resistant cells to explain the copper tolerance of *Silene vulgaris*, *Tussilago farfara* and *Taraxacum officinale*.

THE PROBLEM OF METAL ION SPECIFICITY

Tolerance to one heavy metal does not confer tolerance to another heavy metal, and an intriguing and characteristic feature of metal tolerant plants is their rigid specificity for individual heavy metals (Gregory and Bradshaw 1965; Baumeister 1967, Barker, personal communication, 1967). For example a copper tolerant plant is not zinc tolerant and *vice versa*. This specificity is also apparent in metal tolerant microbial cells (see Tables 1 and 2).

Nevertheless different individual tolerances can occur together and this is correlated with the occurrence of several metals together in toxic quantities in the soil of the original habitat. Plants growing on mine soils of mixed composition are tolerant to all the metals in the soil and combinations of tolerance can occur depending upon the original metal composition of the mine soil (Gregory and Bradshaw 1965). Such combinations of metal tolerances are considered to be multiple tolerances.

In addition (Gregory 1964) demonstrated a correlation between zinc tolerance and nickel tolerance in zinc tolerant populations of *A. tenuis*. No nickel deposits occur in North Wales which seemed to argue that the two metal tolerances are physiologically linked. However, there is no zinc tolerance in nickel tolerant populations collected from nickel mines in South West Germany (Gregory and Bradshaw 1965). Zinc and nickel are closely related in terms of ionic size (zinc 0·69 Å; nickel 0·68 Å) and it was

TABLE I

Heavy metal co-tolerances reported in higher plants (including for comparative purposes Desmids and Mosses)

Plant	Selected against	Plant resistant to: Al	V	Cr	Mn	Co	Ni	Cu	Zn	Y	Ag	Cd	In	Hg	Pb	Reference
Agrostis sp.	Al	×														Clarkson (1966)
Agrostis tenuis	Ni						×		○							Gregory and Bradshaw (1965)
Agrostis tenuis	Cu	○		○	○	○	○	×	○	○	○	○	○	○	○	Gregory (1964) Barker, unpublished data.
Silene vulgaris	Cu							×	○						○	Bröker (1963)
Higher plants	Cu		○	+				×	○							Url (1955)
Mosses	Cu		○	+				×								Url (1956)
Desmids	Cu			+				×								Url (1955)
Agrostis tenuis	Zn	○		○	○	○	○	○	×	○	○	○	○	○	○	Barker, unpublished data
Agrostis tenuis	Zn					○	+	○	×						○	Gregory (1964)
Agrostis tenuis	Zn			○	○	○	+	○	×	○	○	○	○	○	○	Turner and Antonovics, unpublished data
Silene vulgaris	Zn							○	×						○	Bröker (1963)
Agrostis tenuis	Pb	○		○	○	○	○	○	○	○	○	○	○	○	×	Gregory (1964 Barker, unpublished data
Festuca ovina	Pb							○	○						×	Wilkins (1960)

Explanation of Table 1: + co-tolerant, × tolerant; ○ non-tolerant. The absence of a symbol indicates that the metal was not tested.

TABLE 2

Heavy metal co-tolerances reported in micro-organisms

Microorganism	Selected against	Organism resistant to: B	Cr	Mn	Fe	Co	Ni	Cu	Zn	Ag	Cd	Hg	Pb	Reference
Poria vaillantii	Cr	×						+	+					Da Costa (1959)
Thiobacillus thiooxidans	Fe				×				+					Ryder and Colmer (1965)
Saccharomyces cerevisiae	Co					×	○	○		○	○			Ashida (1965)
Saccharomyces cerevisiae	Ni					+	×	○		○	○			Ashida (1965)
Saccharomyces cerevisiae	Cu					○	○	×		○	○			Ashida (1965)
Aspergillus niger	Cu		+					×	+	+				Jurkowska (1952)
Piricularia oryzae	Cu	+						×				○		Yamasaki and Tsuchiya (1964)
Penicillium ochrochloron	Cu							×	+					Bose and Basu (1965)
Ceratostomella paradoxa	Cu							×				+		Sugunakar-Reddy Apparao and Subbaya (1964)
Penicillium notatum	Cu							×				+		Partridge and Rich (1962)
Sclerotinia fructicola	Cu							×				+		Partridge and Rich (1962)
Stemphylium sarcinaeforme	Cu							×				+		Partridge and Rich (1962)
Saccharomyces cerevisiae	Ag					○	○	○		×	○			Ashida (1965)
Saccharomyces cerevisiae	Cd					○	○	+		○	×			Ashida (1965)
Aspergillus niger	Hg					+	+	+	+	+	+	×	+	Ashworth and Amin (1964)
Hypochnus sp.	Hg							+				×		Ashida (1965)
Ceratostomella paradoxa	Hg							+				×		Sugunakar-Reddy, Apparao and Subbaya (1964)
Penicillium notatum	Hg							+				×		Partridge and Rich (1962)
Sclerotinia fructicola	Hg							+				×		Partridge and Rich (1962)
Stemphylium sarcinaeforme	Hg							+				×		Partridge and Rich (1962)
Piricularia oryzae	Hg	○						○				×		Yamasaki and Tsuchiya (1964)

Explanation of Table 2: + co-tolerant, × tolerant, ○ non-tolerant. The absence of a symbol indicates that the metal was not tested.

believed that specificity could be achieved through a control mechanism which was restricted to ions of the same size as zinc. However magnesium (0·65 Å), lithium (0·68 Å) and cobalt (0·70 Å) have ions of similar dimensions to those of zinc and nickel, and indeed cobalt is also like nickel in its chemical and physical properties, but none of these ions could substitute for zinc or nickel in the tolerance mechanism (Gregory 1964; Barker, personal communication, 1967). The tolerance to more than one metal, but where only one of these metals occurs in the original habitat is called a co-tolerance. Co-tolerances are rare and anomalous (see Tables 1 and 2) and more confusion arises as co-tolerances seem to vary depending upon the method of bio-assay (Barker, personal communication 1967).

The remarkable characteristic of metal tolerance is its degree of specificity. This is perhaps not unusual in biological systems, particularly in the field of micronutrient biochemistry. The rigid specificity of enzyme systems for metal ions and the non-replaceability of the essential ion by ions with similar properties is a common feature of '*in vivo*' enzyme systems (Comar and Bronner 1962). However, the particular mechanisms conferring this specificity of metal tolerance remain to be discovered.

CONCLUSIONS

The general pattern that emerges concerning metal tolerance in plants indicates that the poisonous metals may either be excluded from the metabolic system or when present be prevented from exerting an effect because of changes in enzymes which allow normal processes to continue. A significant feature is the role of the cell wall in acting as a heavy metal accumulator in a wide range of different plants. Future studies on this problem may emphasize still further the importance of this cellular fraction in regulating metal toxicities. The whole problem is a very interesting aspect of the ecology and evolution of plants and the comparison of edaphic ecotypes offers an ideal experimental approach for future studies.

ACKNOWLEDGEMENTS

I am greatly indebted to Dr C. Marshall, School of Plant Biology, University College of North Wales, Bangor, for his invaluable advice and criticisms in the preparation of this manuscript; and to the Zinc and Copper Development Associations for financial support.

REFERENCES

ASHIDA J. (1965). Adaptation of fungi to metal toxicants. *A. Rev. Phytopath.* **3**, 153–174.

ASHIDA J., HIGASHI N. and KIKUCHI T. (1963). An electron microscopic study on copper precipitation by copper resistant yeast cells. *Protoplasma* **57**, 27–32.

ASHIDA J. and NAKAMURA H. (1959). Role of sulphur metabolism in copper resistance of yeast. *Pl. Cell Physiol. Tokyo* **1**, 71–79.

ASHWORTH L.J. and AMIN J.V. (1964). A mechanism for mercury tolerance in fungi. *Phytopath.* **54**, 1459–1463.

BAUMEISTER W. (1967). Schwermetall-Pflanzengesellschaften und Zinkresistenz einiger Schwermetallpflanzen. *Angew. Bot.* **50**, 185–204.

BAUMEISTER W. and BURGHARDT H. (1956). Uber den Einfluss des Zinks bei *Silene inflata*, Sm. II. CO_2 Assimilation und Pigmentgehalt. *Ber. dt. Bot. Ges.* **69**, 161–168.

BOLLARD E.G. and BUTLER G.W. (1966). Mineral nutrition of plants. *Ann. Rev. Pl. Physiol.* **17**, 77–105.

BOSE R.G. and BASU S.N. (1965). Effects of copper, zinc and mercury on the growth and cellulolytic activities of *Pencillium ochrochloron* Biourge. *Indian J. exp. Biol.* **3**, 42–44.

BRADSHAW A.D., MCNEILLY T.S. and GREGORY R.P.G. (1965). Industrialisation, evolution and the development of heavy metal tolerance in plants. *Ecology and the Industrial Society* (Ed. G.T. Goodman, R.W. Edwards and J.M. Lambert), pp. 327–343.

BRENCHLEY W.E. (1938). Comparative effects of cobalt, nickel and copper on plant growth. *Ann. appl. Biol.* **25**, 671–694.

BRÖKER W. (1963). Genetisch-Physiologische Untersuchungen uber die Zinkvertraeglichkeit von *Silene inflata*, Sm., *Flora, Jena* **153**, 122–156.

BROWNING B.H., RUSSELL P, KINGSNORTH S.W. and PEERLESS R.J. (1959). Inactivation of organo-mercurial fungicides in groundwood pulp made from logs stored in salt water and the possible role of sulphur compounds. *Nature. Lond.* **183**, 1346–1347.

CARTWRIGHT B. (1966). Studies on copper deficiency in nodulated subterranean clover. Ph.D. Thesis, Univ. Nottingham, Sutton Bonington.

CLARKSON D.T. (1966). Aluminium tolerance in species within the genus *Agrostis*. *J. Ecol.* **54**, 167–178.

COMAR C.L. and BRONNER F. (1962). *Mineral Metabolism*. II. *The Elements*. New York.

DA COSTA E.W.B. (1959). Abnormal resistance of *Poria vaillantii* (D.C. ex Fr) cke, strains to copper-chrome-arsenate wood preservatives. *Nature, Lond.* **183**, 910–911.

DANIEL M.R., DINGLE J.T., and LUCY J.A. (1961). Cobalt tolerance and mucopolysaccharide production in rat dermal fibroblasts in culture. *Expl Cell Res.* **24**, 88–105.

DIEZ-ALTARES C., and BOROUGHS H. (1961). La localizacion intracelular del zinc en las plantas. *Turrialba* **11**, 162.

DIEZ-ALTARES C., and BORNEMISZA E. (1967). The localization of zinc-65 in germinating corn tissues. *Plant and Soil* **26**, 175–188.

DYKEMAN W.R. and DE SOUSA A.S. (1966). Natural mechanisms of copper tolerance in a copper swamp forest. *Can. J. Bot.* **44**, 871–878.

EAGON R.G., SIMMONS G.P. and CARSON K.J. (1965). Evidence for the presence of ash and divalent metals in the cell wall of *Pseudomonas aeruginosa*. *Can. J. Microbiol.* **11**, 1041–1042.

EARLEY E.B. (1943). Minor element studies with soybeans. I. Varietal reactions to concentrations of zinc in excess of the nutritional requirement. *J. Am. Soc. Agron.* **35**, 1012–1023.

ERNST W. (1966). Okologisch-soziologische Untersuchungen an Schwermetallpflanzengesellschaften Sudfrankreichs und des Ostlichen Harzvorlandes. *Flora, Jena* **156**, 301–318.

GREGORY R.P.G. (1964). The mechanism of heavy metal tolerance in certain grass species. Ph.D. Thesis, Univ. Wales.

GREGORY R.P.G. and BRADSHAW A.D. (1965). Heavy metal tolerance in populations of *Agrostis tenuis*. Sibth. and other grasses. *New Phytol.* **64**, 131–143.

GURD F.R.N. and WILCOX P.E. (1956). Complex formation between metallic cations and proteins, peptides and amino acids. *Advanc. Protein Chem.* **11**, 311–427.

HEWITT E.J. (1963). The essential nutrient elements: requirements and interactions in plants. *Plant Physiology—A Treatise*, **III**, Steward, F.C., Ed., Acad. Press, New York.

HORIO T., HIGASHI T and OKUNUKI K. (1955). Copper resistance of *Mycobacterium tuberculosis avium*. II. The influence of copper ion on the respiration of the parent cells and copper-resistant cells. *J. Biochem.* (*Tokyo*) **42**, 491–498.

JURKOWSKA H. (1952). Zdolnosc przystosowania sie Kropidlaka (*Aspergillus niger*) do miedzi. *Acta microbiol. Pol.* **1**, 107–122.

KIKUCHI T. (1965a). Production of hydrogen sulfide from sulfite by a copper-adapted yeast. *Pl. Cell Physiol. Tokyo* **6**, 37–45.

KIKUCHI T. (1965b). Studies on the pathway of sulfide production in a copper-adapted yeast. *Pl. Cell Physiol. Toyko* **6**, 195–209.

KIKUCHI T. (1965c). Some aspects of the relationship between hyper-hydrogen sulphide producing activity and copper resistance of yeast. *Mem. Coll. Sci. Kyoto Univ.* B **31**, 113–124.

MURAYAMA T. (1961a). Studies on the metabolic pattern of yeast with reference to its copper resistance. III. Enzymic activities related to the tricarboxylic acid cycle. *Mem. Ehime Univ.* IIB **4**, 43–52.

MURAYAMA T. (1961b). Studies on the metabolic pattern of yeast with reference to its copper resistance. IV. Characteristics in the tricarboxylic acid cycle. *Mem. Ehime Univ.* IIB **4**, 53–66.

NAKAMURA H. (1962). Adaptation of yeast to cadmium. *V*. Characteristics of RNA and nitrogen metabolism in the resistance. *Mem. Konan Univ. Sci.* **6**, Art 31, 19–31.

NEVINS D.J., ENGLISH P.D. and ALBERSHEIM P. (1967). The specific nature of plant cell wall polysaccharides. *Pl. Physiol.* **42**, 900–906.

NICOLLS O.W., PROVAN D.M.J., COLE M.M. and TOOMS J.S. (1965). Geobotany and geochemistry in mineral exploration in the Dugald River Area, Cloncurry District, Australia. *Trans. Instn Min. Metall.* **74**, 695–709.

NORTHCOTE D.J. (1963). The nature of plant cell surfaces, *The Structure and Function of the Membranes and Surfaces of Cells. Biochem. Soc. Symp.* **22**, 105–125.

PARTRIDGE A.D. and RICH A.E. (1962). Induced tolerance to fungicides in three species of fungi. *Phytopath.* **52**, 1000–1004.

PORTER H., JOHNSTON J., and PORTER E.M. (1962). Neonatal hepatic mitochondrocuprein. I. Isolation of a protein fraction containing more than 4% copper from mitochondria of immature bovine liver. *Biochem. biophys. Acta* **65**, 66–73.

REILLY C. (1967). Accumulation of copper by some Zambian plants. *Nature, Lond.* **215**, 667–668.

REPP G. (1963). Die Kupferresistenz des Protoplasmas höherer Pflanzen auf Kupferzböden. *Protoplasma* **57**, 642–659.

RYDER L.A. and COLMER A.R. (1965). An iron oxidizing bacterium from the effluent of lead and zinc mines. *Proc. La. Acad. Sci.* **28**, 5–11.

SENO T. (1963). Genetic relationship between brown colouration and copper resistance in *Saccharomyces cerevisiae. Mem. coll. Sci. Kyoto Univ.* B **30**, 1–8.

SOMERS E. (1963). The uptake of copper by fungal cells. *Ann. appl. Biol.* **51**, 425–437.

SUGUNAKAR-REDDY M., APPARAO A. and SUBBAYA J. (1964). Adaptation of *Ceratostomella paradoxa* Dade to copper sulfate and mercuric chloride. *Indian J. exp. Biol.* **2**, 211–215.

TURNER R.G. (1967). Experimental studies on heavy metal tolerance. Ph.D. Thesis, Univ. Wales.

TURNER R.G. and GREGORY R.P.G. (1967). The use of radioisotopes to investigate heavy metal tolerance in plants. *Isotopes in Plant Nutrition and Physiology*. IAEA/FAO, Vienna.

URL W. (1955). Resistenz von Desmidiacen gegen Schwermetallsalze. *Wien Sitzungsber.* **164**, 206–230.

URL RW. (1956). Uber Schwermetall -, zumal Kupferresistenz einiger Moose. *Protoplasma* **46**, 768–793.

WILKINS D.A. (1960). The measurement and genetical analysis of lead tolerance in *Festuca ovina. Rep. Scott. Pl. Breed. Stn* 1960, p. 86.

YAMASAKI Y. and TSUCHIYA S. (1964). Studies on drug resistance of the rice blast fungus, *Piricularia oryzae*, Cav. II. Cross resistance of drug resistant strains. *Bull. natn. Inst. agric. Sci. Tokyo* **11**, 53–67.

DISCUSSION (B) ON MECHANISMS OF MINERAL NUTRITION

Recorded by DR D. H. JENNINGS and MR J. F. HANDLEY

There was no doubt that the meeting was very anxious to endorse the need to study halophytes in more detail both from the ecological and from the physiological standpoint. Professor Epstein felt that there was every reason to believe that it would be possible to breed halophytic crop plants in view of the fact that there is no basic biological incompatibility between salinity and plant life. He also strongly supported a plea of Dr Glentworth—though he felt it should be directed at physiologists rather than ecologists—that more study should be made of the wide-ranging halophytic communities such as those in North America and that special attention should be given to the diversity of physiological form which can be found in such communities.

Much of the discussion about the paper of Dr Woolhouse centred around the exact location of the enzymes which he had been studying. Dr Woolhouse said that he had not yet been able to look at his preparations under the electron microscope. On light microscopical criteria, there are no particles attached; furthermore, the preparations do not respire. They are prepared by the usual standard procedures. He did agree that the preparations could be heterogeneous with regard to phosphatase activity, feeling as with Dr Loughman that it need not be specifically associated with the root surface. Dr Woolhouse also agreed that his preparations may also contain an active phosphatase in mucigel material as described by Dr Rovira on the basis of his own studies.

Professor Mengel drew attention to the fact that, although the activity of the phosphatase can be related to active transport, if one assumes that transport is by a phosphorylated carrier, the limiting process is not the splitting-off of the phosphate but the priming reaction for it via the kinase. Furthermore, one would expect this enzyme to be located on the inner side of the membrane.

In the discussion about aluminium toxicity, Professor Russell drew attention to those plants which are so-called aluminium accumulators. Dr Rorison made a plea that we should be clear what we mean here. He pointed out that ecologically we are normally concerned with soils of pH 4·5 and below, where aluminium is toxic to susceptible species. At pH's higher than 5·5, these plants will grow and not suffer from aluminium toxicity. Furthermore there is plenty of aluminium inside the plant under these conditions. Dr Clarkson supported this, producing evidence from his own studies on *Agrostis stolonifera* which had been growing in calcareous grassland and on podsolized brown-earth soils. The plants in the former instance had an aluminium concentration at an order of magnitude higher than in the latter. Part of this was of course due to the larger root systems of plants growing on the calcareous soil.

With regard to the mechanism of aluminium toxicity, Dr Woolhouse drew attention to the work of Amoore on the so-called mitotic complexes which seemed to contain a cytochrome system, which if present could be a possible site for the action of aluminium. Dr Clarkson felt that this finding would fit in with those others which indicate that iron deficiency and aluminium toxicity lead to the same thing—essentially a 'run-down' in the number of mitoses. He felt that, since this was so, it might be worthwhile to try to get a relatively high level of iron or iron complexes into roots to see if one can get roots started again after aluminium poisoning.

There was a wide-ranging discussion about the sequestering in cells of substances so that they are unable to exert a toxic effect. Dr Woolhouse was puzzled as to what happens in a system such as a cell wall when sites within it become saturated with zinc such that zinc comes into equilibrium with the zinc in the solution, which will then enter the cytoplasm. Dr Jennings pointed out that it was important to remember that there are cases where the plant itself is producing a toxic compound. This is the case in the production of fluoracetate in tropical species and what happens here is not known. Dr Jennings felt that it is important to remember that plants can synthesize such materials and tuck them away, so that metabolically speaking they are not effective. He also felt that we have to be careful in assuming that we are going to meet similar mechanisms throughout the plant kingdom. Dr Loughman supported this, pointing out the striking case of herbicides, where for instance 2–4 D can go into a tolerant plant and come out again as 2–4 D and be present in the plant in high concentration. He thought that the important point to establish in such instances was the concentration in the cytoplasm of the substance under consideration and the capacity for its vacuolar accumulation.

Professor Harley raised a subject in part relating to the matter of the previous day's meeting, in which he criticized the outlook of Dr Jefferies with regard to his studies on malic dehydrogenase. Professor Harley considered that it was wrong to talk about the stimulation of an enzyme by a high concentration of an ion. Rather, it was better to think about the whole enzyme system, where one would have to take account of other control mechanisms. From this viewpoint, one would consider the case of tolerance to high calcium as being one in which the enzyme system is susceptible to normal regulation, even though there were abnormally high concentrations of the ion present. Dr Sims in answer to this said that he thought tolerance of enzyme systems to high concentrations of ions was a symptom of the cell being limited in the amount of ATP available for moving ions out of the cell against an electrochemical potential gradient. Dr Jennings urged caution about the idea that all energy for ion movement must come from ATP. In any case, one should draw up an energy balance sheet before coming to conclusions such as those put forward by Dr Sims. Mr Nye pointed out that cells had 'bags of energy' to stop the accumulation of toxic ions. The fact is that the free energy of the complete oxidation of glucose is of the order of 600 Kcal, whilst the free energy required to make a 10-fold change in a mole of ion is only something like 2 Kcal.

Mr Nye also raised the point of the relationship between short-term uptake studies and the site of uptake of nutrients by growing whole plants. Dr Scott Russell felt that there was little virtue in trying to extrapolate from, say, studies with excised roots to what is happening in the whole plant. However, Professor Epstein, although feeling that we knew little about the relationship about which Mr Nye was concerned, did present evidence to show that information about ion uptake by growing plants could be interpreted in terms of experiments which were concerned with determining essentially instantaneous rates of uptake. Professor Epstein drew attention to the studies of Osmond in Australia on the ion relations of *Atriplex*. Plants were grown for a considerable period and their ionic content was determined. The data obtained indicated that there are indeed in these plants two mechanisms for potassium uptake—one highly specific and the other not nearly so specific and operating at higher concentrations. Further, Professor Epstein pointed out, if you analyse whole plants which have been growing in solutions in which there is a certain ratio of potassium to rubidium, the ratios of the two ions in the individual organs are as predicted by the model postulated on the basis of short-term studies in which potassium and rubidium are absorbed by the same mechanism.

AN ECOLOGIST'S VIEWPOINT

A. D. BRADSHAW
Department of Botany, University of Liverpool

The ecologist seeks to understand the relationship of plants to their surroundings and why particular plants grow in particular places. Perhaps not all the contributions to this Symposium have borne this in mind. For this reason it is appropriate to go back and examine the very simple aspects of the problem.

THE IMPORTANCE OF SOIL NUTRIENTS

The ecologist must first satisfy himself that soil nutrient status is an important ecological factor. There is now a wide variety of very powerful evidence that shows this to be true; but it is not all correlated and it is still very possible to underrate the importance of soil nutrient factors. One major example of this is the account given by Tansley (1939) of the ecological factors determining the vegetation of chalk grassland. He gives extreme emphasis to the part played by drought and even includes some drawings of rooting profiles. However, these same grasslands are now being extremely heavily used by the agriculturist. On them are not only growing crops of cereals, but also high-yielding grass species normally found in damp fertile grasslands such as *Phleum pratense*, timothy, *Festuca pratensis*, meadow fescue and *Dactylis glomerata*, cocksfoot. Indeed, such pastures rarely suffer from drought and often do better over summer periods than their equivalents in lowland conditions. All this has been brought about by the use of general fertilizer. The problem facing plants in natural chalk grassland is not shortage of water but an acute shortage of nitrogen, phosphate and potassium.

This shows very clearly that some of the best evidence for the importance of soil nutrient status as a factor determining plant distribution is to be got particularly from the fertilizer usage of the agriculturist on grassland, both natural and sown. It is quite clear that there are very few grasslands in which soil nutrients are not limiting; the exception is perhaps in the high grade ryegrass and clover pastures of one or two highly specific areas in the British Isles (Davies 1960). In certain cases, experiments have been set up in which a range of different fertilizer treatments have been applied to natural grassland for a long period. The most famous of these is the Park

Grass experiment at Rothamsted started in 1856, which now provides quite overpowering evidence for the role of soil nutrient levels (Brenchley 1958). The distribution of any species, for instance *Anthoxanthum odoratum* (Fig. 1), in these plots shows this. The experiment deserves much more attention from ecologists and physiologists than it has received up to now, and Miss Thurston's description is a very proper contribution to this volume. Another less well known but equally impressive experiment, since its age is much less and the differences it shows just as great, is that

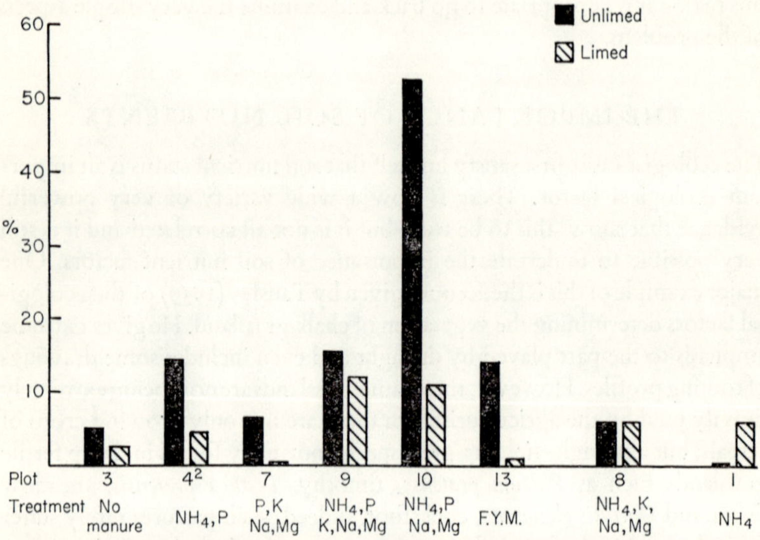

FIG. 1. Percentage of *Anthoxanthum odoratum* in some of the Park Grass experimental plots, Rothamsted (Brenchley 1958).

established by Milton on various hill grasslands near Aberystwyth (Milton 1940). Both of these are still running. But it is not necessary to rely on other people's rather long-term experiments. The same evidence can be obtained from simple experimental plots set out in any selected plant communities visited regularly for teaching purposes. The Bangor plots (Fig. 2) are designed to be fertilized once a year during the course of an annual class visit. Not only can such treatments reveal vast effects on the amount of growth of the species already present, but also on their frequency; new species often appear very quickly. In four years, for instance, the NPK Ca treatment transformed an area of *Eriophorum vaginatum*, cotton grass, bog to a pasture dominated by *Holcus lanatus*, Yorkshire Fog. Such plots are not only illustrative, they are also provocative, since they may produce results which are not necessarily those expected.

The final and most obvious piece of evidence for the importance of soil nutrient factors is the wide variety of communities whose distribution can be related to soil type. There is no better place to see this than in the British Isles. It is well documented, but not necessarily well understood yet. The specific nutrient factors involved and the way they operate do not necessarily turn out the way they are expected. This is well shown by Professor Pigott in salt marsh vegetation; salinity is by no means the only important factor.

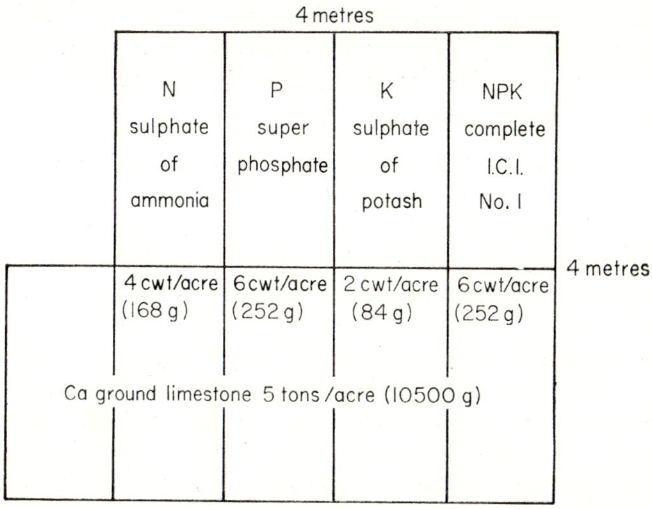

FIG. 2. Plan of small scale manurial plots in natural communities near Bangor.

While we are happy to accept the effects of soil variation in determining major differences between plant communities, it is not clear that we are prepared to accept its effects on a small scale. There is very little work on this. Dr Gittins shows a very beautiful variation in community composition in his ordination of a limestone grassland, which can to a large extent be related to variation in nutrient levels. The distances separating contrasting sites were of the order of twenty metres. But there is no reason why they should not be of the order of twenty centimetres. Snaydon (1962) has shown a highly localized variation which is perhaps the rule rather than the exception in many natural communities. We must also be prepared for local nutrient variation in other ways, for instance, the variation from one depth in a soil profile to another described by Dr Newbould, or from one time in the year to another (Russell 1961, p. 303). The variation may be on a micro-scale. The detailed evidence of microbial distribution (Hill and Gray 1967) clearly points to highly localized variation. But we need not be too sophisticated in looking for such evidence, for the casts of earthworms, the droppings of birds and the death of plants such as thistles and buttercups, must all provide highly localized release of nutrients.

THE PROBLEMS OF CO-HABITATION

Professor Epstein has encouraged us to look outside the *Avena* coleoptile and the barley root. The most remarkable thing we find if we do look outside is that species do not occur in isolation. Potentially or actually, species occur together. This rather obvious, but often forgotten, point which Darwin emphasized so much, leads us to three different characteristics of species which we must examine closely, characteristics which have hardly been mentioned in this symposium.

The characteristics of species which enable them to cope with the general properties of the environment

A group of species occupying any one situation, soil type or habitat will all have to cope with the same environmental factors. So we must look for the qualitative and quantitative similarities of species which provide their necessary adaptation. Such characteristics can effectively be examined in simple experiments, as spaced plants with soil (e.g. Pigott, Rorison, Grime, in this volume) or in nutrient culture (Clymo 1962; Clarkson 1967). These experiments are enormously productive and they produced subtleties of of all sorts; each paper has revealed something new. There is no need to stress this approach here, but to emphasize that we must beware of overlooking the obvious. Pigott and Taylor (1964) have shown the greater importance of phosphate rather than nitrogen in the distribution of nettles, yet the agriculturist is well aware of the limiting part played by nitrogen in nearly all situations and there is good evidence of species differences in nitrogen response (Bradshaw *et al.* 1964). At this moment nitrogen is perhaps the least well investigated soil nutrient, almost certainly because it is difficult to analyse. As we go from one area of the world to another, we must be prepared for particular elements to differ greatly in importance, such as sulphur in the Coast Range of California, and zinc in the Ninety-Mile Desert of Southern Australia.

The characteristics of species in their relationship to other species which cause potentially co-habiting species not to do so

This is the problem of competition, where plants are trying to exploit a resource in short supply. Here we must look for qualitative similarities and quantitative differences. Since all plants are qualitatively similar in their nutrient requirements (they all need the same macro- and micro-nutrients) and since these are so often limiting then it follows that direct competition between species for available supplies must be quite normal. Such competition is complex and is tied with competition for other things, especially light. What little work has been done emphasizes the importance of nutrients. In *Phalaris tuberosa*, canary grass and *Dactylis glomerata* mixtures 75% of the competitive effects were due to nutrients (Donald 1958). Ellenberg (1958) has emphasized the general phenomenon more than many people. In this volume we have the evidence of Dr Rorison and Dr Hackett. Indeed, it is unlikely that the realized niche of any plant species is the same as its fundamental niche, a point well emphasized by zoologists (Hutchinson 1966).

But we have to show that in such relationships it is nutrient factors which are involved. There are a variety of experiments that can be done. Most simply, species can be grown in mixed and pure stands in varying nutrient conditions. The experiments of Mr van den Bergh (this volume and 1968) using an analysis of relative replacement rates, are elegant. Another simpler, but equally impressive experiment involving competition for potassium between *Trifolium subterraneum*, subterranean clover and *Ehrharta calycina*, perennial veldt grass is that of Rossiter (1947). But we need to take the analysis further and discover how such changes in mixtures come about.

α values for uptake activity, and amounts of root growth can both be determined. From this we should be able to make models predicting the growth in mixtures from the behaviour in monoculture and see if they are borne out in practice. The relationship between root growth and performance in mixtures has been demonstrated. For instance, the outcome in mixtures of *Agropyron spicatum* and *Bromus tectorum* is fully explained by the differences in root growth (Harris 1968); and in mixtures of root pruned and normal wheat plants, Newman has shown that nitrogen uptake is related to amount of root (Table 1). But the relationship between uptake activity and performance in mixtures has not yet been demonstrated. It appears from the work of Mr van den Bergh that the actual growth pattern response to nutrient level is more directly important in determining the outcome of competition. But this pattern of response itself may be due to uptake activity.

Relationships are sometimes very complex. In a series of experiments involving mixtures of barley and oats on a variety of soils, de Wit (1960, p. 17) in some cases obtained a 'Montgomery' effect, where the lower

TABLE 1

Effect on uptake of nitrogen of mixing root pruned and normal wheat plants (uptake of N in mg/plant) (Newman, unpublished)

	Alone	Mixed
Unpruned	6·1	10·4
Pruned	5·3	3·9

60% of roots cut in pruned plants.

yielding species in pure stand is yet the better competitor in a mixture. This, however, only happened on some soils and not others. With the advantages of hind-sight, it is a great pity that the nutrient relationship in those experiments were not examined. In some very interesting experiments of Williams and McCown (personal communication) involving competition between *Erodium cicutarium* and *Bromus mollis* for sulphur the effects of differential uptake were visible only in mixtures. The same is true in experiments involving mixtures of *Glycine javanica* and *Panicum maximum*, in the absence of *Rhizobium* (de Wit, Tow and Ennik 1966): the relative nitrogen uptake in the two species determined as % dry matter was quite different when the species were grown in mixtures to what it was when they were grown in pure stands.

The possible unpredictable outcome of competition where nutrients are involved is particularly well shown by investigations on the growth of clover populations with different nutrient responses in natural situations (Snaydon and Bradshaw 1962a). In sand culture, populations from low nutrient habitats showed much less reduction in growth at low nutrient levels than did populations from high nutrient habitats. However, despite the fact that high nutrient populations suffered a greater reduction, they never came to have a yield as low as that of the low nutrient populations. This relationship was retained when the populations were grown as spaced plants under low nutrient upland conditions. However, when the populations were transplanted into a natural *Festuca–Agrostis* sward under low nutrient conditions, the high nutrient populations died extremely rapidly, although the others did not. It would be easy to believe that the death of the lowland populations was due to something else such as grazing or climate. However, when the experiment was repeated with the addition of fertilizer, the growth of the lowland populations was considerable, and the relationship between the two populations was just as it was under spaced conditions (Fig. 3). The nutrient relationship in such experiments must indeed be complex. It would seem that a plant growing at 90% of its maximum yield is in some ways more fit than another plant growing at only 50% of its maximum yield, even if the actual yield of the latter is more

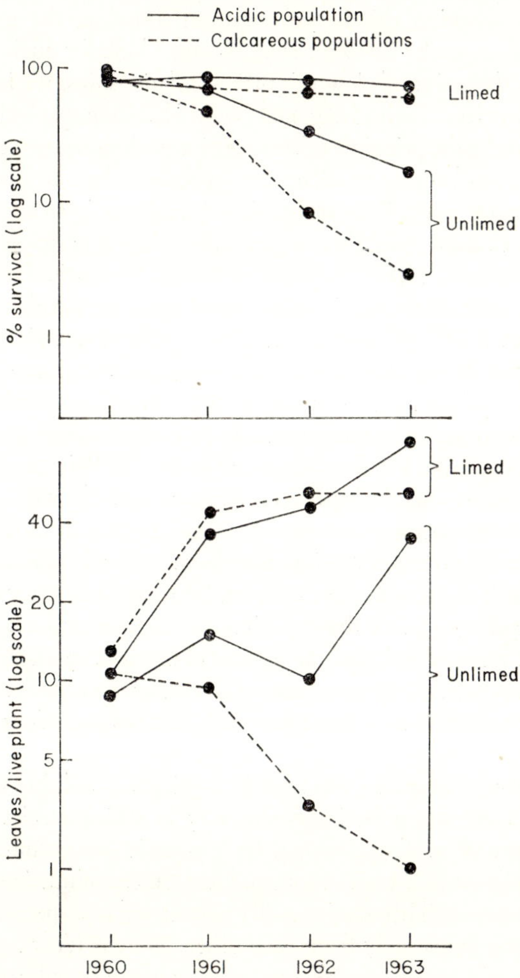

FIG. 3. Survival of clover populations in an upland sward near Aber, Caerns. (data of Snaydon, unpublished).

than that of the former. It is very possible that this is due to the geometric relationships discussed by Mr van den Bergh; but it could also be true that a plant growing at only 50% of its maximum yield is under physiological stress and may have lowered resistance to all the other adverse factors of its environment.

The characteristics of species which permit them to co-habit

Ecologically speaking this is the common situation, the end point of evolution envisaged by Darwin. Here we must look for qualitative differences. From the point of view of nutrient relationships, we know almost nothing. If nutrients are limiting, and two co-habiting species both require them, then under the principle of competitive exclusion one would exclude the other, unless because of niche diversification they do not effectively inhabit the same niche. Even if the niche is really only an abstract hypervolume (Hutchinson 1966) the competitive exclusion principle is very real. To inhabit different niches, species must differ by qualitative differences, in space or time, or in requirement. Zoologists are familiar with niche diversity, but botanists interested in plant nutrition certainly are not, if this symposium is anything to go by. It would be very interesting to find the botanical equivalent to the situation in the American Warblers (Fig. 4).

Timing differences are an important way by which differences in nutrient requirement can arise. Van den Bergh and de Wit (de Wit 1960) have shown that while *Phleum* and *Anthoxanthum* compete directly with one another under continuous summer conditions, in nature they do not. This is due to the fact that in natural conditions they have markedly different timing in their growth. The same has been shown by Khan in mixtures of flax and linseed (Harper 1965). But the experimental analysis of such situations is still lacking: we need to know the pattern of nutrient uptake over a period of time for co-habiting species. The possibility of very marked differences between the species is clear from the work that has been done on agricultural crops.

The occurrence of spatial variation in nutrients in soils has already been mentioned. Here is a most important way in which species could have available distinctly different niches. Dr Tinker in this volume found one centimetre of root to 2·3 ccs of soil, and movement of nutrients only over very short distances. This would readily permit different species to maintain their roots quite separately from each other. More evidence of the spatial relationships of the roots of different species in natural conditions is needed. Professor Russell emphasizes however that the problem will be complex because of the different availabilities and mobilities of different ions.

We know almost nothing about differences in nutrient requirements of species and how they might lead to stable co-habitation. The only differences of this sort which appear to have been investigated are those involving nitrogen and *Rhizobium*. Ennik (de Wit 1960) and de Wit, Tow and Ennik (1966) have shown very clearly the ability of *Lolium perenne* and *Trifolium repens*, and *Glycine* and *Panicum* to co-exist. In the case of *Glycine* and

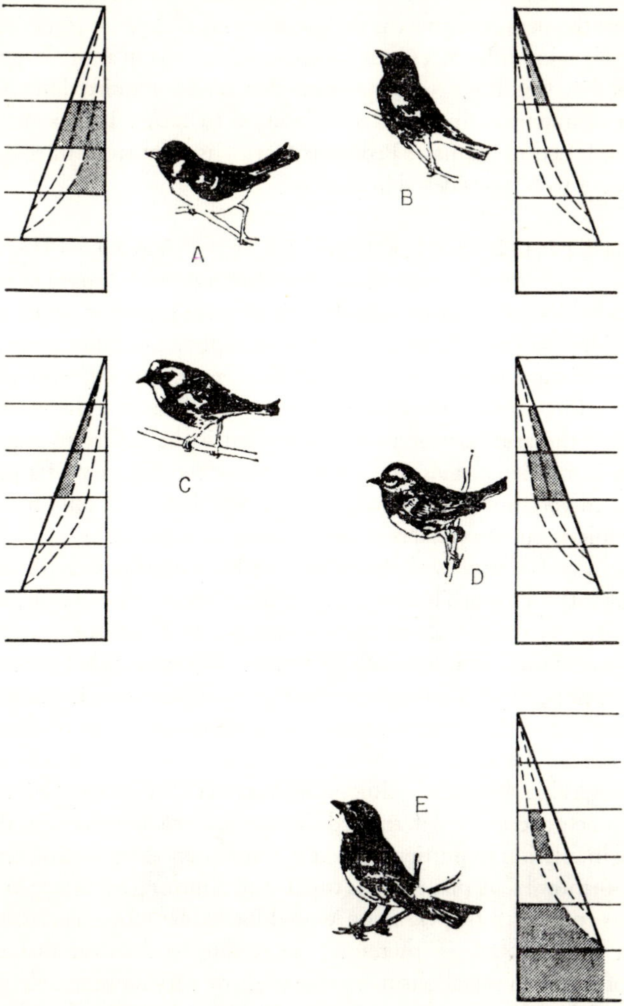

FIG. 4. The contrasting niches of five sympatric American warblers in the spruce forests of Maine.

A. *Dendroica tigrina*. B. *D. castanea*.
C. *D. fusca*. D. *D. virens*. E. *D. coronata*.

The stippling indicates the part of the tree in which the species spends more than half its time when in the tree. (Hutchinson, 1966, from MacArthur.)

Panicum analyses of nitrogen show clearly the co-operative aspect of their inter-relationship. There was not only transfer of nitrogen from *Glycine* to *Panicum*, but the increased nitrogen yield of the *Panicum* was not offset by a decreased yield of *Glycine*. Such experiments are an enormous step forward, but the depressing fact is their scarcity. It is difficult to believe that the sort of major differences in food sources that exist in animals can occur in plants, but we only need to propose that species require different proportions of nutrients for stable co-habitation to ensue. This principle has been established in animals: Professor Harley is very right in suggesting that it needs to be considered in plants.

THE PROBLEMS OF NUTRIENT VARIATION

So far, we have considered nutrient variation in time and space in relation to the co-habitation of species. Such variation has, however, an impact on the individual species. Plants are fixed in one spot and cannot escape from nutrient variation in time. They may also grow into different areas in space, or be forced into such areas by competition. Whatever the exact cause, one part of the armoury of a successful plant is the ability to cope with variation in nutrient levels. Indeed the greater the capacity of a plant to endure such variation, the greater chance it will have, not only to survive in one habitat, but also to spread into a range of habitats.

How can species cope with this problem? They must possess phenotypic flexibility in the sense of Thoday (1953). Either they must be so well buffered with storage systems such as those discussed by Professor Harley that they can maintain their internal characteristics no matter what the external environment is, or they must have the necessary phenotypic plasticity to adjust their metabolic processes to take maximum advantage of changes in environment. The existence of such systems is now well known for morphological and some physiological characters (Bradshaw 1965). They appear to be genetically determined like any other characteristic. But the evidence in relation to nutrient uptake is almost non-existent. Soil nitrogen varies enormously. A plant that is capable of adjusting its nitrogen metabolism to cope with this variation would indeed be more successful than one that could not. It is, therefore, interesting to discover that certain aspects of nitrogen metabolism are able to show very considerable phenotypic plasticity. At the moment interest is mainly concerned with nitrate reductase, whose level appears to be very readily controllable by nitrate (Beevers *et al.* 1965). Other inducible enzyme systems connected with mineral nutrition must also exist. In fact, it would be the absence of inducible systems that would be surprising. This point of view has been argued already

PLATE 1. Boundary between toxic mine and ordinary soil, Flintshire, N. Wales: species that occur on both soils such as *Plantago lanceolata*, *Agrostis tenuis* and *Anthoxanthum odoratum* do so only because they evolve tolerance.

facing p. 425

very clearly in this volume by Jeffreys *et al.* It is clear from their work on *Lemna* that evidence for phenotypic change in uptake systems can be found and it is a subject which will surely receive more attention in the future.

The alternative way in which a species can cope with nutrient variation is by the possession of genetic flexibility; in other words, the ability to change genetically. Species are not static organisms produced by special creation. They are dynamic and evolving: under the influence of natural selection (which can often be extremely powerful) they can change genetically extremely rapidly. There was a time when, although we paid lip service to the concept of the ecotype, we never considered that it had any relevance to problems of nutrient uptake. However, we now know that there is as much variation between populations in their adaptation to different nutrient levels as there is an adaptation to variation in climate and other factors. Some of the genetic change of this sort is very spectacular (Plate 1), e.g. metal tolerance in *Agrostis tenuis* (Gregory and Bradshaw 1965). Such an ability to evolve tolerance is not by any means a special characteristic in a few odd species. Again and again you find it turning up in a new, quite different species, such as *Rumex acetosa*. This capacity to evolve is clearly of great adaptive value in the face of varying nutrient conditions. Without it, none of the species concerned could grow in the sites contaminated with any metals in which they are found. It would be interesting to know whether the reason why a large number of species are not found in such areas is because they do not possess an ability to evolve. It is, perhaps, not without significance that several of the species able to evolve tolerance, e.g. *Rumex acetosa*, are found in serpentine areas where heavy metal toxicity may often be associated. Such evolutionary ability is not confined to spectacular characteristics. Adaptation to quite ordinary common soil nutrients may be evolved, for example to varying phosphate levels in *Trifolium repens* (Snaydon and Bradshaw 1962b); and it can be found in almost any species you care to take (Crossley and Bradshaw 1968). Dr Goodman gives some excellent examples in this volume. Although such evolutionary adaptation is in relation to normal macronutrients which have no toxic effects, it still can be very important in determining whether or not a species can exist in a particular environment. Lowland populations of *Trifolium repens* will not survive in upland conditions where nutrient levels are low. If *Trifolium repens* did not possess the ability to evolve populations with different nutrient requirements, its distribution would be much more restricted. Ecological amplitude has, therefore, not only phenotypic but also genetic components of considerable adaptive significance.

Plants have evolved in natural situations and the characteristics they possess relate to what happens in them. Even if for practical purposes we must resort to experimental studies under controlled laboratory conditions, we must continually go back to these natural situations to see what particular characteristics involving mineral nutrition are of paramount importance.

REFERENCES

BEEVERS L., SCHRADER L.E., FLESHER D. and HAGEMAN R.H. (1965). The role of light and nitrate in the induction of nitrate reductase in radish cotyledons and maize seedlings. *Pl. Physiol.* **40**, 691–698.

VAN DEN BERGH J.P. (1968). An analysis of yields of grasses in mixed and pure stands. *Agric. Res. Rep. Wageningen* **714**, 1–71.

BRADSHAW A.D. (1965). Evolutionary significance of phenotypic plasticity in plants. *Adv. Genet.* **31**, 115–155.

BRADSHAW A.D., CHADWICK M.J., JOWETT D. and SNAYDON R.W. (1964). Experimental investigations into the mineral nutrition of several grass species. IV. Nitrogen level. *J. Ecol.* **52**, 665–676.

BRENCHLEY W.E. (1958). *The Park Grass Plots at Rothamsted* (1856–1949). Rothamsted Experimental Station, Harpenden, Herts.

CLARKSON D.T. (1967). Phosphorus supply and growth rate in species of *Agrostis* L. *J. Ecol.* **55**, 111–118.

CLYMO R.S. (1962). An experimental approach to part of the calcicole problem. *J. Ecol.* **50**, 707–731.

CROSSLEY G.K. and BRADSHAW A.D. (1968). Differences in response to mineral nutrients of populations of ryegrass, *Lolium perenne*, and orchard grass, *Dactylis glomerata*. *Crop Sci.* **8**, 383–387.

DAVIES WM. (1960). *The Grass Crop*. London.

DONALD C.M. (1958). The interaction of competition for light and nutrients. *Austral. J. agric. Res.* **9**, 421–435.

ELLENBERG H. (1958). Bodenreaktion (einschliesslich Kalkfrage). *Encyclopedia of Plant Physiology*. (Ed. W. Ruhland.) Springer-Verlag, Berlin, **IV**, 638–708.

GREGORY R.P.G. and BRADSHAW, A.D. (1965). Heavy metal tolerance in populations of *Agrostis tenuis* Sibth. and other grasses. *New Phytol.* **64**, 131–143.

HARPER J.L. (1965). The nature and consequences of interference amongst plants. *Genetics Today: Proc. 11th Int. Cong. Genet.* London, **2**, 465–482.

HARRIS G.H. (1968). Some competitive relationships between *Agropyron spicatum* and *Bromus tectorum*. *Ecol. Mon.* **37**, 89–111.

HILL I.R. and GRAY T.R.G. (1967). Application of the fluorescent antibody technique to an ecological study of bacteria in soil. *J. Bacteriol.* **93**, 1888–1896.

HUTCHINSON G.E. (1966). *The Ecological Theater and the Evolutionary Play*. Newhaven.

MILTON W.E.J. (1940). The effect of manuring, grazing and cutting on the yield, botanical and chemical composition of natural hill pastures. I. Yield and botanical composition. *J. Ecol.* **28**, 326–356.

PIGOTT C.D. and TAYLOR K. (1964). The distribution of some woodland herbs in relation to the supply of nitrogen and phosphorus in the soil. *J. Ecol.* **52**, suppl., 175–185.

ROSSITER R.C. (1947). The effect of potassium on the growth of subterranean clover and other pasture plants on Crawley sand. *Australian Jour. Council Sci. Indust. Res.* **20**, 389–401.

RUSSELL E.J. (1961). *Soil Conditions and Plant Growth.* 9th Ed., London.

SNAYDON R.W. (1962). Microdistribution of *Trifolium repens* L. and its relation to soil factors. *J. Ecol.* **50**, 133–143.

SNAYDON R.W. and BRADSHAW A.D. (1962a). The performance and survival of contrasting natural populations of white clover when planted into an upland *Festuca/Agrostis* sward. *J. Brit. Grassld. Soc.* **17**, 113–118.

SNAYDON R.W. and BRADSHAW A.D. (1962b). Differences between natural populations of *Trifolium repens* L. in response to mineral nutrients. I. Phosphate. *J. exptl. Bot.* **13**, 422–434.

TANSLEY A.G. (1939). *The British Islands and their Vegetation.* Cambridge.

THODAY J.M. (1953). Components of fitness. *Symp. Soc. Exptl. Biol.* **7**, 96–113.

DE WIT C.T. (1960). On competition. *Versl. landbouwk, Onderz. Ned.* **66** (8), 1–82.

DE WIT C.T., TOW P.G. and ENNIK G.C. (1966). Competition between legumes and grasses. *Versl. landbouwk. Onderz. Ned.* **687**, 1–30.

A SOIL SCIENTIST'S VIEWPOINT

E. W. Russell

Department of Soil Science, University of Reading

In the context of this Symposium, the soil scientist's viewpoint concerns the soil as the home of the roots of the plants growing on the soil; for the well-being of a plant is very largely dependent on the health and vigour of its root system. Plants need roots for three separate reasons; to anchor them into the ground, to take up water and to take up nutrients. Anchorage is of fundamental importance for trees and bushes, and also for many plants growing on blowing sands, but otherwise even quite a poorly developed root system is normally adequate for anchorage. The problems concerned with the efficiency of roots for taking up water from the soil are not the subject of this Symposium, and I will not be considering them further except in so far as they are relevant to the problems concerned with the efficiency of roots for taking up nutrients.

The papers that have been given at this Symposium have shown very clearly how many different factors control the rate at which plant roots can take up nutrients and other elements from the soil, for it must always be remembered that roots have only a limited ability to discriminate between the various ions and other soluble substances which occur in the soil in their neighbourhood. The principal factors include the form in which the element is present in the soil, its concentration in the soil solution, the rate at which the element can move from its labile pool in the soil to the surface of the roots where uptake is taking place, the efficiency with which roots can transfer the element from outside the root to within the root cells probably into the vacuoles of the cortical and endodermal cells, and the rate at which the roots themselves are growing and ramifying through the soil. All these are essential factors in the mineral nutrition of plants, but only some belong to the field of soil science. Thus research on the uptake of minerals by plants must of necessity involve the closest collaboration between the soil scientist and other disciplines, of which plant physiology is obviously one of the most important. It has been one of the most interesting features of this Symposium, from my point of view, that the fields of joint interest between the different disciplines has been highlighted so clearly, and I hope this will augur very well for the practical acceptance of the need for more active inter-disciplinary research on these very important problems of plant nutrition.

THE RESERVOIR OF IONS IN THE SOIL

This Symposium has been concerned principally with the uptake of nutrient elements needed in major amounts—nitrogen, sodium, potassium, magnesium, calcium and phosphorus—although the uptake of aluminium and other heavy metals in toxic amounts has also been an important theme. All these elements are normally taken up either in the form of simple ions or else through the activity of the fungal component of mycorrhizas growing in the soil; and even these fungal hyphae may be taking up these elements as ions from the soil solution. It has come out very clearly in this Meeting that uptake by the root is very dependent on the concentration of ion close to the root surface; and since uptake involves the removal of ions from this region, uptake over periods of days depends on the factors which control the rate of transport of ions from the bulk of the soil into these regions, which in turn depends on the reservoirs of these ions in the soil.

This topic of the ionic reservoirs in the soil was not systematically discussed at this Meeting, although it was mentioned in connection with specific ions by a number of speakers, so I am proposing to make a few comments on it, and will begin with calcium. The amount of calcium ions in the soil solution, though often present in greatest concentration, is usually only a small part of the calcium reservoir in the soil itself. The principal reservoirs are the exchangeable calcium on the clay and humus and, for calcareous soils, calcium carbonate. Humus probably does not contain any calcium, apart from the exchangeable calcium it holds, but undecomposed plant debris does, though because of technical difficulties, little work has been done on the rate at which this calcium becomes transferred to the exchangeable form. Thus for non-calcareous soils the calcium reservoir is almost identical with the labile pool of calcium ions, which in turn is nearly the same as the exchangeable ions, and all the calcium in this pool appears to be equally available for uptake by the plant.

Potassium differs from calcium in a number of ways. Its concentration in the soil solution is usually over ten times lower and may be nearly a hundred times lower than the calcium concentration, and the level of exchangeable calcium is usually much higher than of exchangeable potassium, yet plants frequently contain more potassium than calcium. As with calcium, the labile pool of potassium consists mainly of the exchangeable potassium but, as shown by Professor Arnold, not all the labile pool is equally available to plants for some is held very tightly by the clay. As the vegetation removes potassium, the potassium ion concentration falls in the soil solution and the exchangeable potassium falls, but as the concentration falls close to the clay surface, potassium ions often begin to be transferred from a non-exchangeable to the exchangeable form. As the exchangeable pool becomes depleted, so it becomes partially refilled; and when it has been depleted to a certain level, the rate of uptake of potassium by the vegetation becomes equal to the rate of transfer of potassium from the non-exchangeable form.

Potassium therefore also differs from calcium in non-calcareous soils in that most clay minerals contain very considerable amounts of non-exchangeable potassium whereas they contain very little if any non-exchangeable calcium; and it is a part of this non-exchangeable potassium in the clay fraction that forms this additional reservoir of potassium. Some of this non-exchangeable potassium is fairly rapidly transferred to the exchangeable form in most soils relatively high in exchangeable potassium as the vegetation takes up potassium from the soil; but the greater the amount of potassium removed from the soil, the slower the rate of transfer becomes. Little is known about the rate of transfer in strongly depleted soils, but on the Rothamsted clay loam it appears to be about 25–50 kg per ha annually. This is a rate well in excess of the demands of natural vegatation, so the importance of depletion is usually due to the low concentrations of potassium in the soil solution rather than to an actual shortage of potassium itself. This means that rates of transfer are usually of little importance to the ecologist who is interested in vegetation that is not being harvested or burnt annually, for then the sole source of loss of potassium is that in the water draining out of the area, which is much less than the rate of release in loam and clay soils. It could be of importance to foresters planting up strongly depleted soils to short rotation coniferous forest, for trees must rely on this release for the considerable amounts of potassium they will immobilize in their dry matter during their period of growth.

Potassium also differs from calcium in a further way that may be of considerable importance, for rain washes potassium out of the leaves of the vegetation much more rapidly than calcium, so that there is a rapid circulation of potassium from the soil to the leaf and back to the soil, and since most roots of the vegetation are in the surface layer of the soil, this may be an important mechanism for transferring potassium from mature leaves to the growing points. For as far as I know the importance of this cycle has never been studied in detail, probably because of the very great technical difficulties involved.

Phosphorus presents yet another picture, because it occurs in both inorganic and organic combinations in the soil. The soil solution contains inorganic phosphate ions, principally the $H_2PO_4^-$ ion, except in high pH soils when the HPO_4^{--} ion becomes important, and these ions are certainly important sources of phosphorus for the plant. But the soil solution also contains some soluble organic phosphates, possibly principally inositol phosphates, some of which can be taken up by plant roots, and some appear to be of little value to plants except perhaps if their roots contain mycorrhizas.

Phosphorus differs from the cations previously discussed in that the phosphate ion concentration is very much lower in soils than the potassium, often over a hundred times lower. Further, as Mr Nye showed in his paper, its rate of diffusion in a soil is also very much lower, and is in fact so low the the soil reservoir from which the roots can draw their phosphorus must be within a fraction of a millimetre from the root surface instead of the several centimetres for potassium. Phosphorus is, however, similar to potassium in that as a root removes phosphate from a soil, and the labile phosphate falls, some non-labile phosphorus will be transferred to the labile form. This point has not been examined in the same detail for phosphorus as for potassium so less is known about the relation between phosphate ion concentration and the rate of transfer of phosphate from non-labile to labile forms.

Several speakers have noted that we still know very little about the details of how labile phosphate is held on a soil and the form in which the non-labile phosphate that can be transferred to the labile form occurs. If a soluble phosphate is added to a soil, the phosphate is slowly converted over periods of months and possibly years into forms of decreasing solubility, and the roots of those plants capable of taking up phosphate from dilute solutions can use this phosphate many years after it was applied. In spite of nearly a century of study by chemists and physical chemists, our knowledge of the chemistry of these phosphates is still very limited, probably because they occur as thin films of indefinite chemical composition. In calcareous soils the phosphate is probably present largely as films of hydroxy- and fluor-apatite on the surface of the calcium carbonate crystals or films, and on non-calcareous soils as aluminium and ferric phosphates on films of hydrated aluminium and ferric oxides. But as these films form, the phosphate precipitates probably carry down some other ions present in the soil solution, thus preventing films of definite composition and therefore possessing definite solubility products from forming.

Phosphorus, however, is an important constituent both of plant litter and of humus, so that the phosphate status of the surface soil is very dependent on the rate of decomposition of these two components. Unlike the transfer of inorganic non-labile to labile phosphorus, the rate of production of inorganic from organic phosphates, or the rate of mineralization as it is often called, is not much affected by the concentration of phosphate ions in the soil solution, but is more dependent on the phosphorus content of the organic matter being decomposed. In so far as the rate of mineralization is in excess of the rate of uptake, as it often is in the autumn, mineralization will add to the labile or near-labile pool.

Nitrogen differs from all the elements so far discussed in that almost the whole of the nitrogen reservoir in a soil is in the organic form, with only a very small pool of labile inorganic nitrogen present as ammonium ions, and, in soils that are not too acid, a larger pool of nitrates all present in the soil solution. As Mr Nye pointed out, since nitrates are not absorbed by soil particles their rate of diffusion to the root during periods of uptake is much higher than for potassium; and also since the nitrate concentration is much higher usually than the potassium, a greater proportion of the nitrogen taken up by the vegetation reaches the roots during periods of active transpiration dissolved in the water flowing to the roots. There is a further difference between nitrogen and the ions previously considered in that under some conditions gaseous nitrogen in the soil air is converted into organic forms by a number of nitrogen-fixing organisms or systems; and some of these excrete soluble nitrogen compounds into the soil solution that are readily ammonified. The rate of supply of ammonium ions to the soil, which in turn controls the rate of production of nitrates, thus depends both on the rate of decomposition of plant debris and soil humus and on the rate of supply of soluble nitrogen compounds produced by the nitrogen-fixing systems. Although our knowledge of the rate of production of these substances is very imperfect, due to lack of suitable techniques, it is known that it becomes of increasing importance as the level of nitrate and ammonium ions in the soil decreases, and that for non-symbiotic fixation it is dependent on an adequate source of energy for the organisms.

THE TRANSFERENCE OF NUTRIENTS FROM THE SOIL INTO THE ROOT

The root-soil interface, where this transference takes place, constitutes a field of study where the plant physiologist and the soil scientist will increasingly have to work closer together. Not only must the nutrients be able to reach the root surface but the root must be able to take them up. It is known in practice that many roots will not function properly if the aeration of the soil in their neighbourhood is too poor, and under some conditions, for example, roots will lose potassium to the soil rather than take it up from the soil. Unfortunately this is a field where our knowledge is extremely poor, for we still have surprisingly little really reliable quantitative data on the effects of poor aeration on nutrient uptake.

In the past it has been assumed that plant roots could only function effectively if there was an adequate oxygen concentration outside the root surface, though it was known that marsh plants could grow in anaerobic soils. It was then shown that these marsh plants possess an oxygen pathway from the atmosphere through the leaves and stems of the plant to its roots, and for typical marsh plants the internal diffusion coefficient is sufficiently high to allow oxygen to diffuse out of the root into the soil. Barber was able to show, using ^{15}O, that this pathway is not confined to marsh plants but that it exists, though with an appreciably lower diffusion coefficient, in many plants normally considered to need aerobic conditions. This work has therefore opened up again the whole problem of the effect of aeration around the outside of the root on nutrient uptake.

My own personal view is that we have in the past over-emphasized the need for oxygen by the root. The soil contains many organisms which can function in the absence of oxygen, but when they are functioning in these conditions they are liable to produce products of reduction that may be extremely harmful to root activity. Thus, a typical reduction product is hydrogen sulphide, which can affect root activity at concentrations as low as 10^{-6} M. Butyric and other fatty acids are also sometimes produced in poorly aerated soils in concentrations sufficient to affect root growth, though this is probably rare in natural conditions. However, we still know very little about what substances are produced under these conditions, though circumstantial evidence suggests that some of them may have quite marked effects on root activity. The principal reason that poor aeration is harmful to root activity may well be due to some of these products of reduction being present in the soil at too high a concentration rather than to a direct absence of oxygen.

It is worth emphasizing here that even well-aerated soils may have many pockets of anaerobic soil when they are wet but well drained, for such soils will hold water in all pores finer than about 30 micron diameter; and in many fine-textured soils, such as soils high in fine sand, silt or clay, the individual crumbs making up the soil may have all their pores finer than this. Since oxygen diffuses very slowly through water and since the water in the soil pores is bathing active micro-organisms which are extracting oxygen from the water, oxygen can in fact only diffuse through 1–2 mm of wet soil so any volume of soil further away than this from an air pore will be anaerobic in the wet well-drained condition.

At this point it is worth remembering that not only does the soil close to the root affect the root but that roots also affect the soil close to them. I have already mentioned that oxygen sometimes leaks out of a root into the soil, but roots by extracting water from a soil usually cause air to enter the soil to take the place of the water extracted. Thus, an active root tends to improve the environment in its neighbourhood.

THE GROWTH OF ROOTS INTO SOILS

Plant roots are typically active organs which grow and ramify through the soil at certain times of the year. The faster they grow and the more they ramify the greater is the volume of the soil they can tap for nutrients; and this is especially noticeable for ions that can only diffuse slowly through a soil, such as phosphate for example, but to a lesser extent potassium. Again there is still regrettably little knowledge on the soil factors that influence this growth, and this is another field where I hope the soil scientist and the plant physiologist will be able to co-operate. We know that both organic and inorganic toxins are present in soils which can reduce root growth, and we have heard how aluminium ions and the ions of some heavy metals can have this effect.

The size distribution of the soil pores is a very important factor affecting root growth. It is well known that plants growing on well drained coarse sandy soils develop a deep, very extensive root system, whilst the same plants on a clay soil may only develop a shallow and very restricted one. The effect of a compacted layer in the soil profile can have a very marked effect on the root distribution, for it can prevent any roots entering the layer, so if it is close to the surface, the soil is edaphically very shallow although at first sight it looks deep. Peats over many boulder clays, and mor humus over poor sands commonly cause the vegetation to be shallow rooting; and if such soils are planted to fast growing coniferous trees they form areas where the trees are very liable to windthrow during gales.

CONCLUSION

From the soil scientist's point of view, the mineral nutrition of plants growing in a soil is dependent on the form of the reservoirs of each nutrient in the soil, the rate nutrients diffuse from the reservoirs to the plant roots during uptake, the soil condition which affects the rate of transfer of a nutrient from outside the root surface into the cortical or endodermal cells, and the soil conditions which affect the ability of the plant to grow a deep and extensive root system. The first two of these fall in the field of soil science, but the second two concern the plant physiologist as much as the soil scientist. And it is only when these effects are fully understood that the ecologist can safely explain his observed facts of plant distribution. On the other hand the ecologist is in an excellent position to test the accuracy or the adequacy of the ideas of the soil scientist and the plant physiologist by studying how far they are concordant with his observations in the field.

A PHYSIOLOGIST'S VIEWPOINT

J. L. HARLEY
Department of Botany, University of Sheffield

This conference has been concerned in a broad sense with the extent to which ecological distribution and behaviour may be ascribed to variation in supply, availability, and utilization of nutrients derived from the soil. The subject is one of great complexity which has at its core all the physiological problems concerning the rate and manner of absorption and utilization of nutrients by plants. These are themselves very complex, and experimental work upon them has so far mainly sought to describe general concepts and has not yet spent much effort upon the differences between species, between plants of different ages and stages, or between plants from different environmental conditions. The conference has served to emphasize this deficiency, but it has also included accounts of studies, mostly of a biochemical nature, of different varieties and species. I have been asked to comment in a general way on what has been said, in order to emphasize problems, criticize hypotheses, and summarize conclusions for discussion.

For purposes of experimental investigation physiological research on nutrient absorption has abstracted three broad sets of processes involved in mineral nutrition. These are:

(i) the absorption and accumulation of substances into cells,
(ii) the movement and circulation of substances in the plant body,
(iii) the nutritional and physiological functioning of the substances absorbed.

These three partial sets of processes are not really separable but are integral parts of the whole process of nutrition.

Movement of nutrients in the plants and accumulation in the cells

Professor Weatherly attempted to integrate particularly the first two sets of processes and in doing so produced a brilliant summary which drew together many aspects that had been discussed in other papers, especially emphasizing the need to consider the plant as a whole. But in my view he did somewhat less than justice to the importance of accumulation in cells, which was the subject of Dr Jennings' contribution, and one that may be of ecological importance. As Dr Loughman pointed out, accumulation is a process paralleling and competing with the movement of ions from the external source to the shoot. The rates of the two processes are differentially susceptible to applied changes of conditions.

In considering the potential importance of accumulation ecologically we have to decide whether a plant is mainly or solely utilizing for growth in any ecological surroundings, nutrients which are currently being absorbed by its roots; or whether there is a store, accumulated especially in the roots, which might smooth out the variations in nutrient supply and demand imposed periodically, seasonally, or sporadically by the environment or growth phase.

The estimation of transport index, that is the proportion of a substance absorbed in a given time that is transported to the shoot in the same time, shows significant and sometimes considerable retention in the root system in almost any investigation with whole plants. The extent varies quantitatively with the age and condition of the plant, with the concentration of nutrients supplied before and during any experiment, and with other external factors. The relative proportions transported and accumulated may vary in different parts of the same root or root system (see for instance Weibe and Kramer 1954, Canning and Kramer 1958, Brouwer 1954). Even if one accepts some distortion of these estimates because of the unsterile conditions, as the paper of Dr Barber and other work on the effects of microorganisms would predict, there can be little doubt that accumulation in the roots may be a significant part of total uptake. In the past, laboratory research has perhaps tended to overemphasize accumulation, but it is important not to swing too far in the opposite direction and deny the cells of the root the ability to accumulate nutrients, nor to ignore their capacities in this regard. This is important since periodicity of habitat conditions, especially periodicity of nutrient availability and of root development, occur in ecological surroundings and do not necessarily synchronize with periodicity of growth of the shoot and of reproductive structures.

To emphasize the possible ecological importance of accumulation the functioning of ectotrophic mycorrhizas affords an example which is especially fitting because I have been asked to remedy the absence of reference in the Conference, to this common ecologically important type of absorbing organ.

In ectotrophic mycorrhizas there is what may be termed an extra cortex, the fungal sheath, which encloses the cortex of the host. This layer, like the cortex of the root, has spaces between its hyphal cells through which movement of substances by diffusion or mass flow in solution could take place exactly as described by Weatherley for the root cortex; but if we were to conclude that these spaces constituted the main or only routes of movement of nutrients into the host as far as the endodermis, we would ignore much of the available experimental evidence, and make nonsense of the structural organization of these mycorrhizas which are the nutrient absorbing organs of many of the dominant trees of temperate regions.

The processes of nutrient absorption by ectotrophic mycorrhizas differ little qualitatively from those of uninfected roots but very greatly quantitatively, especially in the extent to which nutrients are accumulated in the cells of the fungal sheath. This fungal layer has much greater powers of accumulation of some nutrients than the cortical cells of the host tissue beneath it, or of those of uninfected roots. It is an extra accumulating zone of great efficiency.

Of course there can be mass flow or diffusional movement through the walls and spaces of the fungal layer as through those parts of the cortex of an uninfected root. For instance, if excised mycorrhizas are kept in

TABLE I

Effect of stripping mycorrhizas on phosphate uptake rate. Figures are relative to the rate of uptake in the intact condition

Concentration mM	0·032	0·32	3·20	32·00
Sheath	1·71	1·69	0·96	1·03
Core	5·50	4·27	2·23	1·30

phosphate solutions for a few hours their behaviour in this regard depends upon the applied concentration. (Table 1). From high concentrations (approximately 30 mM) phosphate passes through the fungal sheath and enters the host tissue about as rapidly as it does into the cortex of host tissue freed by dissection from its sheath; but at 0·032 mM the uptake into the host is increased fivefold by its being freed of the sheath. This difference in the effect of removal of the sheath lies in the extent to which the rate of diffusion of phosphate through the spaces of the sheath exceeds the rate at which phosphate is removed from the diffusion channels by accumulation into the sheath cells. Indeed it is possible to find phosphate concentrations (near 3 mM) where any factor inhibiting the accumulatory process of both fungus and host, e.g. low temperature or dilute metabolic poison, diminishes accumulation into the sheath cells so that the net rate of diffusion of phosphate to the host is stimulated and uptake into the host is not inhibited to the expected degree. In Table 2 low temperature and dilute azide do not greatly affect or even slightly stimulate the rate of phosphate accumulation in the host of mycorrhizas in 3·2 mM phosphate.

TABLE 2

Effects of inhibitors and low temperatures on phosphate uptake by sheath and core of beech mycorrhizas. Figures are percentages of the rate of uptake at normal temperature in absence of inhibitors

Phosphate conc. mM	0·32		3·2		32	
	Sheath	Core	Sheath	Core	Sheath	Core
0·2 mM azide (1)	6·2	34	10·2	101	35	38
(2)	4·9	61	7·4	110	22	50
1·0 mM iodoacetate	46	49·5	34	101	50·6	51·2
2–5°C	17	13	24	69	31	26

In ectotrophic mycorrhizas then, the two circuits of movement through the tissues, that through the living cells and that through the free space are both present, but the latter seems to be of negligible importance in phosphate uptake from ecologically probable concentrations because of the rapid accumulation of phosphate into the sheath cells.

The separation of the two processes of accumulation and movement through the tissues, as discussed by Dr Loughman must take place within the living system including the cortex as well as the endodermis and stelar parenchyma of the root. The process of accumulation which is a general feature of the functioning of root systems is obviously an exceedingly important process in ectotrophic mycorrhizas where it has been further investigated. During absorption of phosphate by beech mycorrhizas from ecologically possible concentrations, 80–90% of that absorbed is accumulated in the fungal layer. The 10–20% that passes to the host during absorption does so as inorganic phosphate by a route by which it equilibrates with only a small part (much less than 1%) of the soluble phosphate in the fungus (Harley, McCready and Brierley 1954; Harley and Loughman 1963; Jennings 1964). In this respect there are great similarities with the process described in barley by Crossett and Loughman (1966) as shown in Fig. 1. Phosphate accumulated in the fungus of mycorrhizas becomes mobilized and may be translocated to the host in conditions of deficient external supply by a mechanism which is oxygen and temperature sensitive (Harley and Brierley 1954). This metabolically linked system of mobilization of phosphate, accumulated during uptake in the root system, has been demonstrated to occur in both mycorrhizal and non-mycorrhizal pine seedlings by

Morrison (1962). In both, phosphate accumulated in the root system is translocated to the shoots if plants are transferred to phosphate deficient conditions, but the greater accumulations in root systems of mycorrhizal pines are mobilized and translocated over a longer period than the lesser accumulations in uninfected systems.

With ions or substances presented to root systems in the soil in higher concentrations than phosphate, or again with ions less avidly absorbed by the cells than phosphate, mass flow and diffusion through the free spaces are likely to be of greater importance as a feature of the uptake process, and

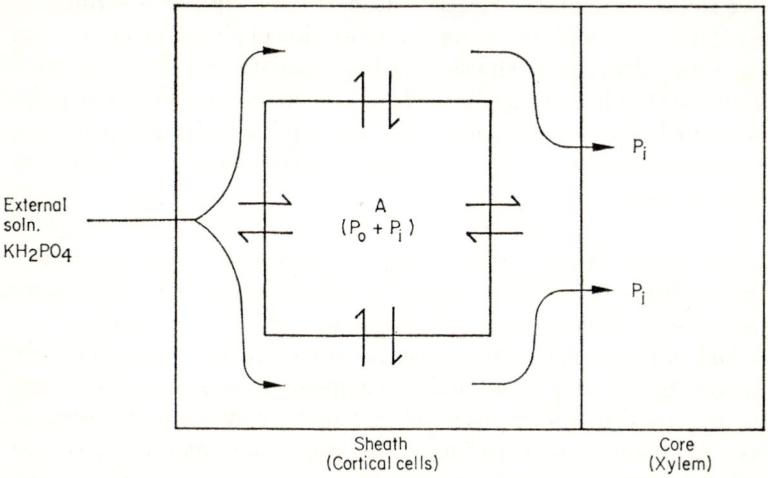

FIG. 1. Diagram of the route of movement of phosphate through the sheath of beech mycorrhizas to the host. The parallel between this and uninfected roots is indicated by the words in brackets.

it is to be expected that ions such as calcium, sodium, or sulphate absorbed in lesser quantities or less avidly by the living cells, would not only penetrate more readily the free space by physical processes, but also be more sensitive to increase and decrease in their rates of absorption to changes in transpiration rate.

Although accumulation in the fungal tissues of ectotrophic mycorrhizas may be more striking than is accumulation in the cortex of other absorbing organs, there is no case for regarding it as unimportant in the latter. Moreover since there will be a variation in the degree of penetration of the free spaces occasioned by variations in transpiration rate, the model which I have proposed will go some way to explaining effects of transpiration on absorption rate.

Before passing on to other topics it is of interest to note that, although, rightly or wrongly the properties of mycorrhizas may be regarded as too specialized to contribute to general discussion, the endotrophic, vesicular-arbuscular, mycorrhizal condition is very widely present in vascular plants (even many of those named in our discussions) as Dr Barber emphasized. Very many recent experiments have shown these to be especially efficient in nutrient uptake, particularly in phosphate uptake (see for instance Nicolson 1968). Nicolson reports that in experiments, by G.D. Bowen and Barbara Mosse, as yet unpublished, the fungus-containing cells were seen especially to accumulate phosphate in short-term experiments. Another, very different example of primary accumulation of phosphate in root cells is given by Jeffrey (1964) who showed that in *Banksia ornata* 30% of the total phosphate absorbed during 24 hours was accumulated as inorganic polyphosphate in the root. He put forward the hypothesis that polyphosphate accumulation might be a common adaptation to long term survival in phosphate deficient soils.

Dr Rorison in his paper made reference to the work, as yet unpublished, by Dr Nassery on comparative phosphate uptake and accumulation in a group of British plants from contrasted habitats. In the course of his work Nassery made estimates of polyphosphates, using the definitions and methods of Jeffrey, in *Deschampsia flexuosa* and *Urtica dioica*. In neither did he find any significant quantity of these compounds, but as Rorison pointed out, he noted that *Deschampsia* readily accumulated soluble inorganic phosphate. In a comparative experiment, *Deschampsia* and *Urtica* were grown in a culture solution very high in phosphate and then transferred to phosphate-free culture. *Urtica* grew for only two weeks using accumulated stores before its growth rate decreased and deficiency symptoms appeared, but *Deschampsia* continued to grow steadily for the full five weeks of the experiment without change of rate or showing any deficiency symptoms, Fig. 2. It is clear that the potential importance of accumulation in the roots requires careful assessment as an ecologically important facet of periodic growth and development, and of existence in habitats where there are periodic or seasonal changes in nutrient supplying power of the soil.

Selective uptake and carrier mechanisms

Dr Epstein emphasized in his very valuable paper the importance of selective absorption in the nutrition of plants and its relevance to problems of ecological adaptation. In particular he demonstrated in a most enlightening manner how research into carrier mechanisms, far from only being a difficult and fascinating pursuit of laboratory physiologists, is of prime importance in the interpretation of the absorptive processes of all plants; and indeed its study may point a way to explanations of differential behaviour of different species under natural conditions. Carriers specific or selective for particular ions, if they vary in their detailed properties, importance or incidence, in ecologically diverse species, may be powerful factors in

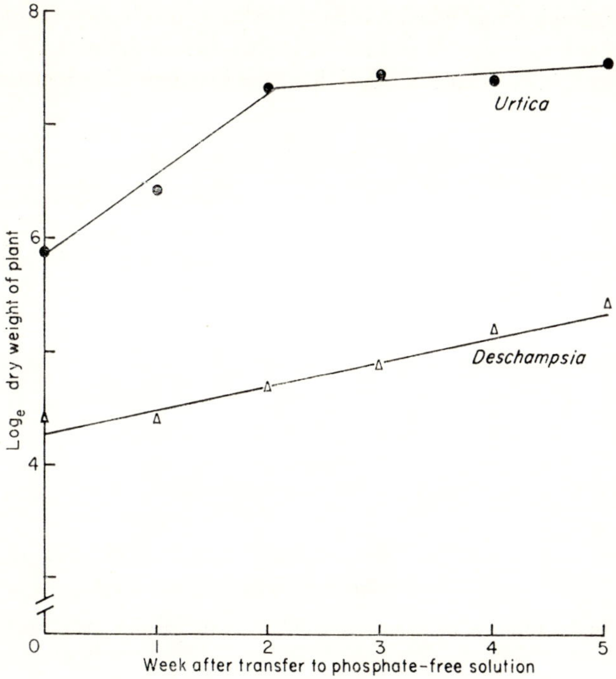

FIG. 2. Relative growth rate of *Urtica dioica* and *Deschampsia flexuosa* in phosphate-free culture after a period in 10^{-3} M phosphate. (Results of H. Nassery.)

the maintenance of internal ionic and nutrient balances which may differ so greatly from those in the external habitat. One of the most important lines of evidence, apart from the observation of selective uptake, which leads to the conclusion that carriers exist, results from the study of changes in the rate of uptake with external concentration. Essentially the curve of uptake rate over a range of low concentrations follows a curve approaching a rectangular hyperbola. This makes Michaelis–Menten kinetics, or analyses derived from it, attractive ways of studying the relationship. The conclusion is reached that a carrier forms a compound with the absorbed ion as a stage in absorption. Such analyses have been immensely productive of experimental problems, but it seems to me that they may be at times not the only profitable procedure. Dr Epstein has interpreted certain departures from the hyperbolic over wide ranges of concentration, as evidence for a dual carrier system. One carrier is thought to function over low concentration ranges and one over high. I should like to give an alternative explanation of such a curve. An exactly similar curve of phosphate uptake against concentration is found when beech mycorrhizas absorb phosphate (Fig. 3). From what I have already said it is clear that there is diffusive penetration of the tissues in this material which increases in extent with concentration.

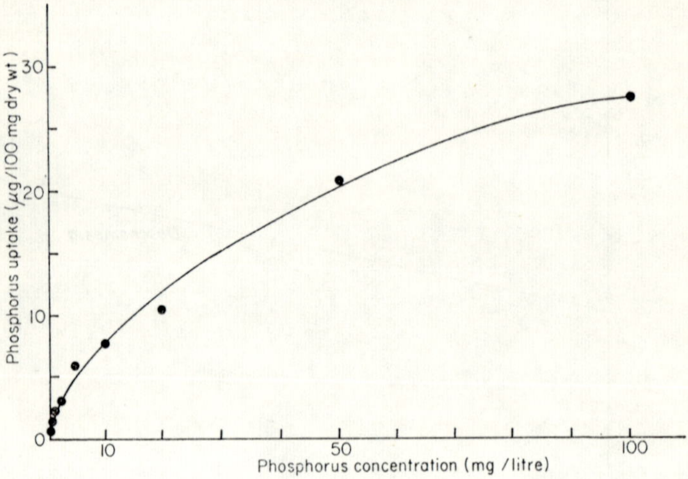

FIG. 3. The absorption of phosphate from different external phosphate concentrations by beech mycorrhizas.

In Table 1 it is shown that detached sheath has little greater uptake than attached sheath from 3 mM phosphate. All hyphal cells are almost equally supplied by diffusive penetration. If we make the assumption that each individual cell responds hyperbolically in uptake to increased concentration, then the observed curve (Fig. 3) is an expected resultant of the summation of all the metabolic uptakes of the individual cells. At each external concentration each cell layer is presented with a concentration which is the resultant of diffusive penetration of the free space less the metabolic removal of solute by the cell layers outside it. Such a model serves to emphasize the two paths through a root cortex, and to explain the effects of transpiration, and the possible increase in concentration in the cortical space of ions not rapidly accumulated in the cells. At the same time it does not deny the validity of the carrier hypothesis in its simpler form. I think this is a discussible topic where the points of view of Weatherley, Loughman, Epstein and Jennings may meet.

In any event, the form of the curve relating uptake by the whole plant to concentration is not a simple one. Its lack of simplicity makes for difficulties in deciding what assumptions should be made by Mr Nye and his colleagues about the evaluation of factor α, the *proportionality coefficient* or *absorbing power* of the root. It seems likely that in dealing with substances avidly absorbed and present in the soil in low concentrations that the best first approximation is that α varies hyperbolically with concentration, but for substances present in high concentrations with lower absorption rates the relationship will be modified significantly by factors which affect physical penetration into the tissues or, to take Dr Epstein's interpretation, by the increasing significance of a second carrier mechanism.

Absorption by root systems

It should be noted whilst we are concerning ourselves with these problems of the possible role of accumulation in the cells of the root system, that they have a bearing on the use of analyses of the change in nutrient content of the shoots as estimates of the total absorption by the plant. Mr Nye and Dr Newbould used such analyses in their experiments from which they concluded that the roots in certain soil horizons, the lower horizons, contributed unexpectedly excessive proportions of the total uptake of nutrients. These results indeed showed that the roots in those horizons contributed disproportionately high amounts to the nutrient supply of the shoot. Therefore to reach the conclusion that they drew, an assumption must also be made that, during any selected time period, the ratios of the quantity which is translocated to the shoot to that which is utilized or accumulated in the root are constant in all parts of the root system. This assumption seems to be contrary to the experimental results of such observers as Brouwer (1954), Weibe and Kramer (1954) and Canning and Kramer (1958) who have shown differences in the ratio of total uptake to quantity translocated in different parts of a single root axis under standardized conditions.

These problems are made even more complex when one considers the results of experiments and observations which show that root systems, their growth and geometry, vary in a single species with nutrient supply and other external conditions. It is a fairly commonplace observation that relative root growth, as measured in the simplest way by the ratio of the dry weight of the root to that of the shoot, is greater in low- than in high-nutrient supply (the extreme starvation and excessive ranges excluded). Again, ratios of root to shoot are commonly lower in low than in high light intensities. Such variations, mentioned by Professor Pigott, which differ in extent between species, seem to merit as detailed an appraisal in respect of their ecological implications as do the relative tolerances of species to other factors, such as heavy metals about which we have had much interesting and valuable information from many speakers. Relative root growth and root geometry and the relationship of these to nutrient absorption, are capable of much more detailed analysis than a simple comparison of ratios, and some work is proceeding on these lines, such as that of Dr Hackett (1968) in Dr Scott Russell's laboratory.

TABLE 3

Estimated nutrient uptakes of forest stands of 100 years' growth. Kg per hectar (after Rennie 1955)

	Ca	K	P
Pines	203	91	21
Other conifers	438	234	41
Hardwoods	879	225	50

I would like to conclude by suggesting that there are other differences between plant species that may be of great ecological importance in their relevance to mineral nutrition. I will take just two examples. It is a very general assumption that oxygen supply to the root is an important factor in nutrient uptake. Dr Loughman has made the points that phosphate absorption may in some degree and under certain circumstances be mediated by the glycolytic system, and also that barley seedlings might become adapted to low oxygen supply. It is clear that these properties, especially if they vary between species, are potentially of ecological importance. Moreover much recent work with various plants has indicated that internal ventilation of the tissues of the root system may be more extensive than has hitherto been assumed (see for instance Brown 1947; Evans and Ebert 1960; Barber, Ebert and Evans 1962; Greenwood 1966). The existence of variations and their relevance to ecological conditions not only of waterlogged areas but also in root competition are perhaps worthy of investigation.

Lastly it may not be too naïve to point out that the different requirements of different species for nutrients is a subject worthy of notice. If one inspects the results of analyses per unit weight of the plant bodies of different species of the same growth form, it is clear that there are remarkable variations (see for instance Baumeister 1958). In Table 3 some of Rennie's (1955) estimates are given for stands of trees. It cannot be doubted that such differences in demand are likely to be relevant to the interpretation of ecological distribution and behaviour.

REFERENCES

BARBER D.A., EBERT M. and EVANS N.T.S. (1962). The movement of ^{15}O through barley and rice plants. *J. Exp. Bot.* **13**, 397–403.

BAUMEISTER W. (1958). Die Aschtenstoffe. *Encyclopaedia of Plant Physiology*, Vol. IV, 5–36.

BROUWER R. (1954). Water and salt uptake by roots of intact *Vicia faba* plants. *Acta. Bot. Neerl.* **3**, 264.

BROWN R. (1947). The gaseous exchange between the root and the shoot of seedlings of *Cucurbita pepo*. *Ann. Bot. Lond.* NS, **11**. 417–437.

CANNING R.E. and KRAMER P.J. (1958). Salt absorption and accumulation in various regions of the root. *Am. J. Bot.* **45**, 378–382.

CROSSETT R.N. and LOUGHMAN B.C. (1966). The absorption and translocation of phosphorus by seedlings of *Hordeum vulgare* L. *New Phytol.* **65**, 459–468.

EVANS N.T.S. and EBERT M. (1960). Radioactive oxygen in the study of gas transport down the root of *Vicia faba*. *J. Exp. Bot.* **11**, 246–257.

GREENWOOD D.J. (1966). Studies on the transport of oxygen through stems and roots of vegetable seedlings. *New Phytol.* **66**, 337–347.

HACKETT C. (1968). A study of the root system of Barley. 1. Effects of nutrition on two varieties. *New Phytol.* **67**, 287–299.

HARLEY J.L. and BRIERLEY J.K. (1954). The uptake of phosphate by excised mycorrhizal roots of the beech. VI. *New Phytol.* **53**, 240–252.

HARLEY J.L. and LOUGHMAN B.C. (1963). The uptake of phosphate by excised mycorrhizal roots of the beech. IX. *New Phytol.* **53**, 92–98.

JEFFREY D.W. (1964). The formation of polyphosphate in *Banksia ornata*, and Australian heath plant. *Aust. J. Biol. Sci.* **17**, 845–854.

JENNINGS D.H. (1964). Changes in the size of orthophosphate pools in mycorrhizal roots of beech with reference to the absorption of the ion from the external medium. *New Phytol.* **63**, 181–193.

MORRISON T.M. (1962). Absorption of phosphate from soils by mycorrhizal plants. *New Phytol.* **61**, 10–20.

NICOLSON T.H. (1967). Vesicular arbuscular mycorrhiza—a universal plant symbiosis. *Sci. Prog., Oxf.* **55**, 561–581.

RENNIE P.J. (1955). The uptake of nutrients by mature forest growth. *Plant and Soil.* **7**, 49–95.

WEIBE H.H. and KRAMER P.J. (1954). Translocation of radioactive isotopes from various regions of roots of Barley seedlings. *Pl. Physiol.* **29**, 342–348.

V. SUMMATION

Discussion

Recorded by Dr C. Marshall

Referring to Professor Bradshaw's comments on the competitive interaction between plants Dr Rovira pleaded for attention to be given also to antagonistic effects on microorganisms. He cited as an example the grass *Aristida polyganza* which grows in soils of low nitrogen status and produces root exudates that are toxic to *Rhizobium*. This prevented the establishment of legumes and so there was no increase in soil fertility and no subsequent invasion by other species. Thus although a species might appear to have little competitive ability it must be realized that significant interactions between higher plants and microorganisms do occur and can have great influence on the competitive situation. Dr Vine considered that this was a good argument for working with the ecosystem as a whole so that all the components of the soil-plant-atmosphere interrelationship could be evaluated for given situations. Using the ecosystem approach one considered the reciprocal effects of the plants upon the soil and the soil upon the plants and this was shown to be of some importance when studying the ecology of strongly leached sites. In a further comment Dr Vine discussed the data on phosphate uptake from different soil depths presented in the paper by Mr Nye and concluded that the ecosystem approach would be valuable in following the leaching and decomposition of phosphorus compounds down the profile and their uptake and transport by plant roots.

From a general ecological viewpoint Mr Greig-Smith was concerned that so little attention had been given throughout the Symposium to the nutritional problems of long lived species. In a forest where the balance of species was not established during early development but at a much later time, very little was known about competition for nutrients among mature plants. Even if rather specialized conditions were required for establishment these were not of great ecological importance as, in the long run, the nutrient conditions affecting the mature plants would be more important. Another problem about which little was known was the frequent lack of correlation between the vegetative composition and the soil nutrient level. Such a situation was evident in the tropical rain forest and needed attention. Lastly Mr Greig-Smith cautioned the biochemists on the dangers of what he termed 'Victorian optimism' in that several speakers seemed to be adopting the attitude that every chemical difference that was found between species must be of definite adaptive value. In reply to this last point Professor Harley, Dr Sims and Dr Jennings defended the work on metabolic differences between species and it was generally felt that the biochemist was well aware of the limitations of his experiments.

In response to Professor Russell's comments on aeration Dr Barber discussed the significance of oxygen transport from the shoot to the root system in a wide range of plants. The use of the radioisotope ^{15}O in rice varieties had shown that the oxygen moved by diffusion and that there was a correlation between the movement of oxygen down to the roots and the development of air spaces. Studies on barley however had shown that the transfer of oxygen from shoot to root under well-aerated conditions was likely to be of little significance but that there was the possibility that transport could increase under less well-aerated soil conditions. This view was supported by Dr Woolhouse for *Nardus stricta* in which he described a marked change in the amount of root intercellular air space in relation to the wetness of the habitat in which the plant was growing. Other adaptive changes at low oxygen tension were mentioned by Dr Crawford in commenting on the metabolism of roots under anaerobic conditions induced by flooding.

Professor Weatherley pointed out that one ought also to consider the effect of low oxygen tension in the root in terms of the whole plant. For a low oxygen tension in the root would lead to an increased resistance to water flow and this could result in a water stress in the shoot. The stomata would then close or partly close and if there was a transpiration dependent fraction in salt transport it would decline. Secondly photosynthesis would be reduced and so the circulation of metabolites in the whole plant would decline. Thus to get an integrated picture of the influence of low soil oxygen tension on growth, the whole plant must be examined.

The problem of root exudates was again raised by Dr Rovira, and he described in some detail work done by Bowen in Adelaide in which he found that the rate of exudation from roots of *Pinus radiata* was greater under phosphorus deficiency than in complete nutrient solution. A technique was described in which the uptake and accumulation of ^{32}P by the root could be followed using a chromatogram scanner (Bowen and Rovira 1967). Dr Barley however queried whether the rate of nutrient uptake was limited by properties of the plant or the soil and suggested that there was considerable evidence to support the latter particularly from experiments in which the uptake of ions in solution phase was compared with uptake from soil. He stated that very great differences could be obtained but concluded that the general problem could only be resolved when more uptake data became available from experiments in which the transpiration rate had been high, as the vast majority of people worked under conditions of low transpiration.

There was considerable discussion on Professor Weatherley's model of ion uptake and transport in the whole plant and it was generally agreed that it was of great use in integrating the transport system as a whole. He confirmed that there could be considerable ion accumulation in cortical vacuoles under certain conditions and that this could predominate against movement across the root. But some experiments showed that such accumulation was minimal and that root mineral content was virtually constant. Mycorrhiza when present could be regarded as an extra cortical system abstracting ions from the mass flow stream and giving back to it in times of shortage so that transport of ions to the shoot would be maintained. Professor Harley stressed that this was of considerable ecological importance.

Professor Epstein questioned Professor Harley's explanation of absorption isotherms in terms of a diffusion model. He felt that it was particularly difficult to reconcile with many experimental findings and restated data from his paper on the similarity of the pattern of ion uptake by different plant tissues. In this case diffusion would imply that the total velocity of the diffusion path and all the limiting things in it were the same in two organs that were totally different in their geometry. This and other evidence suggested that a diffusion based model would not allow adequate explanation. In reply Professor Harley felt sure that the diffusion model was right for the experiments that he had described but agreed that it would not be adequate for all situations. But for any nutrient that was absorbed and accumulated gradually there was likely to be considerable diffusional complication.

The problem of the efficiency of nutrient circulation was raised by Professor Pigott in describing the differing mobilities of phosphorus in *Mercurialis perennis* and *Urtica dioica*. In the latter the transport of phosphorus from old to new leaves was inefficient in that senescing leaves contained high levels of phosphorus. In *Mercurialis* recirculation was more efficient and thus overall less phosphorus would be required for growth. This aspect of metabolism could clearly be of considerable importance for successful growth on soils of low phosphate status. Dr Loughman suggested that as virtually all the phosphorus in the cell was in the vacuole the efficiency of mobilization might be related to the ability to move it across the vacuolar membrane.

Finally, there was some discussion on the occurrence of mycorrhiza in crop species and it was concluded that the roots of most plants had some sort of fungal association of this type.

REFERENCE

BOWEN G.D. and ROVIRA A.D. (1967). Phosphate uptake along attached and excised wheat roots measured by an automatic scanning method. *Aust. J. biol. Sci.* **20**, 369–370.

LIST OF THOSE WHO ATTENDED THE SYMPOSIUM

ALSTON Mr B. Department of Agricultural Chemistry, Queens' University, Belfast.
ANDREWS Miss R.E. Department of Botany, University College of Wales, Penglais, Aberystwyth.
ARMSTRONG Dr W. Department of Botany, University of Hull.
ARNOLD Prof. P.W. School of Agriculture, University of Newcastle upon Tyne.
ASHFORD Miss A.E. Department of Botany, University of Leeds.
ASPREY Prof. G.F. Department of Botany, University College, Cardiff.
ATKINSON Mr D. 22 St. Georges Terrace, Newcastle upon Tyne 2.
BANNISTER Dr P. Department of Botany, University of Glasgow.
BARBER Dr D.A. ARC Radiobiological Laboratory, Letcombe Regis, Wantage, Berks.
BARLEY Mr K.P. Waite Agricultural Research Institute, Glen Osmond, Adelaide, South Australia.
BELL Mr J.N.B. Department of Botany, University of Manchester.
BERGH Mr J.P. VAN DEN. Englaan 2, Wageningen, The Netherlands.
BINSTEAD Mr R. 3 Spanish Road, London, S.W.18.
BOATMAN Dr D.J. Department of Botany, University of Hull.
BOATMAN Dr S.G. ICI Agricultural Division, Jealott's Hill Research Station, Bracknell, Berks.
BRADSHAW Prof. A.D. The Hartley Botanical Laboratories, University of Liverpool.
BRERETON Dr A.J. An Foras Talúntais, Johnstown Castle, Wexford, Ireland.
BREWSTER Mr J.L. Soil Science Laboratory, Department of Agriculture, Parks Road, The University, Oxford.
BROWN Mr A.H.F. The Nature Conservancy, Merlewood Research Station, Grange over Sands.
CARLETON Mr T.J. School of Plant Biology, U.C.N.W., Bangor.
CHADWICK Dr M.J. Department of Biology, The University, Heslington, York.
CHAPPELL Mr H.G. Department of Biological Sciences, College of Technology, Portsmouth.
CLAPHAM Prof. A.R. Department of Botany, The University, Sheffield 10.
CLARKSON Dr D.T. ARC Radiobiological Laboratory, Letcombe Regis, Wantage, Berks.
COTTON Mr J. 27E, Well Street, Exeter.
COURT Miss E. 18 Douglas Road, Goodmayes, Essex.
COWLING Mr D.W. Grassland Research Institute, Hurley, Berks.
CRAWFORD Dr R.M.M. Department of Botany, University of St. Andrews, Fife.
CROOKE Dr W.M. The Macaulay Institute for Soil Research, Craigiebuckler, Aberdeen.
CUNNINGHAM Dr R.K. Ministry of Overseas Development, Eland House, Stag Place, London, S.W.1.

DALE Mr J. The Nature Conservancy, Penrhos Road, Bangor.

DANIELS Mr R.E. Department of Botany, University of Nottingham.

DAVIES Mr W.N.L. Tate & Lyle Research Centre, Westerham Road, Keston, Kent.

DAVISON Dr A.W. Department of Botany, University of Newcastle upon Tyne.

DELAP Dr ANNE V. East Malling Research Station, Maidstone, Kent.

DENBIN Mr J.A. Northfields Farm, Priors Marston, Nr. Rugby.

DERRY Miss S. 102 Earlham Road, Norwich.

DIJKSHOORN Dr W. Institute for Biological and Chemical Research on Field Crops and Herbage, Wageningen, Holland.

DOBSON Mr A.T. Department of Botany, University of Hull.

DUNHAM Mr R.J. St. Catherine's College, Oxford.

EPSTEIN Prof. E. Department of Plant Nutrition, University of California, Davis, California 95616, U.S.A.

EVANS Dr and Mrs G.C. Botany School, Downing Street, Cambridge.

FREIJSEN Dr A.H.J. Biological Station, Oostvoorne, The Netherlands.

FRENCH Mr R.A. Rothamsted Experimental Station, Harpenden, Herts.

FULLERTON Miss H.T. Department of Agricultural Chemistry, University of Glasgow.

GAMBI VERGNANO Dr ORNELLA 17 via Bartolommeo Scala, 50126 Firenze, Italy.

GARDNER Mr G. Department of Botany, The University, Sheffield 10.

GEBRE-EGZIABHER Mr T.B. School of Plant Biology, U.C.N.W., Bangor.

GIGON Mr A. Geobotan. Inst. E.T.H., Zürichbergstr 38, 8044 Zürich, Switzerland.

GILBERT Mr O.L. Department of Botany, University of Newcastle upon Tyne.

GILES Mr B.R. c/o Westfield College, London, N.W.3.

GITTINS Dr R. Institute of Statistics, North Carolina State University, Raleigh, North Carolina 27607, U.S.A.

GLENTWORTH Dr R. The Macaulay Institute for Soil Research, Craigiebuckler, Aberdeen.

GOLDSMITH Dr F.B. Department of Botany, University College, Gower Street, London, W.C.1.

GOODMAN Mr G.T. Department of Botany, University College, Swansea.

GOODMAN Dr J.P. Welsh Plant Breeding Station, Plas Gogerddan, Nr. Aberystwyth.

GORE Mr A.J.P. The Nature Conservancy, Merlewood Research Station, Grange over Sands, Lancs.

GRACE Mr J. Department of Botany, The University, Sheffield 10.

GREENHAM Dr D.W.P. East Malling Research Station, Maidstone, Kent.

GREIG-SMITH Mr P. School of Plant Biology, U.C.N.W., Bangor, Caerns.

GRIME Dr J.P. Department of Botany, The University, Sheffield 10.

GRUBB Dr P.J. Botany School, Downing Street, Cambridge.

GRUE Mr J.D. Department of Botany, University College, Gower Street, London, W.C.1.

GULLIVER Mr R.L. 40 Crosbie Road, Birmingham 17.

GUNARY Dr D. Levington Research Station, Ipswich, Suffolk.

HACKETT Dr C. ARC Radiobiological Laboratory, Letcombe Regis, Wantage, Berks.

HANDLEY Mr J.F. University College, Gower Street, London, W.C.1.

HARDING Mr C.P. Alcuin College, University of York, Heslington, York.

HARLEY Prof. J.L. Department of Botany, The University, Sheffield 10.

HASLAM Dr F. Department of Agricultural Botany, University of Leeds.

HAYNES Mr F.N. Department of Biological Sciences, College of Technology, Portsmouth, Hants.

HEGARTY Mr T.W. Botany School, Downing Street, Cambridge.

HIGGS Mr D.E.B. Department of Chemistry and Biology, Hatfield College of Technology, Herts.

HODGSON Mr J.G. Department of Botany, The University, Sheffield 10.

HOWARD Mrs E.M. River House, Piddinghoe, Newhaven, Sussex.

HUNT Mr R. Department of Botany, The University, Sheffield 10.

HUXLEY Mr T. The Nature Conservancy, 12 Hope Terrace, Edinburgh 9.

JACKSON Mr D.K. Imperial College, Sunninghill, Ascot, Berks.

JAMES Mr D.B. Department of Agricultural Botany, Institute of Rural Science, Penglais, Aberystwyth.

JARVIS Dr P.G. Department of Botany, St. Machar Drive, University of Aberdeen.

JARVIS Mr S.C. Preston Montford Field Centre, Shrewsbury.

JEFFERIES Dr R.L. School of Biological Sciences, University of East Anglia, Norwich.

JEFFREY Dr D.W. 9 Borwick Lane, Warton, Lancs.

JENNINGS Prof. D.H. The Hartley Botanical Laboratories, University of Liverpool.

JONES Mrs H. Department of Botany, U.C.S.W., Cathays Park, Cardiff.

JONES Dr L.H.P. Grassland Research Institute, Hurley, Berks.

KENTZER Mr Z.M. 717 Abbey Lane, Sheffield 11.

KENWORTHY Dr J.B. Department of Botany, University of Aberdeen.

KING Mr T.J. Botany School, South Parks Road, The University, Oxford.

KIRKBY Mr E.A. Department of Agriculture, University of Leeds.

KNIGHT Mr A.H. The Macaulay Institute for Soil Science, Craigiebuckler, Aberdeen.

KØIE Prof. M. Botanical Laboratory, Gothersgade 170, Copenhagen.

LAND Mr J.B. School of Agriculture, Sutton Bonington, Loughborough.

LEE Dr J.A. Department of Botany, University of Manchester.

LIDDLE Mr M.J. 7 Pear Tree Gardens, Aston Cantlow, Solihull, Warwicks.

LLOYD Dr P.S. Department of Botany, The University, Sheffield 10.

LOUGHMAN Dr B.C. Department of Agricultural Sciences, Parks Road, The University, Oxford.

MACDONALD Dr I.R. Department of Plant Physiology, The Macaulay Institute, Craigiebuckler, Aberdeen.

MCLELLAND Miss J.E. Fircroft, Fir Tree Lane, Littleton, Chester.

MCWILLIAM Miss J. 56 College Road, Bangor.

MALLOCH Mr A. St. John's College, Cambridge.

MARSHALL Dr C. School of Plant Biology, U.C.N.W., Bangor.

MARTIN Dr M.H. Department of Botany, University of Bristol.

MARTIN Mr P. Welsh Plant Breeding Station, Plas Gogerddan, Nr. Aberystwyth.

MENGEL Prof. K. 3 Hannover-Kirchrode, Buntweg 8, Germany.

MIDDLETON Mr C.P. Department of Biological Studies, Lanchester College of Technology, Coventry.

MILES Mr J. The Nature Conservancy, Blackhall, Banchory, Kincards.

MORRIS Mr P.J. Biology Section, College of Technology, Oxford.

MOSS Mr P. Department of Soil Fertility/Chemistry, Agricultural Institute, Johnstown Castle, Wexford, Eire.

MOTT Dr C.J.B. Soil Science Department, University of Reading.

MYERSCOUGH Dr P.J. Department of Botany, King's Buildings, University of Edinburgh.

NEWBOULD Dr P. ARC Radiobiological Laboratory, Letcombe Regis, Wantage, Berks.

NEWBOULD Prof. P.J. New University of Ulster, Coleraine, N. Ireland.

NEWMAN Dr E.I. Department of Botany, University College of Wales, Penglais, Aberystwyth.

NORTON Dr G. School of Agriculture, Sutton Bonington, Nr. Loughborough.

NYE Mr P.H. Soil Science Laboratory, Department of Agriculture, Parks Road, The University, Oxford.

OHLROGGE Prof. and Mrs A.J. Department of Agronomy, Purdue University, Lafayette, Indiana, U.S.A.

OXLEY Mr E.R.B. School of Plant Biology, U.C.N.W., Bangor.

PEARSON Dr M.C. Department of Botany, University of Nottingham.

PERCIVAL Mr M.J.L. Botany School, Oxford.

PERKINS Dr D.F. The Nature Conservancy, Penrhos Road, Bangor.

PIGOTT Prof. C.D. Department of Biological Sciences, St. Leonard House, Lancaster.

PRICE JONES Dr D. Jealott's Hill Research Station, Bracknell, Berks.

PRIME Dr C.T. The Chestnuts, Farleigh Common, Warlingham, Surrey.

PROCTER Mr J. Botany School, South Parks Road, The University, Oxford.

RACKHAM Mr O. Corpus Christi College, Cambridge.

RAHMAN Mr M.S. Imperial College, Sunninghill, Ascot, Berks.

RAMZAN Dr M. Department of Agriculture, Parks Road, The University, Oxford.

RATCLIFFE Dr D. Department of Botany, University of Leicester.

READ Mr M.W. Moor House, Garrigil, Alston, Cumbs.

REDDAWAY Mr E.J.F. Department of Botany, Westfield College, London, N.W.3.

REILEY Dr J. School of Plant Biology, U.C.N.W., Bangor, Caerns.

ROBERTS Mr E. P.O. Box 7062, Kampala, Uganda.

RORISON Dr I.H. Department of Botany, The University, Sheffield 10.

ROVIRA Dr A.D. C.S.I.R.O. Division of Soils, Adelaide, S. Australia.

RUSSELL Prof. E.W. Department of Soil Science, University of Reading.

RUTTER Prof. A.J. Department of Botany, Imperial College, London, S.W.7.

RYCHNOVSKA Dr MILENA. Ecological Department, Czechoslovak, Academy of Sciences, Stará 18, Brno, Czechoslovakia.

SALAMAN Miss S. Department of Chemistry and Biology, Hatfield College of Technology, Herts.

SANDERSON Mr J. Paddock House, Ham Road, Wantage, Berks.

SANKHLA Dr N. Botanisches Institut der Universität, 6 Frankfurt a. M. Siesmayerstr. 70, Germany.

SCOTT RUSSELL Dr R. ARC Radiobiological Lab., Letcombe Regis, Wantage, Berks.

SHORROCKS Dr V.M. Borax Consolidated Ltd., Borax House, Carlisle Place, London, S.W.1.

SIMS Dr A.P. School of Biological Sciences, University of East Anglia, Norwich.
SMITH Dr C.J. Department of Agricultural Botany, University of Reading.
SNAYDON Dr R.W. Agricultural Botany Department, University of Reading.
SOUTHERN Mr H.N. Animal Ecology Research Group, Botanic Garden, High Street, Oxford.
SPENCE Dr D.H.N. Department of Botany, University of St. Andrews, Fife.
SPRUCE Miss E.M. Department of Botany, The University, Sheffield 10.
STEWART Mr G. Department of Botany, University of Bristol.
STEWART Dr J.W.B. Department of Soil Science, University of Saskatchewan, Saskatoon, Canada.
STEWART Mr W.S. Department of Botany, University of Glasgow.
SUTTON Dr C.D. Levington Research Station, Ipswich, Suffolk.
SUTTON Mr F. Department of Botany, The University, Sheffield 10.
TABOR Mr B. 132 Wellington Road, Manchester 14.
TAYLOR Dr K. Department of Botany, University College, Gower Street, London, W.C.1.
TEATHER Mr D.C.B. Department of Biology, University of London Goldsmith's College, London, S.E.14.
THAKE Mrs B. 327 City Road, London, E.C.1.
THORP Mr T.K. Department of Botany, University of Glasgow.
THURSTON Miss J.M. Rothamsted Experimental Station, Harpenden, Herts.
TINKER Dr P.B. Department of Agriculture, The University, Parks Road, Oxford.
TOMLINSON Mr R.W. Department of Botany, University of Hull.
TOMLINSON Dr T.E. Jealott's Hill Research Station, Bracknell, Berks.
TURNER Dr BRENDA J. Department of Geography, University of Leicester.
TURNER Dr R.G. Shell Research Ltd., Woodstock Agricultural Research Centre, Sittingbourne, Kent.
URQUHART Mr C. Department of Botany, University of Birmingham, Research Gardens, Winterbourne, Birmingham 15.
VAIDYANATHAN Dr L.V. Soil Science Laboratory, Department of Agriculture, The University, Parks Road, Oxford.
VINE Dr H. 15 Holmwood Drive, Leicester.
WADSWORTH Dr R.M. Department of Botany, University of Reading.
WAGGITT Mr P.W. Childs Hall, University of Reading.
WATERS Mr S.J.P. Department of Botany, Bedford College, Regent's Park, London, N.W.1.
WEATHERLEY Prof. P.E. Department of Botany, St. Machar Drive, University of Aberdeen.
WEBB Mr M.J. Department of Biological Sciences, St. Leonard House, The University, Lancaster.
WESTLAKE Mr D.F. River Laboratory, E. Stoke, Wareham, Dorset.
WHITE Dr D.J.B. Department of Botany, University College, London, W.C.1.
WHITE Mr J. Department of Botany, University College, Dublin 4.
WHITMORE Dr T.C. Forest Research Institute, Kepong, Selangor, Malaya.
WHITTON Dr B.A. Department of Botany, University of Durham.
WILKINS Dr D.A. Department of Botany, University of Birmingham.
WILLIAMS Dr J.T. Lanchester College of Technology, Coventry.
WILLIS Dr A.J. Department of Botany, University of Bristol.

WILSON Mr J. 76 Main Street, Sedbergh, Yorks.
WILSON Mr J.B. Department of Botany, University College of Wales, Penglais, Aberystwyth.
WOODFORD Dr E.K. Grassland Research Institute, Hurley, Nr. Maidenhead, Berks.
WOOLHOUSE Dr H.W. Department of Botany, The University, Sheffield 10.
WOOLHOUSE Mr R. Department of Botany, University of Manchester.

AUTHOR INDEX

Bold figures refer to pages on which full references appear; n indicates that the reference appears in a footnote.

SUBJECT INDEX